Trends in Mathematical Economics

Alberto A. Pinto • Elvio Accinelli Gamba
Athanasios N. Yannacopoulos
Carlos Hervés-Beloso
Editors

Trends in Mathematical Economics

Dialogues Between Southern Europe
and Latin America

 Springer

Editors
Alberto A. Pinto
Matemática
Universidade do Porto
Porto, Grande Porto, Portugal

Athanasios N. Yannacopoulos
Statistics
Athens University Economics Business
Athens, Attiki, Greece

Elvio Accinelli Gamba
Economía
Universidad Autónoma de San Luis Potosi
San Luis Potosí
San Luis Potosí, Mexico

Carlos Hervés-Beloso
Ciencias Económicas y Empresariales
Universidade de Vigo
Vigo, Pontevedra, Spain

ISBN 978-3-319-32541-5   ISBN 978-3-319-32543-9   (eBook)
DOI 10.1007/978-3-319-32543-9

Library of Congress Control Number: 2016944142

Printed on acid-free paper

This Springer imprint is published by Springer Nature
The registered company is Springer International Publishing AG Switzerland

*Alberto Adrego Pinto dedicates this volume
to Maria Barreira Pinto
Elvio Accinelli dedicates this volume
to Maria del Huerto Bettini*

Preface

This book includes selected papers that have been presented or discussed in the following conferences held in 2014: the 3rd International Conference Dynamics Games and Science III—DGS III, the 1st Hellenic-Portuguese Meeting on Mathematical Economics, AUEB, Athens, Greece, and XV Jornadas Latinoamericanas de Teoría Económica (JOLATE), Guanajuato, México.

The 3rd International Conference Dynamics Games and Science III—DGS III, on the occasion of the 50th birthday of Alberto A. Pinto, aims to bring together world top researchers and practitioners. DGS III represents an opportunity for MSc and PhD students and researchers to meet other specialists in their fields of knowledge and to discuss and develop new frameworks and ideas to further improve knowledge and science. DGS I was realized in 2008 at the University of Minho, in honor of Mauricio Peixoto and David Rand, and DGS II was realized in 2013 at the Calouste Gulbenkian Foundation, Lisbon.

The main purpose of the Hellenic-Portuguese Meeting on Mathematical Economics is to bring together researchers and students into a unique event to discuss and foster the spread of mathematical methods for game theory and economics in different countries particularly Portugal, Greece, and Spain. This meeting is organized every year and takes place in these countries looking to develop contacts and networks with Latin American researchers and students in the area of mathematical economics and game theory.

JOLATE is an annual meeting of the Latin American Association of Economics (ALTE). The main objective of ALTE is to provide a framework to promote and spread mathematical methods and research results in economic theory in Latin America. ALTE is involved in supporting activities related to economic theory at very different levels such as basic research, application, and education. The association has built up a Latin American network including universities and research centers in Argentina, Brazil, Chile, Colombia, Mexico, and Uruguay.

ALTE organizes the JOLATE meeting, a scientific conference that annually joins an increasing number of researchers and practitioners of mathematical economics methods, to contribute to the diffusion of their work and to the development of interactions between them to encourage potential future joint collaborations as well.

JOLATE meetings have taken place in many different places in Latin American. The Universidad Nacional del Sur in Bahia Blanca, Argentina, organized the first one in 1999. Since then, other host universities were Universidad Nacional de San Luis (Argentina), Universidad de la República (Uruguay), Universidad Autónoma San Luis Potosí (México), Universidad de Chile (Chile), Instituto de Matemática Pura y Aplicada (IMPA, Brasil), Universidad EAFIT (Colombia), Universidad de los Andes (Colombia), and Centro de Investigaciones Matemáticas (CIMAT, México).

With this volume, the editors not only contribute to the advancement of research in these areas but also inspire other scholars around the globe to collaborate and research in these vibrant, emerging topics.

San Luis, Argentina   Alejandro Neme
                                                                                                Jorge Oviedo

Acknowledgments

The editors of this volume would like to thank all authors for their contributions which reflect the diversity of areas within mathematical economics developed, particularly, in Latin America and southern Europe. We also recognize the invaluable work of the reviewers whose comments and suggestions have largely benefited the edition of this volume.

We thank Robinson Nelson dos Santos, Associate Editor, Mathematics, Springer-Verlag, São Paulo, and Susan Westendorf, Project Coordinator, Springer Nature, for invaluable suggestions and advice and for assistance throughout this project.

Alberto Adrego Pinto would like to thank LIAAD INESC TEC and to acknowledge the financial support received by the ERDF (European Regional Development Fund) through the Operational Programme Competitiveness and Internationalization (COMPETE 2020) within project "POCI-01-0145-FEDER-006961" and by the national funds through the FCT (Fundação para a Ciência e a Tecnologia) (Portuguese Foundation for Science and Technology) as part of project UID/EEA/50014/2013 and within project "Dynamics, optimization and modelling" with reference PTDC/MAT-NAN/6890/2014. Alberto Adrego Pinto also acknowledges the financial support received through the Special Visiting Researcher Program [Bolsa Pesquisador Visitante Especial (PVE)] "Dynamics, Games and Applications" with reference 401068/2014-5 (call: MEC/MCTI/CAPES/CNPQ/FAPS), at IMPA, Brazil.

Elvio Accinelli acknowledges the financial support received through the project "Trends in Mathematical Economics Dynamics and Game Theory with Applications to the Economy," supported by the special program of CONACYT (México) "Estancias Sabática en el extranjero," with reference 264820, and through the project "Imitación, Bienestar, Crecimiento y Trampas de Pobreza," CONACYT with reference 167004.

Carlos Hervés-Beloso acknowledges the support by ECOBAS (Xunta de Galicia. Project AGRUP2015/08).

A. N. Yannacopoulos would like to thank Athens University of Economics and Business for its support of the meetings when they took place in Greece, as well as all the participants, who have honored us with their contributions to the meetings and this volume.

Contents

List of Contributors

Grisel Ayllón Aragón Tecnologico de Monterrey, Mexico City, Mexico

David Cantala El Colegio de Mexico, Mexico City, Mexico

Enrique R. Casares Departamento de Economia, Universidad Autonoma Metropolitana Unidad Azcapotzalco, Mexico City, Mexico

Rafael Delfin-Vidal Departamento de Actuaría, Física y Matemáticas, Universidad de las Américas Puebla, Puebla, Mexico

Matías Fuentes Escuela de Economía y Negocios, Centro de Investigación en Economía Teórica y Matemática Aplicada, Universidad Nacional de San Martín, Buenos Aires, Argentina

Patricia Lucia Galdeano Facultad de Ciencias Físico Matemáticas y Naturales, Departamento de Matematicas, Universidad Nacional de San Luis, San Luis, Argentina

Adriana Gama-Velázquez El Colegio de México, Mexico City, Mexico

Bodil Olai Hansen Department of Economics, CBS, Frederiksberg, Denmark

Oliver Juarez-Romero CIMAT, Guanajuato, Gto., Mexico

Sheri Markose Economics Department, University of Essex, Colchester, UK

Samuel Gil Martin Facultad de Economia, Universidad Autónoma de San Luis Potosí, San Luis Potosi, Mexico

Ernesto Mordecki Mathematics Center, School of Sciences, Universidad de la República, Montevideo, Uruguay

Federico de Olivera Mathematics Center, School of Sciences, Universidad de la República, Montevideo, Uruguay

Departmento de Matemática, Federico Garcia Lorca entre Pastori y Goya, CeRP del Sur, Atlántida, Uruguay

William Olvera-Lopez CIMAT, Jalisco S/N, Valenciana, C. P. 36240 Guanajuato, Guanajuato, Gto, México

San Luis Potosí, SLP, Mexico

Leobardo Plata UASLP, San Luis Potosí, SLP, Mexico

Luis Guillermo Quintas Facultad de Ciencias Físico Matemáticas y Naturales, Departamento de Matematicas, Universidad Nacional de San Luis, San Luis, Argentina

Departamento de Matematicas, IMASL (UNSL-CONICET), Universidad Nacional de San Luis, San Luis, Argentina

Franco Robledo Facultad de Ingeniería, Universidad de la República, Montevideo, Uruguay

Guillermo Romero-Meléndez Departamento de Actuaría, Física y Matemáticas, Universidad de las Américas Puebla, Puebla, Mexico

Joss Sánchez-Pérez Facultad de Economía, UASLP, San Luis Potosí, Mexico

Facultad de Economía, UASLP, Av. Pintores s/n, Col. B. del Estado, San Luis Potosí, Mexico

Francisco Sanchez-Sanchez CIMAT, Guanajuato, Gto., Mexico

Martin Shubik Department of Economics, Yale University, New Haven, CT, USA

Horacio Sobarzo El Colegio de Mexico, Centro de Estudios Economicos, Mexico City, Mexico

Andrés Sosa Mathematics Center, School of Sciences, Universidad de la República, Montevideo, Uruguay

William D. Sudderth School of Statistics, University of Minnesota, Minneapolis, MN, USA

Andrianos E. Tsekrekos Department of Accounting and Finance, School of Business, Athens University of Economics and Finance, Athens, Greece

Mich Tvede Newcastle University, Newcastle upon Tyne, Tyne and Wear, UK

Martín Varela Facultad de Ingeniería, Universidad de la República, Montevideo, Uruguay

Omar Viera Facultad de Ingeniería, Universidad de la República, Montevideo, Uruguay

Stylianos Z. Xanthopoulos Department of Mathematics, University of the Aegean, Karlovassi Samos, Greece

Athanasios N. Yannacopoulos Department of Statistics, School of Information Sciences and Technology, Athens University of Economics and Business, Athens, Greece

Chapter 1
Breaking the Circular Flow: A Dynamic Programming Approach to Schumpeter

Martin Shubik and William D. Sudderth

Abstract Starting with a simple Robinson Crusoe economy, then adding in sequence one, then many random variables, we consider the effect of an innovation in the means of production. We then consider a many-agent economy that utilizes money. The success of the innovation for Crusoe depends on the availability of physical goods, his decisions, and chance. The success of innovation in a money-utilizing, many-person economy depends on financing and the locus of financial control, as well as the amount of resources invested and on one or more random events. The coordination and guidance problems posed by the latter are orders of magnitude more difficult than the former. Utilizing a parallel dynamic programming approach, we present models for which the insights of Schumpeter are consistent with the observations of general equilibrium but involve a complex vista of a dynamic economy with finance and incomplete markets and a recognition of the coordination problems irrelevant to general equilibrium theory. Our simple mathematical models illustrate the breaking of the circular flow of income. Here we concentrate on the case where there is only one opportunity for innovation and consider the conditions for the emergence of a new equilibrium. When innovation may take place at any period, the outcome to any individual becomes path dependent. History counts and financial guidance is critical. We limit our modeling of the financial structure to a central bank.

Keywords Cost innovation • Schumpeter • Circular flow • Strategic market games

JEL Classification: C73, D24, G32

M. Shubik
Department of Economics, Yale University, Box 208281, New Haven, CT 06520-8281, USA
e-mail: martin.shubik@yale.edu

W.D. Sudderth (✉)
School of Statistics, University of Minnesota, 224 Church St. S.E., Minneapolis, MN 55455, USA
e-mail: sudde001@umn.edu

© Springer International Publishing Switzerland 2016
A.A. Pinto et al. (eds.), *Trends in Mathematical Economics*,
DOI 10.1007/978-3-319-32543-9_1

1.1   Context and Circular Flow

We construct simple models that achieve a formal mathematization of a fundamental insight that Schumpeter had over a 100 years ago on the need to break the circular flow of finance required in a closed economy in equilibrium when there is the possibility of innovation. Our concern is to be able to illustrate the relationship between real assets and money and debt, noting also that the aspects of banking and who controls the financing become significant at even the most basic level of theory. This requires investigating the nature of the cash flows and how the amount of money, credit, and prices change even in greatly simplified models of innovation.

A literature search indicates that "the breaking of the circular flow" has been hardly treated in Anglo-American theorizing. Yet we believe it to be of considerable significance in both the reconciliation of the Schumpeterian approach to Walrasian economics and in going beyond Walrasian equilibrium to develop a basic theory of dynamics.

The work on Schumpeterian theory done primarily in Italy presents somewhat richer models, highly complementary with those here (cf. Dosi et al. 1988, 2013; Caiaini et al. 2013). They use simulation methods and a macroeconomic approach showing the relationship with both Keynes (1936) and Minsky (1986).

The success or failure of an innovation in production is here modeled as a random event with the probability of success being a function of the amount of real resources invested in an attempt to innovate.

1.1.1   The Evolution of Control

We begin with a study of Robinson Crusoe, who as a solitary individual does not need finance.[1] His optimization problem has constraints imposed by real resources and his production technology. A mass economy faces problems in coordination far beyond those of Crusoe. The introduction of a fiat money provides a means of exchange where much of the control of issue is in the hands of the government and a private banking system.

In a mass market, Crusoe's optimization is replaced with a similar type of optimization for a small family firm, but with more financial constraints imposed by money and prices and constraints created by the presence of many individuals and possibly more commodities available in the markets.

---

[1] Although he may find accounting useful as an aide memoire, and with a stretch of the imagination, could set up a virtual market to calculate virtual prices for himself.

By fixing default rules and monetary issue rules, a government can bound the price system from below and above in an economy utilizing fiat money. In general, the price levels in a system with uncertainty cannot be uniquely specified.[2]

1.1.2 The Circular Flow and Equilibrium

In a modern economy, much of economic activity calls for the use of money and credit, both for decentralization and control. Money, credit, and financial institutions provide the link between statics and equilibrium and dynamics and disequilibrium.[3]

General equilibrium deals precisely with equilibrium states. In spite of its elegance and abstraction, as was noted by Koopmans (1977), general equilibrium theory is preinstitutional. Because the economic world is highly complex and multivariate, radical simplification is called for in the mathematization of the models studied. When process models of general equilibrium are mathematically formulated, even the convergence to equilibrium from positions out of equilibrium in simple dynamic models may be difficult to establish. In contrast, the literature on innovation is always process oriented. There are several simulations of these processes, but the predominant approach to understanding innovation is via the essay, often bolstered with empirical studies analyzed statistically.

Although originally written over a 100 years ago, Schumpeter's work on *The Theory of Economic Development* (Schumpeter 1934) provided an insightful description (in essay form) of a plausible dynamic process involving the interaction of the financial and physical processes of the economy intermixed with the sociopsychological factors of optimism and pessimism. No formal mathematical model was developed. Many years later, Schumpeter (1939) produced two volumes on *Business Cycles* attempting to fit several centuries of innovation into Juglar, Kitchin, and Kondriateff cycles. These provide an encyclopedic tour of innovations but little new light on cycles.

In the last 20–30 years, there has been a surge in the writing on innovation, as is evinced in the works of Nelson and Winter (1982), Dosi et al. (1988), Nelson (1996), Lamoreaux and Sokoloff (2007), Baumol (2002), Bechtel et al. (1996) and many others. Beyond these works, an understanding of the analogy between economic innovation and biological mutation is growing.

---

[2]Prices will depend on details of initial conditions and asset structure as well as default and issue conditions.

[3]A work which is in considerable agreement in spirit but different in technique is that of Godely and Lavoie (2007) heavily devoted to a balance sheet and transaction flow model of the monetary and financial control system of a modern economy. This work utilizes simulations and is far closer to applied macroeconomic problems. It also stresses Kaldor's concern with the tendency of economic theorizing to gloss over the difficulties inherent in differentiating stocks from flows.

1.1.3 Types of Innovation

The study of innovation cannot be approached monolithically. There are at least four distinct types of innovation, namely:

- radically new product innovation;
- engineering variation of current product;
- distribution, network, information, and communication innovation;
- organization, cost reduction, or other process innovation influencing efficiency.

In terms of uncertainty, they are highly different. The most difficult to handle by conventional economic analysis are radical product and network innovations. Both the production procedures and the demand acceptance are unknown. The subjective probabilities for success, if any, may be cooked up by stretched analogy with other products and networks that have succeeded or failed and only can be quantified for the purpose of the construction of imaginary or pro forma financial statements used to persuade potential investors.

More or less standard product variation fits reasonably well into the current theory of oligopolistic competition. The large firms selling, say, refrigerators have products that are close to being identical. It is the job of marketing and the production engineers to have a spice shelf full of technically known modifications or additions that can help to differentiate the product. Costs and demand can be reasonably estimated for such innovations. Innovation can also fit into a modified model of a competitive market, as has been shown by Boldrin and Levine (2008). The cost innovation discussed here can be considered in competitive markets, especially when one takes into account that the appropriation by others of new ideas, industrial secrets, and expertise is by no means instantaneous.

By far, the most prevalent form of innovation in most modern economies is process innovation involving organization and frequently reducing costs of production by orders of magnitude. New inventions call for expensive prototypes. Even if the market for the new product is clearly present, over the first few years, especially with mass market possibilities, there is a considerable focus on unit cost reduction. The prototype is highly expensive, and the first batch for sale, though cheaper than the prototype, is usually produced at nowhere near the intended cost.

1.1.4 Property Rights, Information, and Appropriation

> Drive for show, but putt for dough.
> Old golf saying

The modeling and analysis of innovation are replete with difficulties. In much of the mythology of purely competitive markets, adjustments usually take place immediately. In fact, in a dynamic system, profits are made by innovators having the lead ahead of the myriads of time lags in the diffusion of information and expertise.

The time it takes for an industrial secret to leak and the delays and barriers caused by legal, accounting, and tax considerations are considerable.

Virtually everything is permeable at some point. Thus, patent protection must be looked at as a time delay device and other barriers to entry as delay devices. Law cases are often brought merely as time delay instruments.

In Crusoe's world, none of these details exist.

1.2 How to Finance Innovation

In a modern economy, there are many different ways in which innovation is financed. They depend on many empirical details concerning the nature of the money and credit, transactions costs, knowledge, liquidity, evaluation ability, attitude toward risk, laws, taxation, and other factors. In a complex economy such as that of the United States, many different specialists may be involved. They include inventors, their families and friends, entrepreneurs, venture capitalists, large and small firms, bankers, and the government.

Among the many ways to finance, we note five forms of financing. They are financing by:

- The owners with their own and family resources
- The owners utilizing a capitalist or an investment banker
- The firm utilizing retained earnings
- The firm using a capital market
- The firm borrowing from (and/or subsidized by) government

In current United States practice, much financing for cost innovation is either self-financed by the firm's management and/or owners or an arrangement between a firm and its financiers. Government may encourage innovation and may subsidize the firms rather than be a direct investor.

Crusoe is not bothered with these institutional details. For him, innovation involves physical goods and his ideas and ability, not finance or complex ownership and expertise conditions.

1.3 Models with Cost Innovation

Assume that the probability of the success of an improvement in the efficiency of production (which in a monetary economy can be interpreted as a cost reduction) and its size can be estimated reasonably well. To be specific, we suppose that from the initial production function f for Crusoe, a new improved production function, say g, is obtained with probability $\xi(k)$ after a successful innovation. Here the probability $\xi(k)$ of the improvement is an increasing function of the resources k invested in innovation. With probability $1 - \xi(k)$, the innovation fails and the production function is unchanged. (For a given investment, the improvement may be two-dimensional, there being a trade-off between the size of the improvement and its probability of success for a given investment. For simplicity, we consider the one-dimensional cross section where the improved function g is given and the function $\xi(k)$ is the probability of success.) We assume that $\xi(0) = 0$ so that an investment of zero corresponds to no attempt at innovation.

In our models, we assume that at the start of the game there is the opportunity for innovation. In essence, the first move is a strategic decision to take or reject a gamble to try to improve efficiency. The innovation is modeled as a random event whose value depends on the size of investment.

1.4   Robinson Crusoe in a Nonmonetary Economy

Consider the simple very well-known model in which a single agent produces a good for his personal consumption. Suppose the agent begins with $q \geq 0$ units of the good, puts i units into production, and consumes the remaining $x = q - i$ thereby receiving $u(q - i)$ in utility. The agent begins the next period with $f(i)$ units of the good, and the game continues. (Both the utility function u and the production function f are assumed to be concave, nondecreasing on $[0, \infty)$ with $f(0) = 0$.) The value of the game $V(q)$ to Robinson Crusoe is the supremum over all strategies of the payoff function

$$\sum_{n=1}^{\infty} \beta^{n-1} u(x_n),$$

where x_n is the amount of the good consumed in period n and $\beta \in (0, 1)$ is a discount factor. For this model without the possibility of innovation, the value function V satisfies the Bellman equation

$$V(q) = \sup_{0 \leq i \leq q} [u(q - i) + \beta V(f(i))].$$

Assume that there is a unique positive input i_1 such that $f'(i_1) = 1/\beta$. (This is certainly the case if, as is often assumed, f is strictly concave, $f'(0) = \infty$, and $\lim_{i \to \infty} f'(i) = 0$.)

Theorem 1 (Karatzas et al. (2006)). *If the initial value of the good is q_1, then an optimal strategy is to input i_1 in every period. Consequently,*

$$V(q_1) = \frac{1}{1 - \beta} \cdot u(q_1 - i_1).$$

Thus, the stationary equilibrium in Crusoe's economy has an amount q_1 of goods produced and an amount $x = q_1 - i_1$ consumed in every period.

1.4.1   Innovation by Robinson Crusoe

Assume now that our single agent with goods q is allowed to input i for production and invest j in innovation, where $0 \leq i \leq q$, $0 \leq j \leq q - i$. The agent consumes the remainder $q - i - j$. The innovation is successful with probability $\xi(j)$ resulting in an improved production function g, where $g(q) \geq f(q)$ for all q with strict inequality for some q. The innovation fails, and the production function is unchanged with probability $1 - \xi(j)$. Let V_1 be the value function for the game with production function f without innovation as in the previous section, and let V_2 be the value function for the game with the improved production function g. Then the value function V of the game with innovation satisfies

$$V(q) = \sup_{\substack{0 \leq i \leq q \\ 0 \leq j \leq q - i}} [u(q - i - j) + \beta\{\xi(j)V_2(f(i)) + (1 - \xi(j))V_1(f(i))\}].$$

Let $\psi(i, j)$ be the function of i and j occurring inside the supremum. For an interior optimum, we must have the Euler equations:

$$\frac{\partial \psi}{\partial i} = \frac{\partial \psi}{\partial j} = 0.$$

To find a solution to Crusoe's innovation problem, we must calculate the values of V_1 and V_2 where the quantity of goods is the amount $f(i)$ yet to be determined. Theorem 1 only gives an expression for the value at one equilibrium point, which is different for the two production functions f and g.

1.4.2   A Risk-Neutral Crusoe

If the agent is risk neutral, then, when there is no innovation, there is a simple description of the optimal strategy at every value of q.

Theorem 2. *Assume that $u(x) = x$ for all x. Then an optimal strategy is to input q if $q \leq i_1$ and to input i_1 if $q > i_1$. For $q \geq i_1$, the value of the game is*

$$V(q) = q - i_1 + \frac{\beta}{1 - \beta} \cdot (q_1 - i_1).$$

Proof. A player with goods $q > q' \geq 0$ can always consume $q - q'$ and then play from q'. Hence,

$$V(q) \geq q - q' + V(q').$$

Consider now $q \leq i_1$ and a strategy that inputs $i < q$. The best possible return from such a strategy is

$$q - i + \beta V(f(i)).$$

But an input of q gives a best return of

$$\beta V(f(q)) \geq \beta \cdot [f(q) - f(i) + V(f(i))]$$
$$\geq \beta \cdot [f'(q)(q - i) + V(f(i))]$$
$$\geq q - i + \beta \cdot V(f(i))$$

since $f'(q) \geq f'(i_1) = 1/\beta$. So it is optimal to input q when $q \leq i_1$.

Now suppose that $q > i_1$. Since $u' = 1$, the Euler equation reduces to $f'(i) = 1/\beta$ or $i = i_1$. The appropriate transversality condition is trivially satisfied since $q_n = q_1$ for all $n \geq 2$. It is easy to check that the strategy is interior and therefore optimal.

Consider next the innovation problem of the previous section for our risk-neutral agent with $u(x) = x$.

Assume that $f'(i_1) = 1/\beta = g'(i_2)$ where $0 < i_1 < i_2 < \infty$. Then by Theorem 2, $V_1'(q) = V_2'(q) = 1$ for $q \geq i_2$. Thus, if $f(i) \geq i_2$, we have

$$\frac{\partial \psi}{\partial i}(i, j) = -1 + \beta\{\xi(j)f'(i) + (1 - \xi(j))f'(i)\}$$
$$= -1 + \beta f'(i),$$

$$\frac{\partial \psi}{\partial j}(i, j) = -1 + \beta \xi'(j)\{V_2(f(i)) - V_1(f(i))\}.$$

Hence, in this case, the solutions to the Euler equations are

$$i^* = (f')^{-1}(1/\beta) = i_1 \quad \text{and} \quad j^* = (\xi')^{-1}(1/\beta[V_2(f(i^*)) - V_1(f(i^*))]).$$

To illustrate the solution, we calculate it below for a very simple example.

1.4.2.1 A Numerical Example

Assume that the initial production function is $f(i) = 2\sqrt{i}$ and $\theta = 0.1$ so that, after a successful innovation, the production function is $g(i) = 2.2\sqrt{i}$. Set $\beta = 0.95$. Solve

$$f'(i_1) = 1/\beta \quad \text{and} \quad g'(i_2) = 1/\beta$$

to get

$$i_1 = 0.9025, \qquad i_2 = 1.092$$

and

$$q_1 = f(i_1) = 1.9, \qquad q_2 = g(i_2) = 2.299.$$

For $q \geq i_2 > i_1$, it follows from Theorem 2 that

$$V_2(q) - V_1(q) = \frac{i_1 - i_2}{1 - \beta} + \frac{\beta}{1 - \beta}(q_2 - q_1) = 3.791.$$

Assume now that the probability of successful innovation from investing j is $\xi(j) = j/(1+j)$. As noted above, the first Euler equation has the solution $i^* = i_1 = 0.9025$ so that $f(i^*) = f(i_1) = q_1 = 1.9$. Since $1.9 > i_2 > i_1$,

$$V_2(f(i^*)) - V_1(f(i^*)) = 3.791$$

and the solution to the second Euler equation is $j^* = (\xi')^{-1}[1/(0.95)(3.791)] = 0.8977$. Thus, $\xi(j^*) = 0.8977/1.8977 = 0.473$ is the probability that the innovation is successful.

We can use the formula from Theorem 2 to calculate

$$V_2(f(i^*)) = V_2(1.9) = 23.741,$$

and

$$V_1(f(i^*)) = V_1(1.9) = 19.95.$$

These values together with the values for i^* and j^* can be substituted in the formula for the value of the game with innovation to get $V(q) = q + 18.86$ for $q \geq i^*$. The value of the game without innovation can also be calculated as $V_1(q) = q + 18.05$, which shows that it is slightly worth innovating in this instance.

1.4.3   A Risk-Averse Robinson Crusoe with Proportional Production

Many of the interesting features of investment call for the consideration of risk-averse individuals. In general, it is not possible to achieve the easy adjustment to a stationary state that exists with a risk-neutral Robinson Crusoe. However, analytic solutions are available when the utility function has constant elasticity and production is directly proportional to the input.

We take $u(x) = \log x$ and $f(i) = \alpha i$, where α is a positive constant. [The full class of constant elasticity utilities is considered in a nice article of Levhari and Srinivasan (1969).] Thus, the Bellman equation is

$$V(q) = \sup_{0 \leq i \leq q} [\log(q - i) + \beta V(\alpha i)].$$

The Euler equation for an interior solution $i = i(q)$ takes the form

$$\frac{1}{q - i(q)} = \frac{\beta \alpha}{\alpha i(q) - i(\alpha i(q))}.$$

The solution is $i(q) = \beta q$ and does not depend on α. Thus, the optimal plan is for Crusoe to input βq for production whenever he holds q units of the good. Under this plan, Crusoe's successive positions are

$$q_1 = q, \; q_2 = (\alpha\beta)q, \; \ldots, \; q_n = (\alpha\beta)^{n-1}q, \; \ldots,$$

and the optimal return is

$$V(q) = \sum_{n=1}^{\infty} \beta^{n-1} \log(q_n - \beta q_n) = \sum_{n=1}^{\infty} \beta^{n-1} \log((\alpha\beta)^{n-1}(1 - \beta)q).$$

Using properties of the log function and geometric series, we can rewrite the return as

$$V(q) = \frac{\log q}{1 - \beta} + \frac{\log(1 - \beta)}{1 - \beta} + \frac{\beta}{(1 - \beta)^2}[\log \alpha + \log \beta].$$

1.4.3.1 Innovation by a Risk-Averse Robinson Crusoe

Consider now the situation of an agent who begins with the utility $u(x) = \log x$ and production function $f(i) = \alpha i$ as in the previous section and contemplates the possibility of an innovation leading to an improved production function $g(i) = (1 + \theta)\alpha i$.

Let V_1 and V_2 be the original value function, and that after a successful innovation, then the value function $V_1(q)$ is given by the formula of the previous section and $V_2(q)$ is given by the same formula with the constant α multiplied by $1 + \theta$. Thus,

$$V_2(q) = V_1(q) + \frac{\beta}{(1 - \beta)^2} \log(1 + \theta),$$

and the final term above represents the value to Crusoe of having the improved production function. The value function V for the game with innovation can now be written as

$$V(q) = \sup_{\substack{0 \le i \le q \\ 0 \le j \le q-i}} [\log(q-i-j) + \beta\{\xi(j)V_2(\alpha i) + (1-\xi(j))V_1(\alpha i)\}]$$

$$= \sup_{\substack{0 \le i \le q \\ 0 \le j \le q-i}} \left[\log(q-i-j) + \beta\left\{V_1(\alpha i) + \xi(j)\frac{\beta}{(1-\beta)^2}\log(1+\theta)\right\}\right].$$

The Euler equations for an interior solution $i = i(q), j = j(q)$ can be obtained by letting $\psi(i,j)$ be the function inside the supremum and setting its two partial derivatives equal to zero. Here is the result:

$$\frac{1}{q-i-j} = \frac{\beta}{(1-\beta)}\frac{1}{i} = \frac{\beta^2}{(1-\beta)^2}\log(1+\theta)\xi'(j).$$

The first equation can be solved for i to get

$$i = \beta(q-j).$$

This expression for i can then be substituted back in to obtain

$$\xi'(j) = \frac{1-\beta}{\beta^2\log(1+\theta)}\cdot\frac{1}{q-j}.$$

This equation can be solved explicitly if, as in Sect. 1.4.2.1, $\xi(j) = j/(j+1)$. In this case, the equation above for j becomes a quadratic. Using $\beta = 0.95, \theta = 0.1$ as in Sect. 1.4.2.1 and setting $q = 2$, the positive root of this quadratic equation is $j^* = 0.57$, and, for this value, the chance of a successful innovation is $0.57/1.57 = 0.36$.

1.4.4   Innovation Over Many Periods

We close with observations on two more models where Crusoe may have repeated attempts at innovation until success. The first is a direct extension of the model solved above. The difference is that, after a failed attempt at innovation, Crusoe is free to try again if he has the resources to do so. Another extension of the basic model of individual innovation would permit multiple attempts at further innovation even after a success.

1.4.5 Lessons from Crusoe's Innovation

From a viewpoint of economics, the Crusoe models have been simple; but layering on the complexities starting with the first nonmonetary individualistic models points to the transition from a simple pure technology and preference-driven real goods control problem to a related but far greater control problem in an enterprise economy with money.

The specific observations from Crusoe's economy are:

1. The one-person growth model has been well known for many years along with its equilibrium properties. That the system will converge to an equilibrium if initial conditions are not in equilibrium is well known. However, the *duration of the transient length to equilibrium may be of any length*, and this duration is highly dependent on parametric details. Any evaluation of the success of innovation depends on this transient.
2. The only way innovation may take place is by Crusoe voluntarily giving up the use of his own physical assets. This contrasts with the forced savings scenarios available in a monetary economy.
3. The problem of the separation between ownership and control does not appear in Crusoe's world. It arises in a multiperson economy.
4. In spite of being amenable to dynamic programming or continuous-time methods, these mathematical techniques are limited in being able to compute solutions. Simulation and computational techniques are called for.

1.5 Finance and Innovation

The specific "value added" to the topics of innovation, guidance, control, and ownership attempted here is to bridge the conceptual and mathematical gaps between general equilibrium theory and Schumpeter's writings on innovation. In the past 20–30 years, there have been considerable writing and empirical work on innovation and the economic and behavioral questions that it raises. See, for example, Lamoreaux and Sokoloff (2007), Day et al. (1993), Nelson (1996), Nelson and Winter (1982), Shubik (2010), and in particular the essay of Day (1993). The work here is aimed at being complementary with these but aimed specifically at trying to characterize mathematically via a dynamic programming formulation of strategic market games, the monetary aspects of innovation eventually including ownership, financial control, and coordination features of a market economy.

1.5.1 Physical and Financial Assets, Innovation, and Equilibrium

We now examine some of the problems of the interaction between ownership and control for a small firm that acts as a pricetaker in a large monetary economy. We consider models with many independent agents whose actions determine prices.

There are two features of the investment decision in a monetary economy we deal with and contrast with their Crusoe counterparts.[4] They are:

1. Equilibrium in a closed monetary economy prior to the knowledge that innovation is feasible;
2. Innovation in a closed monetary economy with only short-term assets investigating the need for the expansion of money and credit.

The first topic has been dealt with previously in Karatzas et al. (2006) and Shubik and Sudderth (2011). As with Crusoe, some sufficiently tractable examples are provided that can be solved analytically.

A topic of interest for further research would be the study of the effect of repeated innovation on the distribution of firm size and investment.

1.6 The Closed Economy as a Sensing, Evaluating, and Control Mechanism

Prior to considering the formal closed models with innovation, several general items that supply context are covered.

1.6.1 Individual or Representative Agents?

When there is no uncertainty, models utilizing representative agents and models with independent agents solved for type-symmetric noncooperative equilibria give the same equilibrium results. When there is any exogenous uncertainty present, this is no longer generally true. With independent agents, uncertainty is not necessarily correlated. However, with a representative agent, uncertainty is implicitly correlated for all members of the class. In our models, agents can act independently, but the only randomness occurs at the first stage with the probability of a successful innovation. Kirman (1992) provides a perceptive discussion of the dangers in using

---

[4]We have also dealt elsewhere (Shubik and Sudderth (2011)) with equilibrium in an open monetary economy with innovation. An open economy model ignores detailed feedbacks to small individuals. It serves to study partial equilibrium possibilities.

representative agents. We agree with his observations and stress that the assumptions concerning the correlation of behavior among individuals are extremely strong and must only be utilized in an ad hoc manner with care and specific justification.

1.6.2 On Money, Credit, Banks, and Central Banks

In institutional fact, the definition and measurement of the money supply is difficult at best. The distinctions between money and credit are not always clear. Here we utilize a ruthless simplification in order to highlight the distinction between money and credit and to be able to stress economic control. Consider money to be paper gold or some form of blue chip in which payments are made. Credit is a contract between two entities A and B, in which individual A delivers money at period n_1 in return for an IOU or a promise from B to repay an amount of money to A at period n_2. An individual may be a natural person or a legal person such as a firm, a bill broker, a bank, a credit-granting clearing house, or a central bank.

We may consider two ways to vary the money supply. The first and simpler is that the central bank is permitted to print it. Another way to vary the money supply is to accept the IOU notes of commercial banks as money. Say they are red chips, in contrast with the central bank's blue chips. They are accepted in payment on a 1:1 basis with blue chips. A reserve ratio controls the amount a bank can issue; thus, for any k units of red chips issued, a bank must hold one unit of blue chips.[5]

As we wish to maintain as high a level of simplification as possible in order to illustrate the breaking of the circular flow, we select the simpler structure. The banking system is considered as one and called the central bank. It has funds above its reserves[6] that it can lend and it can pay interest on deposits.[7]

1.7 The Separation of Management and Ownership

The next level of complexity above the single type of agent utilizes two types of agents: managers of the firms and stockholder-owners. (In the first model below, there is also a class of *saver agents* who subsist on the returns from their bank deposits.) The economy can be interpreted as a fully defined game of strategy where

---

[5]The justification for the acceptance of reserve ratio banking is in the dynamics along with acceptance of fiat [see, for example, Bak et al. (1999)].

[6]Central bank reserves in a fiat money economy are a creation of law and possibly economic theology. Mathematically, they are just societal rules of the game or an algorithm stating how the central bank can create money. They specify its strategy set. In actuality, the strategy set is also bounded by political pressures.

[7]In general, central banks do not accept deposits from natural persons, but for modeling simplicity here we permit them to do so.

there is a finite measure of firms and of stockholder-owners whose overall actions will influence prices. By assuming that we limit the solution to a type-symmetric noncooperative equilibrium, all agents of each type, even though independent, will employ a strategy common to their type.

1.8   A Closed Economy Prior to Innovation: The Circular Flow of Money Illustrated

The model presented in this section is based on the work of Karatzas et al. (2006) without innovation. It will be extended in the next section to a model with innovation in order to consider the disequilibrium aspects of innovation on the money supply. Our stress so far has been on nonmonetary models of Crusoe as an innovator. From here on, the emphasis is on simple closed economies or macroeconomic models.

The underlying model is that of a "cash-in-advance"[8] market economy with a continuum of firms $\phi \in J = [0, 1]$ that produce goods, all of which must be put up for sale and a continuum of stockholder agents $\alpha \in I = [0, 1]$ who own the firms and purchase these goods for consumption. The agents hold cash and bid for goods in each of a countable number of periods $n = 1, 2, \ldots$. The firms hold no cash[9] and must borrow from a single outside bank to purchase goods as input for production in every period. The bank is modeled as a strategic dummy that accepts deposits and offers loans at a fixed interest rate ρ. In addition to the owner agents, there may be a continuum of *saver agents* $\gamma \in K = [0, 1]$, each of whom holds cash, bids in every period to buy goods for consumption, and subsists entirely on his/her savings. These agents can be thought of as retirees or private capitalists.[10]

The firms are in general corporate, they do not own themselves. They have (at some ultimate level) natural person stockholders who are also consumers. Directly or indirectly, they depend on at least four sets of decisionmakers for debt (and some equity or options) financing. They are the passive savers, the financiers, the commercial banks, and the central bank. Without having to elaborate further, it should be evident that in any dynamic setting, the coordination problem is considerable. In the mathematical model below, we grossly simplify the financial

---

[8]The term "cash-in-advance" is misleading when combined with a finite grid size where no attention is given to how long the time interval is meant to be. The key item of importance is the recognition that individuals **form prices.** They are only given prior prices, and how these are to be utilized is a matter of behavioral specification.

[9]This reflects the payment of the 100 % dividend, the timing of which is irrelevant in a perfect credit rating competitive economy.

[10]In a less Draconian abstraction, the difference between retirees and capitalists is not merely age, but expertise. The role of competent financing as a perception and evaluating device cannot be overstressed.

sector, ignoring the financiers, collapsing the commercial banks and central bank into one, and having the passive savers save in the aggregate bank, while the firms borrow only from this bank.

The firms in this first closed model have no opportunity to innovate and carry no long-term debt. Each firm ϕ begins every period n with goods q_n^ϕ that are to be sold in the market. The total amount of goods offered for sale is defined by

$$Q_n = \int q_n^\phi \, d\phi. \tag{1.1}$$

Each firm ϕ also borrows cash b_n^ϕ from a central bank, with $0 \leq b_n^\phi \leq (p_n q_n^\phi)/(1+\rho)$, where p_n is the price of the good in period n and $\rho > 0$ is the interest rate. There is no demand function in this model, and the prices are formed endogenously as will be explained below.

The firm spends the cash b_n^ϕ to purchase the amount of goods $i_n^\phi = b_n^\phi/p_n$ as input for production and begins the next period with an amount of goods

$$q_{n+1}^\phi = f(i_n^\phi).$$

Here $f(\cdot)$ is a production function, which satisfies the usual assumptions. During period n, each firm ϕ earns the (net) profit

$$\pi_n^\phi = p_n q_n^\phi - (1+\rho)b_n^\phi,$$

since it must pay back its (short-term) loan with interest. The goal of the firm is to maximize its total discounted profits[11]

$$\sum_{n=1}^{\infty} \left(\frac{1}{1+\rho}\right)^{n-1} \cdot \pi_n^\phi.$$

In a given period n, the total profits generated by all the firms are

$$\Pi_n = \int \pi_n^\phi \, d\phi.$$

The profits Π_n are distributed to the owner agents in equal shares at the end of the period.

The owner agents are now considered. A typical owner agent α holds money m_n^α at the beginning of each period n. The agent bids an amount of money a_n^α with $0 \leq a_n^\alpha \leq m_n^\alpha + \Pi_n/(1+\rho)$, which buys him an amount $x_n^\alpha = a_n^\alpha/p_n$ of goods

---

[11]In institutional fact, a large firm has a considerable constituency of customers, employees, the government, and others, as well as the owners.

for immediate consumption. Any extra money an owner agent has is deposited and earns interest at rate ρ. The agent begins the next period with cash

$$m_{n+1}^{\alpha} = (1 + \rho)\left(m_n^{\alpha} - a_n^{\alpha}\right) + \Pi_n.$$

Each agent α seeks to maximize his total discounted utility

$$\sum_{n=1}^{\infty} \beta^{n-1} u(x_n^{\alpha}),$$

where u is a concave increasing utility function and $0 < \beta < 1$ is a given discount factor.

Also considered is a typical saver agent γ, who holds m_n^{γ} in cash at the start of period n. The saver bids an amount c_n^{γ} of cash with $0 \le c_n^{\gamma} \le m_n^{\gamma}$, which buys him a quantity $y_n^{\gamma} = c_n^{\gamma}/p_n$ of goods, and starts the next period with

$$m_{n+1}^{\gamma} = (1 + \rho)\left(m_n^{\gamma} - c_n^{\gamma}\right)$$

in cash. If $v(\cdot)$ is his utility function, with the same properties as $u(\cdot)$, the saver agent's objective is to maximize the total discounted utility

$$\sum_{n=1}^{\infty} \beta^{n-1} v(y_n^{\gamma}).$$

The total amounts of money bid in period n by the owner agents, the firms, and the saver agents are

$$A_n = \int a_n^{\alpha} \, d\alpha, \qquad B_n = \int b_n^{\phi} \, d\phi \qquad \text{and} \qquad \Gamma_n = \int c_n^{\gamma} \, d\gamma,$$

respectively. The price p_n is formed as the total bid over the total production

$$p_n = \frac{A_n + B_n + \Gamma_n}{Q_n}.$$

An equilibrium is constructed as follows. Suppose that all owner agents begin with cash $M_1^A = m^A > 0$, all saver agents begin with cash $M_1^{\Gamma} = m^{\Gamma} \ge 0$, and all firms begin with goods $Q_1 = q > 0$. Thus, the total amount of cash $M_1 = M_1^A + M_1^{\Gamma}$ across agents is equal to

$$m = m^A + m^{\Gamma},$$

and the proportion of money held by the saver agents is

$$\nu = \frac{m^{\Gamma}}{m} = \frac{m^{\Gamma}}{m^A + m^{\Gamma}}, \qquad \text{with} \qquad 0 \le \nu < 1.$$

Suppose that the bids of the agents and firms are

$$a_1 = am, \qquad b_1 = bm, \qquad c_1 = cm,$$

that is, proportional to the total amount of cash, so that the price is also proportional to this amount:

$$p_1 = p(m) = \frac{(a+b+c)m}{q}.$$

Then the profit of each firm is

$$\Pi_1 = p_1 q - (1+\rho)b_1 = (a+c-\rho b)m,$$

the cash of each owner agent at the beginning of the next period is

$$M_2^A = (1+\rho)\left(m^A - am\right) + \Pi_1,$$

and the cash held by each saver agent is

$$M_2^\Gamma = (1+\rho)\left(m^\Gamma - cm\right).$$

Thus, the total amount of cash held by all agents at the beginning of the next period is

$$M_2 = M_2^A + M_2^\Gamma = (1+\rho - \rho(a+b+c))m = \tau m,$$

where we have set

$$\tau = 1 + \rho - \rho(a+b+c).$$

Define

$$r = \frac{(1+\rho)(1-\beta)}{\rho}. \qquad (1.2)$$

The following theorem was established in Karatzas et al. (2006).

Theorem 3. *Suppose that there exists i^* with $f'(i^*) = (1+\rho)/\beta$. Then there is an equilibrium for which, in every period: each firm inputs i^*, produces $q^* = f(i^*)$, and bids the amount $b_n = b^* M_n$; each owner agent bids $a_n = a^* M_n$; and each saver agent bids $c_n = c^* M_n$. Here*

$$a^* + b^* + c^* = r, \qquad b^* = \frac{r}{q^*} \cdot i^*, \qquad c^* = (1-\beta)v \qquad (1.3)$$

and $M_n = M_n^A + M_n^\Gamma$ is the amount of cash held across agents in period n.

Furthermore, in each period n every owner agent consumes the amount $x^* = (1 - \frac{\rho v}{1+\rho})q^* - i^*$, *every saver agent consumes the amount* $y^* = (\frac{\rho v}{1+\rho})q^*$, *whereas every firm makes* $\pi^* M_n$ *in profits, with* $\pi^* = r - (1 + \rho)b^*$.

It is shown in Karatzas et al. (2006) that, in the equilibrium of Theorem 3, the consumption and total discounted utility of the owner agents are decreasing functions of ρ. Such agents prefer as *low* an interest rate as possible. Similarly, the firms also prefer an interest rate as close to zero as possible, in order to maximize their profits. But the situation of the saver agents is subtler: under certain configurations of the various parameters of the model (discount factor, production function, utility function), they prefer as *high* an interest rate as possible, whereas under other configurations, they settle on an interest rate $\rho^* \in (0, \infty)$ that uniquely maximizes their welfare.

We note that the presence of bank deposits at a positive rate of interest enables the creation of a group of individuals who live off the earnings of their money. Thus, even in this simple model, a conflict arises over setting the interest rate with the firms and entrepreneurs pressing the central bank for a lower rate and the pensioners for a higher rate.

Let

$$\tau^* = 1 + \rho - \rho(a^* + b^* + c^*).$$

Then money and prices inflate (or deflate) at rate τ^* in the equilibrium of Theorem 3. We also have $a^* + b^* + c^* = r$, so that the Fisher equation $\tau^* = \beta(1 + \rho)$ holds.

Remark 1. By setting $v = 0$ in Theorem 3, we obtain an economy with only producer firms and owner-consumer agents.[12] We similarly dispense with saver agents in the models below. This will be useful in illustrating the basic problems with the circular flow and money supply with innovation in a simple context. Also, we take $\tau = \beta(1 + \rho) = 1$ so that there is no inflation.

1.9 Innovation in an Asset-Poor Economy: Breaking the Circular Flow

As in the previous models, we aggregate all goods in the model of this section into a single perishable consumable that is utilized in consumption or production or consumed in innovation. There is no capital stock, such as steel mills. There is no "fat" in the economy; resources for innovation must come directly out of consumption resources.

---

[12]Of course, the proportion v has to be strictly less than one; for otherwise, there is no one to engage in productive activity, own the firms, or receive their profits, and the model unravels.

1.9.1   The Meaning of an Asset-Poor Economy

In actuality, a modern economy is rich in real durable assets with a time profile of durables of many ages that are consumed only in production, not consumption. Gross domestic product may be split into consumption and investment. If we consider around 70 % in consumption, then we note that at market prices, the value of real assets such as steel mills, automobile factories, houses, automobiles, machinery, land, and other consumer durables are priced probably between 5 and 10 times the value of consumption.[13] None of these items are meaningfully placed directly in the utility functions of the individuals. Furthermore, it is the services of consumer durables that are ultimately valued and not the durables themselves. This is even truer of items such as steel mills. In the models considered so far, we have not indicated that the presence of this large mass of assets owned by individuals may be such that the loss or exchange of a small percentage of these assets while pursuing innovation will hardly change the consumption of the owners of large amounts of real assets.

In a poor country, the amount of available assets relative to consumption will be much smaller than in a rich one. We consider in this section the extreme simplifying case where innovation must come directly out of consumption. This makes it easier to be specific about the breaking of the circular flow of capital and the match between real assets and money.

In essence, innovation is nothing other than the execution of an idea for a new process to rearrange and employ existing assets in a different manner.[14] It is a breaking of equilibrium that in a rich country calls for an alternative use for productive assets but does not directly cut down heavily on current consumption. In contrast, in an asset-poor economy, an immediate sacrifice in consumption is called for.

1.9.2   Innovation in an Asset-Poor Economy

We consider a model with a class of identical manufacturers; a class of identical, individual consumers, who also own the firms; and an outside or central bank.

There are several possible models that depend on who is in control of the firm and who finances the innovation. Here we assume the managers are in control, the owners are passive, and the central bank is willing to create new money to make investment loans. Many variants are found in a modern economy; however,

---

[13]These are crude approximations based on the Statistical Abstract of the United States for GNP, amount and age of capital, and Cobb–Douglass production.

[14]Bankruptcy in a basic way is similar to innovation in the sense that it involves a nonequilibrium redeployment of assets.

the model selected serves adequately to illustrate the problems with financing and innovation and decrease in purchasing power of the owner-consumers brought about by the creation of new credit.

1.9.2.1 A Model with Managerial Control and Central Bank Lending

As in the model of Sect. 1.8, there is a continuum of firms $\phi \in J = [0, 1]$. Each firm ϕ begins each period n with goods in process q_n^ϕ to be sold in the market and borrows cash b_n^ϕ from the central bank to purchase goods $i_n^\phi = b_n^\phi / p_n$ as input for production. Each firm ϕ begins in period 1 with no long-term debt, but may borrow an amount of money c^ϕ from the bank to purchase goods $j^\phi = c^\phi / p_1$ to be used in innovation. The interest on this long-term debt must be paid in every period, and the short-term loan b_n^ϕ must be paid back with interest at the end of each period n. In general, the long-term rate ρ^* might differ from the short-term rate, but it is sufficient and simpler to assume that they are equal to a common value $\rho > 0$. In order that a firm be able to meet its debt obligations, the bid b_n^ϕ is restricted to lie in the interval $[0, (\hat{p}_n q_n^\phi - c^\phi \rho)/(1 + \rho)]$, where $\hat{p}_n$ is the bank's estimate of the price p_n in period n. (In a rational expectations equilibrium, $\hat{p}_n = p_n$.) The bank may also impose an upper limit E on the long-term loan c^ϕ.

As in the model of Sect. 1.5.1, all firms begin in period 1 with the same production function f_1, and thus, a firm ϕ will begin period 2 with goods $q_2^\phi = f_1(i_1^\phi)$. However, a successful innovation results in the improved production function f_2. Thus, in periods after the first, there are two types of firms—those of type 1 that failed in the attempt at innovation and continue with production function f_1 and the type 2 firms that succeeded and have f_2.

The (net) profit π_n^ϕ of a firm ϕ in period n is the income from its sales in the period minus its interest payments:

$$\pi_n^\phi = p_n q_n^\phi - (1 + \rho)b_n^\phi - \rho c^\phi.$$

Each firm ϕ seeks to maximize its total discounted profits:

$$\sum_{n=1}^{\infty} \left(\frac{1}{1+\rho}\right)^{n-1} \cdot \pi_n^\phi.$$

The total profit in period n of all the firms is the integral

$$\Pi_n = \int \pi_n^\phi \, d\phi$$

and is paid to the consumer-owners in equal shares at the end of the period, as is explained below.

Because we again look for a type-symmetric equilibrium, we assume that all firms begin period 1 with the same quantity $q_1 > 0$ of goods, and we often omit the superscript ϕ below. When all firms begin in the same state, make the same bids b_1 and c, and earn the same profit $\pi_1 = p_1 q_1 - (1 + \rho)b_1 - \rho c$, the total profit and total goods in period 1 simplify to

$$\Pi_1 = \int \pi_1 \, d\phi = \pi_1, \qquad Q_1 = \int q_1 \, d\phi = q_1.$$

Suppose $W(q_1)$ is the overall value of the game to a firm. Let $W_1(q_2, c)$ be the value to a firm beginning period 2 with goods q_2 after a failed investment of c, and let $W_2(q_2, c)$ be the corresponding value after a successful investment. Let $\xi(c/p_1) = \xi(j)$ be the probability of success when $c/p_1 = j$ is invested in innovation. Then the value functions satisfy the following optimality equations:

$$W(q) = \sup_{\substack{0 \leq b \leq \frac{\hat{p}q - \rho c}{1+\rho} \\ 0 \leq c \leq E}} \left[pq - (1 + \rho)b - \rho c + \frac{1}{1+\rho} \cdot \left\{ \left(1 - \xi\left(\frac{c}{p}\right)\right) \right. \right.$$

$$\left. \left. \cdot W_1\left(f_1\left(\frac{b}{p}\right), c\right) + \xi\left(\frac{c}{p}\right) \cdot W_2\left(f_1\left(\frac{b}{p}\right), c\right)\right\} \right], \tag{1.4}$$

where

$$W_k(q, c) = \sup_{0 \leq b \leq \frac{\hat{p}q - \rho c}{(1+\rho)}} \left[pq - b(1 + \rho) - \rho c + \frac{1}{1+\rho} \cdot W_k(f_k(b/p), c) \right] \tag{1.5}$$

for $k = 1, 2$.

For simplicity, we have suppressed super- and subscripts above and will often do so below as well. In both (1.4) and (1.5), the notation $\hat{p}$ is for the bank's estimate of the price for goods in the period, whereas p denotes the price actually formed as will be explained below.

In every period $n \geq 2$ after the first, there will be two types of firms, those called type 1 which have failed in the attempt at innovation and must continue with the production function f_1 and those called type 2 which have succeeded and henceforth have the improved production function f_2. There will be a fraction $\epsilon = \xi(c/p_1)$ of firms of type 2 and $\bar{\epsilon} = 1 - \epsilon = 1 - \xi(c/p_1)$ of type 1 in all periods after the first.

In seeking a type-symmetric solution, we will assume that at the beginning of periods $n \geq 2$, all firms of type 1 (respectively type 2) will hold the same quantity of goods q_n^1 (respectively q_n^2) and earn the same profit π_n^1 (respectively, π_n^2) in the period. Thus, the total profit and totals goods in period n are given by

$$\Pi_n = \bar{\epsilon}\pi_n^1 + \epsilon\pi_n^2, \qquad Q_n = \bar{\epsilon}q_n^1 + \epsilon q_n^2.$$

In addition to the firms, there is also a continuum of consumer-stockholder agents $\alpha \in I = [0, 1]$. Each agent α begins every period n with cash m_n^α and bids $a_n^\alpha \in [0, m_n^\alpha]$ to purchase goods a_n^α / p_n for immediate consumption. The agent deposits the excess cash $m_n^\alpha - a_n^\alpha$ in the bank and gets back $(1 + \rho)(m_n^\alpha - a_n^\alpha)$ at the end of the period.

The accounting profit D_n of the bank in period n consists of its earnings from the loans made to the firms less the interest paid on the deposits of the owners. Thus,

$$D_n = \rho \cdot \left[\int b_n^\phi \, d\phi + c - \int (m_n^\alpha - a_n^\alpha) \, d\alpha \right]. \tag{1.6}$$

For this model, we assume that the profit of the bank, like that of the firms, is paid to the owners in equal shares at the end of the period. (This assumption and a possible alternative are discussed in Sect. 1.9.2.3.) Thus, an owner agent α begins period $n + 1$ with cash

$$m_{n+1}^\alpha = (1 + \rho)(m_n^\alpha - a_n^\alpha) + \Pi_n + D_n. \tag{1.7}$$

The value function V for an owner satisfies

$$V(m) = \sup_{0 \leq a \leq m} \left[u\left(\frac{a}{p}\right) + \beta V((1 + \rho)(m - a) + D + \Pi) \right], \tag{1.8}$$

where u is a concave, nondecreasing utility function and we have again suppressed super- and subscripts.

The price p_n in each period n is formed as the ratio of the total cash bid in the goods market to the total amount of goods for sale. In the type-symmetric case, the prices are given by

$$p_1 = \frac{a_1 + b_1 + c}{q_1}, \qquad p_n = \frac{a_n + \bar{\epsilon} b_n^1 + \epsilon b_n^2}{\bar{\epsilon} q_n^1 + \epsilon q_n^2}, \quad n \geq 2.$$

If $m_1 = m$, then by (1.7)

$$m_2 = (1 + \rho)(m - a_1) + \Pi_1 + D_1$$
$$= (1 + \rho)(m - a_1) + p_1 q_1 - (1 + \rho)b_1 - \rho c + \rho \cdot [b_1 + c - (m - a_1)].$$

Now $p_1 q_1 = a_1 + b_1 + c$. Substitute this into the previous equation, and simplify the result to see that $m_2 = m + c$. A similar calculation shows that $m_n = m + c$ for all $n \geq 2$. Thus, in this model, the money supply has an initial increase because of the long-term loan in the first period and then remains constant.

1.9.2.2 Stationary Equilibrium and the Question of Convergence

A *stationary equilibrium* for the economy of the previous section is an equilibrium in which bids, prices, and the quantities of goods and money remain constant. The economy experiences a shock due to innovation in the first period after which there is always a fixed fraction $\bar{\epsilon}$ of type 1 firms and ϵ of type 2 firms. We cannot expect to have a stationary equilibrium until sometime after the first period. Under some additional regularity assumptions, there does exist a type-symmetric stationary equilibrium for the economy as it is configured after the initial shock.

Assume now that the production functions f_1, f_2 and the utility function u are strictly concave and continuously differentiable and that the production functions satisfy the condition:

$$f_k(0) = 0, \ f_k'(0) = \infty, \ \lim_{x \to \infty} f_k'(x) = 0, \ k = 1, 2.$$

Suppose as above that there is a fraction $\bar{\epsilon}$ of type 1 firms having production function f_1 and holding goods q_1, a fraction ϵ of type 2 firms having production function f_2 and holding goods q_2, and a continuum of consumer-owner agents $\alpha \in [0, 1]$ each with cash m.

Consider the Bellman equation (1.5), and, for $k = 1, 2$, let

$$\psi_k(b_k) = pq_k - (1 + \rho)b_k - \rho c + \frac{1}{1 + \rho} W_k(f_k(b_k/p), \tilde{p}).$$

Recall that $\tilde{q}_k = f_k(b_k/p)$, so $\psi_k(b_k)$ is the expression inside the supremum in (1.5). Standard arguments show that

$$\frac{\partial W_k}{\partial q_k}(q_k, p) = p.$$

Consequently, the Euler equations take the form

$$\psi_k'(b_k) = -(1 + \rho) + \frac{1}{1 + \rho} \cdot \frac{1}{p} \cdot f_k'(b_k/p) \cdot \tilde{p} = 0$$

This holds if and only if

$$f_k'(b_k/p) = (1 + \rho)^2 \cdot \frac{p}{\tilde{p}}. \tag{1.9}$$

In stationary equilibrium, there will be a fixed price p for goods so that $p = \tilde{p}$ and

$$f_k'(b_k/p) = (1 + \rho)^2, \ k = 1, 2.$$

The input of every type k firms is in every period $i_k^* = (f_k')^{-1}((1 + \rho)^2)$ with output $q_k^* = f_k(i_k^*)$.

The Euler equation for a consumer-owner takes the form

$$\frac{1}{p}u'\left(\frac{a}{p}\right) = \frac{\beta(1+\rho)}{\tilde{p}}u'\left(\frac{\tilde{a}}{\tilde{p}}\right) = \frac{1}{\tilde{p}}u'\left(\frac{\tilde{a}}{\tilde{p}}\right),\tag{1.10}$$

where $\beta(1+\rho) = 1$ by assumption and $\tilde{a}$ and $\tilde{p}$ are the agent's bid and the price in the next period. But in stationary equilibrium, $a = \tilde{a}$ and $p = \tilde{p}$. So the only condition on the optimal bid a^* is that $0 \le a^* \le m$.

Let $Q^* = \bar{\epsilon}q_1^* + \epsilon q_2^*$ be the total production when firms of type k input i_k^* for $k = 1, 2$. Now in order to purchase i_k^*, firms of type k must bid $b_k^* = pi_k^*$. Thus, the price must satisfy

$$p = \frac{a^* + \bar{\epsilon}b_1^* + \epsilon b_2^*}{\bar{\epsilon}q_1^* + \epsilon q_2^*} = \frac{a^* + \bar{\epsilon}pi_1^* + \epsilon pi_2^*}{\bar{\epsilon}q_1^* + \epsilon q_2^*},$$

or equivalently

$$\frac{a^*}{p} = \bar{\epsilon}(q_1^* - i_1^*) + \epsilon(q_2^* - i_2^*),$$

which means that the owner agents consume all the goods produced by the firms that are not used by the firms as input for production of goods for the next period.

Let $p = m/Q^*$ so that

$$b_k^* = pi_k^* = \frac{m}{Q^*} \cdot i_k^*, \quad k = 1, 2$$

and

$$a^* = \frac{m}{Q^*} \cdot [\bar{\epsilon}(q_1^* - i_1^*) + \epsilon(q_2^* - i_2^*)] < m.$$

Observe also that, for $k = 1, 2$,

$$q_k^* = f_k(i_k^*) = \int_0^{i_k^*} f_k'(x)\, dx \ge f_k'(i_k^*) \cdot i_k^* = (1+\rho)^2 \cdot i_k^* > (1+\rho) \cdot i_k^*.$$

Thus, the quantities $q_k^* - (1+\rho)i_k^*$, $k = 1, 2$ are strictly positive. Now the conditions on the bids b_k^* that

$$b_k^* \le \frac{pq_k^* - \rho c}{1+\rho}$$

can be rewritten as

$$\rho c \le pq_k^* - (1+\rho)b_k^* = \frac{m}{Q^*} \cdot (q_k^* - (1+\rho)i_k^*).$$

By assumption, the long-term debt c cannot exceed the bound E. Thus, the inequality above will hold if

$$\frac{E}{m} \leq \frac{q_k^* - (1+\rho)i_k^*}{\rho Q^*}.$$

Theorem 4. *If the ratio E/m is sufficiently small, then there is a stationary equilibrium such that, in every period, each firm of type k inputs i_k^*, produces $q_k^* = f_k(i_k^*)$, and bids $b_k^* = \frac{m}{Q^*} \cdot i_k^*$; each owner-consumer agent bids $a^* = \frac{m}{Q^*}[\bar{\epsilon}(q_1^* - i_1^*) + \epsilon(q_2^* - i_2^*)]$. Furthermore, in every period, every owner-consumer agent consumes the amount of goods $\bar{\epsilon}(q_1^* - i_1^*) + \epsilon(q_2^* - i_2^*)$, and every firm of type k makes the profit $\pi_k^* = \frac{m}{Q^*} \cdot (q_k^* - (1+\rho)i_k^*) - \rho c$.*

Proof. The bids a^* and $b_k^*, k = 1,2$ satisfy their Euler equations, and the appropriate transversality condition is trivial because, by stationarity, the payoffs are the same in every period.

We suspect that there is a theorem showing that, possibly under some additional conditions, there is convergence to stationary equilibrium for this simple model. However, even if this is true, convergence may be slow, and a general analytic solution to the model with innovation seems unlikely. Some simple examples for which convergence is fast are in Sect. 1.9.3.

1.9.2.3   The Modeling of Central Bank Profits

In the model of Sect. 1.9.2.1, it is assumed that the amount ρc of long-term interest is part of the accounting profit D_n [defined in (1.6)] of the central bank and is paid in each period to the consumer-owner agents [see (1.7)]. This is one of several fairly natural models each with different financial, economic, and political implications. One possibility is to neutralize the money as it comes in, leaving a deflationary trend in place. Other alternatives are for the bank to subsidize some groups of agents with this income or spend it to buy resources (such as foreign aid subsidies for purchases in the economy, or the destruction of government purchases of resources for a foreign war). As many institutional variants can be defined, the choice among them depends on the questions to be answered and their empirical relevance.

In order to define the minimal viable model, we have collapsed five banking functions into a single institution. They are:

1. Financing circulating capital or goods in process,
2. Accepting consumer savings,
3. Making short-term consumer loans,
4. Making long-term investment banking loans,
5. Varying the money supply.

Here we have chosen a model with only three types of agents: the firms, the consumer-owners, and a banking system. This seems to be the minimal number necessary to build a playable game that illustrates the phenomenon of breaking the circular flow of capital.

1.9.3 Two Simple Examples

In this section, equilibria are calculated for two very simple examples. In both examples, the production functions f_1 and f_2 are defined as follows:

$$f_1(i) = \begin{cases} 2i, & 0 \leq i \leq 1, \\ 2, & 1 < i, \end{cases}$$

and

$$f_2(i) = \begin{cases} 4i, & 0 \leq i \leq 1/2, \\ 2, & 1/2 < i. \end{cases}$$

Note that the maximum production level is 2 for both production functions, but that f_2 is more efficient and attains the maximum with an input of 1/2, whereas f_1 requires an input of 1. For both examples, we assume that $\rho = \rho^* = 0.05$ and take $\beta = 1/1.05$.

The first example treats a consumer-producer who labors in isolation to produce goods for his personal consumption and has the opportunity to innovate. The second example contrasts the first with the situation in a monetary economy with many firms and owner-consumers.

1.9.3.1 Robinson Crusoe Revisited

Consider first the situation of Robinson Crusoe equipped with the production function f_1 and without the opportunity to innovate. Suppose that Crusoe begins with a quantity of goods $q > 0$, selects an amount i, $0 \leq i \leq q$ to put into production, and consumes the remaining $q - i$ resulting in a utility of $u(q - i)$. He then begins the next period with goods $\tilde{q} = f_1(i)$ and continues the game.

Let $V_1(q)$ be the value of this one-person game to Crusoe. As in Sect. 1.4, V_1 satisfies the Bellman equation:

$$V_1(q) = \sup_{0 \leq i \leq q} [u(q - i) + \beta V_1(f_1(i))].$$

For simplicity, we assume that Crusoe is risk neutral with utility function $u(q) = q$. It is not then difficult to check that a stationary equilibrium has $q = 2$ and $i = 1$ at every stage of the game. Thus,

$$V_1(2) = \sum_{n=1}^{\infty} \beta^{n-1} u(1) = \frac{u(1)}{1 - \beta} = \frac{1}{1 - 1/1.05} = 21.$$

Similarly, if Crusoe begins with the production function f_2, a stationary equilibrium has $q = 2$ and $i = 1/2$ with value

$$V_2(2) = \sum_{n=1}^{\infty} \beta^{n-1} u(2 - 1/2) = \frac{3/2}{1 - 1/1.05} = 31.5.$$

Next, assume that Crusoe begins with $q = 2$ and the production function f_1, but has the opportunity to invest a portion of his goods in an attempt at innovation. Suppose further that the opportunity to innovate can be represented by a binary lottery ticket that can be obtained by utilizing $j = 1/2$ units of input material. The ticket is such that with probability $1/2$ the innovation succeeds and Crusoe has the production function f_2 thereafter, but also with probability $1/2$ it fails and Crusoe must continue with f_1. Let $V = V(2)$ be the value of this new game.

Now Crusoe can reject the investment opportunity and continue with his original production function f_1 thereby earning $V_1(2) = 21$ or make the investment and receive in expectation

$$\sup_{0 \le i \le 1.5} \left[u(1.5 - i) + \beta \left\{ \frac{1}{2} V_1(f_1(i)) + \frac{1}{2} V_2(f_1(i)) \right\} \right].$$

The optimal choice for the input is again $i = 1$, and the quantity above equals

$$u(1/2) + \frac{1}{1.05} \left\{ \frac{1}{2} V_1(2) + \frac{1}{2} V_2(2) \right\} = \frac{1}{2} + \frac{1}{2.1} \{21 + 31.5\} = 25.5.$$

Since $25.5 > 21$, it pays the nonmonetary Crusoe to innovate. A smaller value for the discount factor β, say $\beta = 0.8$, would go against innovation.

1.9.3.2 A Simple Monetary Economy

The following is an example of the model with many firms and consumer-owners that was presented abstractly in Sect. 1.9.2. The resource base per capita is the same as in the previous example, but consumers now find themselves in an economy that uses fiat money.

Let $m = 1$ be the amount of money held initially by the consumers, and suppose that the firms begin with goods $q = 2$ and the production function f_1. Assume first that the firms do not attempt to innovate. The optimal input for the firms is 1 unit of goods. Thus, if the price of goods is p, the firms borrow and then bid $b = p$ thereby obtaining $i = b/p = 1$ as input in order to produce $\tilde{q} = f_1(1) = 2$ for the next period. The (short-term) loan to the firms is financed by the deposit of $m - a = b$ of the owner-consumers. So the owners bid $a = m - b = m - p$ and

$$p = \frac{a+b}{q} = \frac{m-p+p}{q} = \frac{m}{q} = \frac{1}{2}.$$

The economy is in stationary equilibrium, and each period the firms earn the profit

$$\pi = pq - (1+\rho)b = \frac{1}{2} \cdot 2 - 1.05 \cdot \frac{1}{2} = 0.475$$

with a total discounted return of

$$W_1(2) = \sum_{n=1}^{\infty} \left(\frac{1}{1+\rho} \right)^{n-1} \cdot \pi = 9.975. \tag{1.11}$$

The consumers, like Crusoe in the previous example, are assumed to be risk neutral with utility function $u(q) = q$. In each period, they receive in utility $u(a/p) = u(1) = 1$ with a total discounted utility of

$$V_1(1) = \sum_{n=1}^{\infty} \beta^{n-1} u(1) = 21. \tag{1.12}$$

Now suppose that the firms have the opportunity to innovate. The physical aspects of the economy will be the same as for Crusoe in the previous example, but prices and money will now play a role.

By investing 1/2 unit of goods, each firm can, independently of the others, purchase a lottery that with probability 1/2 results in the improved production function f_2 for the firm. But, also with probability 1/2, the attempt fails thus causing the firm to continue with f_1. The question for the managers of the firms is whether they can improve upon the return achievable without making the attempt at innovation.

To answer this question, assume that the firms do purchase the lottery. Suppose that the price of goods in the first period is p. The firms will need to bid $b + c = p + p/2 = 1.5p$ in order to purchase 1 unit of goods as input for production and 1/2 unit for the innovation attempt. The short-term loan of $b = p$ is again financed by the consumer-owners who bid a and deposit $m - a = b = p$ as before. However, the bid $c = p/2$ is financed by a long-term bank loan which must be repaid over the infinite future in payments of ρc in every period. The price of goods in the first period is then

$$p = \frac{a+b+c}{q} = \frac{m-p+p/2+p}{2} = \frac{1+p/2}{2}.$$

So the price is $p = 2/3$, and $b = 2/3, a = 1 - p = 1/3 = c$. The firms earn in the first period the profit

$$\pi = pq - (1 + \rho)b - \rho c = \frac{2}{3} \cdot 2 - 1.05 \cdot \frac{2}{3} - 0.05 \cdot \frac{1}{3} = 0.6167.$$

The owner-consumers receive in the first period

$$u(a/p) = a/p = \frac{1/3}{2/3} = 1/2.$$

In all subsequent periods, the unsuccessful firms called type 1 with production function f_1 bid $b_1 = p$ in order to input 1 unit of goods, while the successful firms called type 2 with production function f_2 bid $b_2 = p/2$ in order to input $1/2$. As before, these short-term loans are financed by the owner-consumers, who now hold cash $m + c = 1 + 1/3 = 4/3$. So they deposit

$$4/3 - a = \frac{1}{2}b_1 + \frac{1}{2}b_2 = \frac{3}{4}p.$$

Hence, the price in periods after the first satisfies

$$p = \frac{a + \frac{1}{2}b_1 + \frac{1}{2}b_2}{q} = \frac{\frac{4}{3} - \frac{3}{4}p + \frac{3}{4}p}{2} = 2/3,$$

that is, the price equals $2/3$ in every period. (One should not expect constant prices in general. This example was constructed to make for a simple analysis.) Notice that because of the constant price and the constant derivative $u' = 1$, the Euler equation (1.9) is satisfied at every stage.

In periods after the first, the type 1 firms have the profit

$$\pi_1 = pq - (1 + \rho)b_1 - \rho c = \frac{2}{3} \cdot 2 - 1.05 \cdot \frac{2}{3} - 0.05 \cdot \frac{1}{3} = 0.6167,$$

type 2 firms make

$$\pi_2 = pq - (1 + \rho)b_2 - \rho c = \frac{2}{3} \cdot 2 - 1.05 \cdot \frac{1}{3} - 0.05 \cdot \frac{1}{3} = 0.9667,$$

and owner-consumers receive

$$u(a/p) = a/p = \frac{5/6}{2/3} = 5/4.$$

The total expected value to a firm is

$$W = W(2) = \pi + \frac{1}{1+\rho}\left\{\frac{1}{2}\sum_{n=1}^{\infty}\left(\frac{1}{1+\rho}\right)^{n-1}\pi_1 + \frac{1}{2}\sum_{n=1}^{\infty}\left(\frac{1}{1+\rho}\right)^{n-1}\pi_2\right\}$$

$$= 0.6167 + \frac{1}{1.05}\left\{\frac{1}{2}\cdot 21\cdot .6167 + \frac{1}{2}\cdot 21\cdot 0.9667\right\} = 16.4507.$$

Since $16.4507 > 9.975$, the innovation lottery is good for the firms. (The increase in profits is, in part, due to the inflated price of goods.)

The total expected utility for an owner-consumer is

$$1/2 + \beta\sum_{n=1}^{\infty}\beta^{n-1}\cdot\frac{5}{4} = 25.5,$$

which is greater than 21. So the lottery is good for consumers also.

1.10 Summary Remarks

The simple models of this essay serve to reflect mathematically the meaning of Schumpeter's breaking of the circular flow of capital in a closed economy and to illustrate the nature of the cash flows in innovation. In the models here the funds for innovation come from a central bank. Historically, both private and public resources have been involved in innovation.

We believe that mathematical models are, at best, of highly limited scope in applications of the social sciences to everyday life, but they are critical in the arduous task of providing a sound logical structure to help in understanding the dynamics of an economy within any society. Our approach here has attempted to provide some of the insight we need to be able to recognize that there is a unified approach that encompasses economic theory from Cournot, Jevons, Walras, and many others to Schumpeter, Keynes, and their successors. This involves at least six steps of increasing complexity and diversity:

1. General equilibrium theory presents a timeless, preinstitutional, parameter-free basic abstraction of the conditions required for the existence of an efficient price system where individual optimization requires no more information than the existence of these prices.
2. The intellectual cost of general equilibrium was to cut out dynamics. An intermediate step between general equilibrium and full dynamics can be taken by concentrating on minimal process models of an economy provided with a context that includes initial conditions and a law of motion for the dynamics that depends on the actions of the agents and random events. In the simple models studied here, the only randomness occurs at the first stage of play. Such games

are sufficiently simple with minimal information conditions (one information set per player) that the behavioral assumption of Nash noncooperative behavior, or in this simple instance, rational expectations, merits consideration.[15]

3. Once the models suggested above are considered for two or more time periods, the proliferation of special cases, information conditions, and parametric requirements become astronomical. Ad hoc building of the rules of the game can in general be justified, but the behavioral conditions as to what constitutes a solution become critical as was noted in Kirman's relevant critique (Kirman 1992).

4. Dynamic programming methods have been applied by Bewley (1982) to explore the inventory theory properties of the storage of money under uncertainty, then by Lucas (1996) and associates, promoting direct application of low-dimensional rational expectations, primarily representation models to macroeconomics. Similar methods have been used by Karatzas et al. (2006), Karatzas et al. (1994), Thompson and Shubik (1959) and others to help provide a modeling basis for microeconomics dynamics and testable experimental games (Angerer et al. 2010).

5. The limitations of low-dimensional rational expectations are so painfully clear that as we attempt to draw nearer to the application of multistage models to industrial organization and macroeconomic application, the need to resort to simulation, gaming, and explicitly ad hoc structural and behavioral assumptions becomes critical as is manifested in the work of Dosi et al. (2013) and others. Ad hoc studies of specific markets and interacting groups of firms are called for when there is no substitute for knowing the business and structural and behavioral assumptions are specifically laid out.

6. There is no royal road to economic dynamics, but a careful deconstruction of the assumptions about structure and behavior concentrating on context and obedience to the sociopolitical environment and the laws of physics takes us closer to being able to reconcile elegant economic abstractions with a mutating, organic, almost biological growth that reflects socio-politico-economic actuality.

References

Angerer, M., Huber, J., Shubik, M., Sunder, S.: An economy with personal currency: theory and experimental evidence. Ann. Finance 6(4), 475–509 (2010)
Bak, P., Nørrelykke, S.F., Shubik, M.: Dynamics of money. Phys. Rev. E 60(3), 2528–2532 (1999)
Baumol, W.J.: The Free-Market Innovation Machine. Princeton University Press, Princeton (2002)
Bechtel, S.D., et al.: Managing innovation. Deadelus 125(2), 147–166 (1996)
Bewley, T.F.: An integration of equilibrium theory and turnpike theory. J. Math. Econ. 10, 233–268 (1982)

---

[15]The full justification and fleshing out of these comments requires a book length manuscript and such an almost completed manuscript exists at this time enlarging on all the points noted here (Shubik and Smith 2016).

Boldrin, M., Levine, D.K.: Perfectly competitive innovation. J. Monetary Econ. **55**(3), 435–453 (2008)

Caiaini, A., Godin, A., Lucarelli, S.: A stock flow consistent analysis of a Schumpeterian innovation economy. Metroeconomica **65**(3), 397–429 (2013)

Day, R.H.: Bounded rationality and the co-evolution of market and state. In: Day, R.H., Eliasson, G., Wilborg, C. (eds.) The Markets for Innovation, Ownership and Control. North-Holland, Amsterdam and New York (1993)

Day, R.H., Eliasson, G., Wilborg, C.: The Markets for Innovation, Ownership and Control. North-Holland, Amsterdam and New York (1993)

Dosi, G., Freeman, C., Nelson, R., Silverberg, G., Soete, L.: Technical Change and Economic Theory. Pinter, London and New York (1988)

Dosi, G., Fagiolo, G., Napolitano, M., Rovertini, A.: Income distribution, credit and income policies in an agent-based Keynesian model. J. Econ. Dyn. Control. **37**, 1748–1767 (2013)

Godely, W., Lavoie, M.: Monetary Economics: An Integrated Approach to Credit, Money, Income, Production and Wealth. Palgrave, MacMillan, New York (2007)

Karatzas, I., Shubik, M., Sudderth, W.: Construction of stationary Markov equilibria in a strategic market game. Math. Oper. Res. **19**, 975–1006 (1994)

Karatzas, I., Shubik, M., Sudderth, W.: Production, interest, and saving in deterministic economies with additive endowments. Econ. Theory **29**(3), 525–548 (2006)

Keynes, J.M.: The General Theory of Employment, Interest and Money. MacMillan, London ((1936) 1957). Reprint

Kirman, A.P.: Whom or what does the representative agent represent? J. Econ. Perspect. **6**, 117–136 (1992)

Koopmans, T.C.: Concepts of optimality and their uses. Am. Econ. Rev. **67**(3), 261–274 (1977)

Lamoreaux, N., Sokoloff, K.L. (eds.): Financing Innovation in the United States: 1870 to the Present. MIT Press, Cambridge (2007)

Levhari, D., Srinivasan, T.N.: Optimal savings under uncertainty. Rev. Econ. Stud. **XXXVI(2)**, 153–163 (1969)

Lucas, R.W.: Nobel lecture: monetary neutrality. J. Pol. Econ. **104**, 661–682 (1996)

Minsky, H.: Stabilizing an Unstable Economy. Yale University Press, New Haven, CT (1986)

Nelson, R.: The Sources of Economic Growth. Harvard University Press, Cambridge (1996)

Nelson, R.R., Winter, S.G.: An Evolutionary Theory of Economic Change. Harvard, Belknap, Cambridge (1982)

Schumpeter, J.A.: The Theory of Economic Development. Harvard University Press, Cambridge (1934). Original in German 1911

Schumpeter, J.A.: Business Cycles. Mcgraw-Hill, London (1939)

Shubik, M.: Innovation and equilibrium. In: Papadimitriou, D., Wray, L.R. (eds.): The Elgar Companion to Hyman Minsky, pp. 153–168. Edward Elgar Publishing, Northampton, MA (2010)

Shubik, M., Smith, E.: The Guidance of a Enterprise Economy. M.I.T. Press, Cambridge (2016)

Shubik, M., Sudderth, W.: Cost innovation, Schumpeter and equilibrium. Part 1. Cowles Foundation Discussion Paper 1786 (2011)

Thompson, G.L., Shubik, M.: Games of economic survival. Nav. Res. Logist. Q. **6**(2), 111–123 (1959)

Chapter 2
A Review in Campaigns: Going Positive and Negative

Grisel Ayllón Aragón

Abstract In this review, we go through the negative campaign literature to highlight the importance of economic inclusion in the topic. Psychology and political science have been studying the phenomena for the last 20 years; however, economic models have not achieved the goal of interpreting and forecasting the strategical behavior of candidates. Voting models can be classified in two: spatial and probabilistic models. Most of the work done consider the possibility of doing either positive or negative campaign, but not both. We look forward to awake curiosity and better understanding of the subject, starting with a discussion of the definition of a *"negative campaign."*

Keywords Campaigns • Negative campaign • Political economy • Spatial • Probabilistic models • Incentives campaigns • Incentives • Spatial competition • Probabilistic voting • Voting • Negative messages • Positive messages • Credibility

2.1 Introduction

Political campaigns have changed in the past two decades, not only by the means of communication but in the type of messages sent to voters. Commonly, political economy has taken the fact that a campaign is useful to send messages about the policy the candidate would apply if going into office. The most striking, and at the same time, the simplest result we have, is Down's model (1957) where the candidate which announces the policy closest to the median voter's best option will win the election. This result turns out like this, as we assume single-peak preferences of each voter, that is, all voters have an ideal policy within a known policy's space, and any alternative will decrease his welfare as it goes away from their best personal policy option. The immediate consequence of this model is the prediction that radical

G.A. Aragón (✉)
Tecnologico de Monterrey, Campus Ciudad de Mexico, Calle del Puente 222, Col. Ejidos de Huipulco, C.P. 14380, Mexico City, Mexico
e-mail: grisel.ayllon@itesm.mx

© Springer International Publishing Switzerland 2016
A.A. Pinto et al. (eds.), *Trends in Mathematical Economics*,
DOI 10.1007/978-3-319-32543-9_2

candidates will never win and that, in the long run, all political parties will converge their promises to the median voter's ideal policy.

We could go deeper into the means by which candidates communicate with the possible voters, but in this review, we are interested in describing the modeling about the type of messages that they send in a campaign. Empirical works have shown that spots, announcements, or any kind of advertisement is no longer done with the objective of promoting a specific characteristic of the candidate. Messages concerning about their opponents have increased; however, it is too risky to say that we are dealing with an increment in negative campaigning. There is no consensus among economists about the definition of a "negative campaign." In fact, it might be that the case where campaigning focuses on messages about the contenders on the lack of personal abilities, results obtained in a sphere different from the issues discussed in the campaign, or on the real position of the policies in debate on the current campaign. In any case, the negative messages are directed to harm the desirability and credibility of the candidates.

Not only in the USA has the expenditure in campaigns increased in the last years but it is a worldwide phenomenon. In the 1996 US presidential elections, 6 % of the expenditure in Clinton's campaign was for negative advertising versus a 70 % of negative campaign conducted by Bob Dole. UNDP has run surveys in different Latin–American countries to evaluate the level of negative campaigning and the impact of mass media in voters. It was shown that, in Mexico, 11 % of the spots were conducting a negative message for the opponent candidate in the 2006s presidential elections, and the messages increased as the campaign got closer to an end. Martínez and Aguilar (2013) highlighted that time is a crucial aspect in the campaigns, as the length of it can create spaces to impact at a larger scale into swinger voters. Hence, it is a fact that candidates not only expose their political position to the voters, but they also talk about the opposition. This information is transmitted to voters, so they can create their expectations about the true position of the candidates.

The definition itself of negative campaigning differs from author to author. Harrington and Hess (1996) bound the definition of advertisement directed to change the perception of a candidate's ideology. Mattes (2007) defines it as the set of actions encouraged by political actors who talk about one's opponent through the character or valence dimension. The classification of the messages might be subject to the criteria of the researcher, and we do not find clear parameters to distinguish one from the other. There is an emerging type of negative campaign called *contrasting campaign*. This is a subclass of negative campaigns, whose principal strategy is to reveal or say something contrasting what a candidate has said of himself and what it has been really doing. This type of campaign is having a great impact in countries where explicit defamation and injury is forbidden in the campaign, such as in Mexico.[1] In the 2006 Mexican presidential elections the winner candidate Felipe Calderón was 8 % behind López Obrador 1 month before

---

[1] The electoral law (Código Federal de Instituciones y Procedimientos Electorales) forbids any kind of negative campaign. Its Article 38 says: "Abstain from any expression involving diatribe, libel, slander, libel, defamation, or demeaning to the citizens, public institutions or other political parties and their candidates, particularly during election campaigns and political propaganda that is used in an electoral period."

elections. Calderón started a campaign[2] where he affirmed that Obrador was a "danger" to Mexico, as he promised to reduce expenses and create economic growth but instead had duplicated the debt in Mexico City, where he was the governor. Another spot first shows how Obrador is asking for tolerance; afterward he appears in another image contemptuously calling the current president. Felipe Calderón pronounced about the "true" policies that Obrador would implement if running into office versus what he promised to achieve. Even though Calderon did not give any further proofs about his statements, he achieved to win the presidential election. This campaign lasted 180 days, long enough to have an impact in the voters' intention. However, legislation has changed and campaigns last 95 days nowadays and any attempt of defamation is punished. The effects of this new law have not been yet so clear. At least, in the last presidential election, the winner candidate had since the beginning a clear advantage and he was not beaten.

This review is intended to go over the different views and approaches done about negative campaigns. There exists a vast literature in negative campaigning in psychology and political science but not in economics. Few formal models introduce the concept of a negative campaign, and the existing literature has some drawbacks. We are going to present a brief view of the first two sciences about the theme, and then we will go through the main papers done in the economics field.

2.2   Negative Campaigns in Different Fields

Psychology studies give three arguments for the existence of negative campaigning: (1) it stimulates attention to and awareness of the campaign, (2) campaigns may arouse anxiety which stimulates interest, and (3) negative campaign might be a sign of a close race, which is directly related with the marginal utility of going to vote in some cases. People are very aware of negative information; they attend to it more, think about it more, and remember it better, and it is more powerful in shaping our impressions of things (Hodges 1974). These authors argue that emotions are more important than beliefs in predicting voting choice. However, they are aware that there is fragmented evidence suggestion in the degree of effectiveness in the type of messages that candidates send. However, the mainstream in psychology proposes that a negative message is received deeper in the voters' minds than a positive one. Skowronski et al. (1989) showed that given equal amounts of positive and negative messages about the characteristics of a person, the overall impression formed is skewed toward the negative, and Richey et al. (1967) argued that negative data are more persistent over time. Some of these authors have done experiments to prove

---

[2]Calderón posted 60 h of such TV spots in the last phase of the campaign and sent 40 million e-mails with messages contrasting the promises of his contender. A deep analysis about such election processes is due to the political scientist Sergio Aguayo [see Aguayo (2010)].

how a negative message is more effective than any other; however, they do not prove how it impacts the voting intention. With the lack of a control group, these findings are descriptive but intriguing and motivating to model the candidates' strategies.

Political science has focused in the study of the type of messages, frequency, and possible causation of the phenomena. They have done a very tough job of quantification of the spots, tweets, and any kind of political advertisement. They have discussed the effectiveness of sending a negative signal, in terms of voting. They assume that candidates choose the campaign strategy they believe will give them the best chance to win (Bartels 1993); however, they do not model or demonstrate how the incentive mechanism would work. For them, candidates who expect to lose may attack harder the reputation of its contender, as they have "nothing to lose." Stevens (2009) reconsiders the validity of empirical and theoretical research on negative campaigning through advertising, since he states that most studies have only evaluated the role of volume of negative ads as beneficial for voters. He asserts that proportion is also important since voters can be exposed to a large volume of ads, but they can be relatively more exposed to a certain ad so that relative exposure or proportion can have different and offsetting effects. The rationale behind his thesis implies that the current evidence might be inconclusive and ambiguous in terms of the nature of negative campaigns.

2.3   Modeling Negative Campaigns: An Economic Approach

The economic literature has focused mainly in the signaling of the position of the candidates' ideal policies; however, there are not so many models that have captured the strategical point of view of the candidates, where they can have different types of alternatives in their strategies' space. Most part of the models are games where candidates can send only one kind of messages to convince the voters. Political economy literature has developed two different approaches: spatial and probabilistic voting models. In the following paragraphs, we are going to number and state the main idea of some of them.

Polborn et al. (2006) develop a model in which candidates can either send a positive or a negative message to the voters. The candidates have unknown qualities that will be signaled by the decision between doing positive or negative campaigning; however, they cannot do both. Soubeyran (2009) provides sufficient conditions for the existence and uniqueness of a symmetric Nash equilibrium where candidates have as strategies: to attack and to defend. He considers that each candidate has a transformation function which allows them to overcome the attacks and turn them into a positive effect in the probability of winning. His main question, further from being how to win when you can do negative campaign, is how these campaigns affect voter turnout. Aragonés et al. (2007) analyze the conditions under which candidates' reputations may affect voters' beliefs over what policy will be implemented by the winning candidate of an election. They use a dynamic game where candidates can promise a policy different from their ideal point even though the true ideal policies are observable. Rational voters will believe the promises which will be implemented in the future as long as the reputation has value for the candidates. Callander et al. (2007) provide a model where candidates are willing to lie about their policy's intentions creating an effect on other candidates' behavior, changing the nature of political campaigns.

Harrington and Hess (1996) consider that political campaigning is not as simple since candidates may sometimes want to reveal the position of the rival and vice versa. For them, the allocation of resources between positive and negative campaigning is crucial for candidates, since engaging in each type of campaign is costly. The model consists of a one-shot game of spatial competition; candidates have two dimensions: issue and valence. At the initial phase of the game, voters have initial perceptions of candidates' personal and policy attributes, and candidates must allocate resources of any of these. They restrict the concept of advertising only to the ideology space, that is, only actions related to the candidate's issues. They show that candidates who are weak on the character dimension (personal attributes) will be the ones who will engage into negative campaigning, while stronger candidates on the valence dimension will engage into a rather positive campaign. Chakrabarti (2007) extends the model developed by Harrington and Hess by including negative campaigning through the valence dimension as well as on the issue dimension. The effect of introducing valence advertising would mean that candidates denigrate or criticize the opponent on the character attributes. When a candidate's valence index is relatively high, the type of campaign the candidate will engage in will be based on valence issues, whereas if the valence index is low, the candidate will conduct an ideological or issue campaign. Mattes (2007) defines negative campaigning as actions encouraged by political actors who talk about one's opponent through the valence dimension. Each candidate inherits initial positions on both ideology and valence; they compete by revealing the location of one candidate on only one dimension. Voters decide depending on prior information about the candidates and also on information given by each candidate during the campaign phase. The main proposition of the model is that a candidate will engage in negative campaigning if the voters value candidates on a valence basis dimension. These models are embraced in a Hotelling–Downs framework, where the median voter will be the pivotal one. They are not so interested in modeling how they get to an "ideal" policy, but how they impact the vote intention or how they make face to it.

Skapedras et al. (1995) explored the incentives candidates have to engage in negative campaign. They show that, with a two-candidate election, the front runner will tend to skew his messages in a positive way. And in a three-candidate competitions, they recognize two features: (1) when there is a candidate with a very low support, he will only have a positive campaign; (2) the negative campaign will only be directed to the strong candidates.

Another kind of economic models to study elections is based on probabilistic voting models. A basic assumption in these models is that voters are uninformed, and it is not clear who the pivotal voter is. Brueckner et al. (2013) develop a probabilistic voting model with negative campaigning that extends the evidence of other spatial competition models like those of Harrington and Hess (1996) and

Chakrabarti (2007). According to them, probabilistic voting models allow stochastic outcomes from voters' decisions due to a valence effect which affects all other voters making this model preferable to the classical framework. Under this model, the best strategy a candidate can undertake is to locate near the median of voter's preferred policy position without bearing in mind where the opposition is located. Candidates must decide two variables: stated position and the level of expenditure in negative campaign seeking to maximize their expected utility function (the probability of winning times the benefit of holding office minus the cost of ideological divergence). For them, centrist candidates will engage in negative campaigning regardless of the nature of their ideological positions (fixed or chosen), whereas more extreme candidates devote their resources on positive campaign spending.

2.4 Final Remarks

The use of game theory to model the campaigns has faced different difficulties and ambiguous results. It is a fact that candidates do not engage only in one type of campaign. However, the definition of negative campaigning is still not so clear. This can be harmful for the empirical works which try to test the theoretical models. However, using the concept of contrast campaign can be of a great use. If we do not need to classify the type of message sent but only quantify if the message corresponds to his own characteristics of their contenders', then it will be easier to test the theoretical framework.

Different efforts have been done to explain the increasing volume in negative advertisement. Some authors have focused in the difference between attacking in a valence level or the ideological position. Beyond this issue, economists should set a base model to identify the mechanism and the decision variables. Afterward, a second question could be the dimension in which campaigns are more effective. Furthermore, if a candidate is attacked, the final objective is one: harming credibility to have less share of the mass of voters.

Instead of talking of a negative campaign in terms of saying a non-desirable characteristic of the contender, we could look at the promises and pronouncements that candidates announce to convince voters. Promises are all the messages done to promote themselves and inform the voters about their policy intentions. Pronouncements are those messages sent to voters about the policy intentions of the contenders. We cannot separate these decision variables: they create a reputation effect in the sender as well as in the contender. The threat of losing reputation, hence, credibility for the next period, can help to elicit the true policy's positions of the candidates. Candidate's promises are their real policies if the opportunity cost is high enough. Therefore, the existence of pronouncements can be a tool to prevent the candidates to promise the implementation of the median voter's ideal policy.

Another natural question is not only who makes and how much negative campaign, but does it work?

Acknowledgements I would like to thank Enriqueta Aragonés for her useful comments and to Manuel Veléz for his passion on research.

References

Aguayo, S.: La transición en México: 1910–2010, 1st edn. Fondo de Cultura Económica, Mexico (2010)

Aragonés, E., Palfrey, T., Postlewaite, A.: Political reputations and campaign promises. J. Eur. Econ. Assoc. **5**, 846–884 (2007)

Banks, S.J.: A model of electoral competition with incomplete information. J. Econ. Theory **50**, 309–325 (1990)

Bartels, L.: Messages received: the political impact of the media exposure. Am. Polit. Sci. Rev. **87**, 267–285 (1993)

Brueckner, J., Lee, K.: Negative campaigning in a probabilistic voting model. CESifo Working Paper Series 4233, CESifo Group Munich (2013)

Callander, S., Wilkie, S.: Lies, damned lies, and political campaigns. Games Econ. Behav. **60**, 262–286 (2007)

Chakrabarti, S.: A note on negative electoral advertising: denigrating character vs. portraying extremism. Scott. J. Polit. Econ. **54**, 136–149 (2007)

Downs, A.: An economic theory of political action in a democracy. J. Polit. Econ. **65**, 135–150 (1957)

Grofman, B., Skaperdas, S.: Modeling negative campaigning. Am. Polit. Sci. Rev. **89**, 49–61 (1995)

Harrington, J., Hess, G.: A spatial theory of positive and negative campaigning. Games Econ. Behav. **17**, 209–229 (1996)

Hodges, B.H.: Effect on valence on relative weighting in impression formation. J. Pers. Soc. Psychol. **30**, 378–381 (1974)

Martínez, M., Aguilar, R.: Campañas electrorales en M éxico y una visión a Centroamérica, 1st edn. Porrúa, Mexico (2013)

Mattes, K.: Attack politics: who goes negative and why? Paper presented at Midwest Political Science Association, Chicago, IL (2007)

Polborn, M., Yi, D.: Informative positive and negative campaigning. Q. J. Polit. Sci. **1**, 351–371 (2006)

Richey, M.H., McClelland, L., Shimkunas, A.M.: Relative influence of positive and negative information in impression formation and persistence. J. Pers. Soc. Psychol. **6**, 322–326 (1967)

Skowronski, J.J., Carlston, D.E.: Negativity and extremity biases in impression formation: a review of explanations. Psychol. Bull. **105**, 131 (1989)

Soubeyran, R.: Contest with attack and defense: does negative campaigning increase or decrease voter turnout? Soc. Choice Welf. **32**, 337–353 (2009)

Stevens, D.: Elements of negativity: volume and proportion in exposure to negative advertising. Polit. Behav. **31**, 429–454 (2009)

Chapter 3
On Lattice and DA

David Cantala

Abstract We present an application where, in a matching market, Preferences of one side of the market evolve all along the process of the sequential version of the deferred acceptance (DA) algorithm, producing an agenda-dependent and stable outcome. We also provide an example where agents stable matching within the set of achievable matchings. The motivation for this application is simply to show that the original DA algorithm is more versatile than suggested by Hatfield and Milgrom(2005).

Keywords Deferred acceptance • Matching with contracts

3.1 Introduction

We present an application where, in a matching market, preferences of one side of the market evolve all along the process of the sequential version of the deferred acceptance (DA) algorithm, producing an agenda-dependent and stable outcome. We also provide an example where agents of the offering side of the market do not agree on which is the best/worst stable matching within the set of achievable matchings. The motivation for this application is simply to show that the original DA algorithm is more versatile than suggested by Hatfield and Milgrom (2005).

Gale and Shapley (1962) introduce the marriage game and the DA algorithm. They consider a model where there are two sets of players, girls on the one hand and boys on the other hand; girls have preferences over boys, and vice versa. The problem consists in matching girls with boys—one girl with one boy—at a stable assignment. Stability is a natural normative criteria: a matching is stable if partners are acceptable to one another; moreover whenever a girl (boy) prefers a boy (girl) to her (his) match, he (she) prefers his (her) mate to her (him). They establish a first theorem: such a matching always exists since the DA algorithm always produces a stable matching.

D. Cantala (✉)
El Colegio de Mexico, Camino Al Ajusco #20, Tlalpan Pedregal de Santa
Teresa, 10740 Mexico City, Mexico
e-mail: dcantala@colmex.mx

© Springer International Publishing Switzerland 2016

43

A.A. Pinto et al. (eds.), *Trends in Mathematical Economics*,
DOI 10.1007/978-3-319-32543-9_3

The principles of the DA algorithm are the following:

1. fix the side of the market, girls or boys, which makes offers;
2. at each step of the algorithm, single agents on the offering side of the market make offers to the favorite prospect to whom she did no make an offer yet;
3. agents on the accepting side consider her/his mate and the new proposals, and tentatively accept her/his favorite prospect;
4. the process generates a sequence of tentative matchings, where agents on the receiving side never regret to have turn down an offer. It ends when all agents on the offering side either are matched or have made offer to all acceptable prospects.

Roth and Sotomayor (1990) extend the problem to a setting where one side of the market, firms, can hire many agents on the other side of the market, workers, who can work for at most one firm. Thus, firms have preferences defined over subsets of agents. They define the substitutability condition, which captures the idea that the individual quality of workers contributes more to the value of a group of workers than their complementarity. If preferences of firms are substitutable, again, the existence results obtained for one-to-one markets naturally extend to the many-to-one settings.

Gale and Shapley (1962) also establish that, whenever there are many stable matchings, the opposition of interests between both sides of the market is sharp: one matching is the girls best/boys worst stable matching; another one is the girls worst/boys best stable matching. We call this result Theorem 2. It naturally triggers a branch in the literature, pioneered by Knuth (1976), dedicated to the study of lattice structures in matching markets. All these papers, however, use preference orders in their definition of lattices. Blair (1988) establishes the lattice structure under substitutability but with unnatural least upper and lower bounds. Other references on the lattice structure in many-to-one and many-to-many markets and more restricted preferences are Baiou and Balinski (2000), Alkan (1999, 2001, 2002), Alkan and Gale (2003), Martínez et al. (2001), Echenique and Oviedo (2004, 2006), Fleiner (2003).

Hatfield and Milgrom (2005) introduce the matching with contracts model so as to underline analogies between many-to-one matching markets, the labor market studied in Kelso and Crawford (1982), and package biddings from Milgrom (2004) and Cramton et al. (2006) The strategy followed by the authors consists in introducing a two-part model.

First, they introduce contracts, which are either doctor-hospital pairs or firm-worker-wage triplets. We formalize the former case and recall the main ingredients of their model[1]: the set of doctors is D, and the set of hospitals is H. The set of contracts is X, $X \equiv D \times H$. A contract $x \in X$ is bilateral $x_D \in D$, $x_H \in H$. The preferences of a doctor d is $\succ_d$, its associated chosen set $C_d(X')$. The preferences of

---

[1] A reader unfamiliar to the model might read first Hatfield and Milgrom (2005), Hatfiel and Kojima (2008) and Aygün and Sönmez (2013).

any hospital h over subset of doctors is $\succ_h$; the chosen set of hospital h is $C_d(X')$; aggregating for all doctors, we get $C_D(X') = \cup_{d \in D} C_d(X')$. The rejected set is $R_D(X') = X' - C_D(X')$. Similarly for hospitals we define $C_H(X') = \cup_{h \in H} C_h(X')$ and $R_H(X') = X' - C_H(X')$.

A set of contracts X' is a *stable* allocation if:

1. $C_D(X') = C_H(X') = X'$ and
2. there exists no hospital h and set of contracts $X'' \neq C_H(X')$ such that

$$X'' = C_h(X' \cup X'') \subset C_D(X' \cup X'').$$

Hatfield and Milgrom (2005) show the following : if $(X_D, X_H) \subset X^2$ is a solution to the system of equations

$$X_D = X - R_H(X_H)$$

and

$$X_H = X - R_D(X_D),$$

then $X_H \cap X_D$ is a stable set of contracts and $X_H \cap X_D = C_D(X_D) = C_H(X_H)$. Conversely, for any stable collection of contracts X', there exists some pair (X_D, X_H) satisfying the system of equations such that $X' = X_H \cap X_D$.

Second, they define an order $\geq$ over $X \times X$. This order does not depend on preferences; specifically the order is

$$\left((X_D, X_H) \geq (X_D', X_H') \right) \Leftrightarrow \left(X_D \supset X_D' \text{and} X_H \subset X_H' \right).$$

Adapted to the matching with contract environment, the DA is generalized by iterating the following function $F : X \times X \to X \times X$

$$F_1(X') = X - R_H(X')$$
$$F_2(X') = X - R_D(X')$$
$$F(X_D, X_H) = (F_1(X_H), F_2(F_1(X_H))) .$$

The operator is isotone on the lattice $(X \times X, \geq)$, which is why, by Tarski's fixed-point theorem, the process converges to a fixed point: that is, a stable contract. Depending on the initial set of contracts, the fixed point might be in particular the smallest or the highest fixed point in the lattice.

The approach is claimed to be more general than the one by Gale and Shapley (1962) in the sense that the order used by Hatfield and Milgrom (2005) relies on inclusion of sets, rather than preference orders. When agents prefer choosing on larger sets, one recovers the result previously mentioned, Theorem 2, that an outcome matching is the favorite stable matchings for one side of the market is the worst one for the other side of the market is recovered as a particular case.

The argument about the further generality of the matching with contract approach, however, is not fair. Certainly Theorem 2 is a particular case of matching with contracts, but it is also a particular result among DA applications. As we show in the next section, the DA also operates in environments where agents on one side of the market do not agree on which is the best/worst achievable stable matching given the primitives of the model. Our approach complements Echenique (2012) who challenges the generality of matching with contracts approach by considering a problem where wage is part of the description of a contract and showing that bargaining over contracts is equivalent to bargaining over wages, when contracts are substitutes.

3.2 The Leader–Follower Model

Think of a decentralized settings where offers are emitted sequentially and doctors' employment record affects their value/productivity: having worked for hospitals which are leaders in their field brings valuable experience and information that follower hospitals wish to acquire. To capture this real-life feature, we model a matching market where offers are emitted sequentially and index by $t = 0, 1, 2, 3,$ ... the tentative matchings entailed by the DA algorithm.

We assume that hospitals are ordered on a leadership ladder: the lower the i, the index of an hospital, the more her leadership, h_1 being the absolute leader and h_H being the absolute follower. Thus, preferences of hospitals over doctors evolve with tentative matchings until they take a hiring decision over doctors, i.e., until they accept or reject an offer from the doctors.

We denote $\succ_h^0$ the preferences of firm h at $t = 0$. For all doctors d, let i_d^t be index of the hospital with lower index that accepted her offer until t and i_d^h this index when hospital h receives an offer from d. Let Y_h^t be the subset of doctors that have made an offer to hospital h before t and N_h^t the subset of doctors that did not do any offer.

Preferences of hospital h_i follow a *leader–follower pattern* if, for any period t:

Case 1 $d_1, d_2 \in N_{h_i}^t$

$$
d_1 \succ_{h_i}^t d_2 \Leftrightarrow
\begin{cases}
i_{d_1}^{t-1} < i_{d_2}^{t-1} \text{ if } i_{d_1}^{t-1} < i \text{ or} \\
d_1 \succ_{h_i}^0 d_2 \text{ if } (i_{d_1}^{t-1} > i \text{ and } i_{d_2}^{t-1} > i) \text{ or } i_{d_1}^{t-1} = i_{d_2}^{t-1} < i.
\end{cases}
$$

Case 2 $d_1 \in Y_{h_i}^t, d_2 \in N_{h_i}^t$

$$
d_1 \succ_{h_i}^t d_2 \Leftrightarrow
\begin{cases}
i_{d_1}^{h_i} < i_{d_2}^{t-1} \text{ if } i_{d_1}^{h_i} < i \text{ or} \\
d_1 \succ_{h_i}^0 d_2 \text{ if } (i_{d_1}^{h_i} > i \text{ and } i_{d_2}^{t-1} > i) \text{ or } i_{d_1}^{h_i} = i_{d_2}^{t-1} < i.
\end{cases}
$$

Case 3 $d_1 \in N_{h_i}^t, d_2 \in Y_{h_i}^t$

$$d_1 \succ_{h_i}^t d_2 \Leftrightarrow \begin{cases} i_{d_1}^{t-1} < i_{d_2}^{;h_i} \text{ if } i_{d_1}^{t-1} < i \text{ or} \\ d_1 \succ_{h_i}^0 d_2 \text{ if } (i_{d_1}^{t-1} > i \text{ and } i_{d_2}^{;h_i} > i) \text{ or } i_{d_1}^{t-1} = i_{d_2}^{;h_i} < i. \end{cases}$$

Case 4 $d_1, d_2 \in Y_{h_i}^t$

$$d_1 \succ_{h_i}^t d_2 \Leftrightarrow d_1 \succ_{h_i}^{t-1} d_2.$$

We assume that hospitals' preferences over single doctors follow a leader–follower pattern and that preferences of hospitals over subsets of workers are responsive to preferences over individual doctors: the preferences $\succ_h$ are responsive if for all doctors d, d' and subsets of doctors $s \in D \backslash \{d, d'\}$ we have that:

1. $s \cup \{d\} \succsim_h s \cup \{d'\}$ if and only if $d \succsim_h d'$, and
2. $s \cup \{d\} \succ^i s$ if and only if $d \succ^i \varnothing$.

We denote by $q = (q_1, \ldots, q_H)$ the vector of quotas associated to each hospital and $Ch(S, \succ_h, q_i)$, the subset of $S \cup \{\varnothing\}$ of cardinality at most q_i preferred by h_i as for $\succ_{h_i}$. We adapt now the DA to this setting.

The deferred acceptance algorithm with evolving preferences (DAEP)

Consider a market $(D, H, q, \succ^0, P)$, μ^0 is the empty matching, set $t = 1$.

Main iteration

Pick randomly one unmatched doctor d, who makes an offer to h, her top alternative as for $\succ_d^{t-1}$ within the set of hospitals to which she did not make any offer.

If d belongs to $Ch(\mu^{t-1} \cup \{d\}, \succ_h^{t-1}, q_i)$, d is matched to h at μ^t; if any, the doctors in μ^{t-1} who do not belong to $Ch(\mu^{t-1} \cup \{d\}, \succ_h^{t-1}, q_i)$ are unmatched at μ^t, other assignments are unaffected.

Preferences of doctors are updated by all hospitals.

If there are no unmatched doctors or all unmatched doctors have emitted all acceptable offers, the tentative matching is the outcome matching, else, follow the main iteration.

Let π be the order in which doctors have been selected to make offers. Our main theorem states that the outcome matching of the DAEP is stable.

Theorem 3. *Consider a market* $(D, H, q, \succ^0, P)$; *if hospitals report preferences which follow a leader–follower pattern, for any order π in which doctors have been selected to make offers, the outcome matching of DAEP is stable.*

Proof. The outcome matching is individually rational since doctors only make offers to acceptable hospitals and hospitals accept subset of doctor belonging to their choice, thus individually rational when it is accepted, and no individual doctor becomes unacceptable once it is considered acceptable.

We observe that the respective ranking of doctors that did not make offer to any hospital h only improves with respect to doctors who already made an offer to h; thus if a doctor is rejected by an hospital at iteration t, her respective position with respect to those doctors accepted by h at t does not vary afterwards and possibly decreases with respect to doctors who still did not make an offer to h at t. Thus, since preferences of hospitals are responsive, no hospital never regrets to have fired a doctor, which guarantees the fact that the offer process stops and the stability of the outcome matching. The argument is true for all orders π. □

An alternative proof consists in establishing that the DAEP outcome is the outcome of the DA algorithm or of the generalized DA (Hatfield and Milgrom 2005) applied to preferences of hospitals at the last stage of DAEP; we opt for the previous one to show that the DA principles apply to a model with evolving preferences.

Interestingly, the outcome matching of DAEP depends on the order in which doctors make offers to hospitals, as shown in Example 1.

Example 1. The set of hospitals is $H = \{h_1, h_2, h_3\}$, the set of doctors is $D = \{d_1, d_2, d_3\}$, quota is 1 for all three hospitals, and preferences of hospitals at date 0 and doctors are

$$
\begin{array}{ccc}
\succ_{h_1}^0 & \succ_{h_2}^0 & \succ_{h_3}^0 \\
d_1 & d_3 & d_2 \\
d_2 & d_2 & d_3 \\
d_3 & d_1 & d_1
\end{array}
\quad \text{and} \quad
\begin{array}{ccc}
\succ_{d_1} & \succ_{d_2} & \succ_{d_3} \\
h_1 & h_1 & h_1 \\
h_2 & h_3 & h_3 \\
h_3 & h_2 & h_2
\end{array}.
$$

At the beginning of the algorithm, $i_{d_1}^0 = i_{d_2}^0 = i_{d_3}^0 = \varnothing = Y_{h_1}^0 = Y_{h_2}^0 = Y_{h_3}^0$ and $N_{h_1}^0 = N_{h_2}^0 = N_{h_3}^0 = D$.

Order 1

Step 1　Suppose d_1 is picked; she makes an offer to h_1, which is accepted; thus

$$
\mu_1 = \left\{ \begin{array}{ccccc} d_1 & d_2 & d_3 & \varnothing & \varnothing \\ h_1 & \varnothing & \varnothing & h_2 & h_3 \end{array} \right\},
$$

$$
i_{d_1}^0 = 1, i_{d_2}^0 = i_{d_3}^0 = \varnothing,
\quad
\begin{array}{ccc}
\succ_{h_1}^1 & \succ_{h_2}^1 & \succ_{h_3}^1 \\
d_1 & d_1 & d_1 \\
d_2 & d_3 & d_2 \\
d_3 & d_2 & d_3
\end{array}.
$$

Step 2　Suppose d_2 is picked; she makes an offer to h_1, which is rejected; thus $\mu_2 = \mu_1$, $\succ^2 = \succ^1$.

Step 3　Suppose d_2 is picked; she makes an offer to h_3, which is accepted; thus

$$
\mu_3 = \left\{ \begin{array}{cccc} d_1 & d_2 & d_3 & \varnothing \\ h_1 & h_3 & \varnothing & h_2 \end{array} \right\},
$$

$$i^0_{d_1} = 1, i^0_{d_2} = 3, i^0_{d_3} = \varnothing, \quad \begin{array}{ccc} \succ^3_{h_1} & \succ^3_{h_2} & \succ^3_{h_3} \\ d_1 & d_1 & d_1 \\ d_2 & d_3 & d_2 \\ d_3 & d_2 & d_3 \end{array} \ .$$

Step 4   Suppose d_3 is picked; she makes an offer to h_1, which is rejected; thus $\mu_4 = \mu_3, \succ^4 = \succ^3$.

Step 5   Suppose d_3 is picked; she makes an offer to h_3, which is rejected; thus $\mu_5 = \mu_4, \succ^5 = \succ^4$.

Step 6   Suppose d_3 is picked; she makes an offer to h_2, which is accepted; the outcome matching is

$$\mu_{\text{order1}} = \left\{ \begin{array}{ccc} d_1 & d_2 & d_3 \\ h_1 & h_3 & h_2 \end{array} \right\} \ .$$

Order 2

Step 1   Suppose d_3 is picked; she makes an offer to h_1, which is accepted; thus

$$\mu_1 = \left\{ \begin{array}{ccccc} d_1 & d_2 & d_3 & \varnothing & \varnothing \\ h_3 & \varnothing & \varnothing & h_1 & h_2 \end{array} \right\} ,$$

$$i^0_{d_3} = 1, i^0_{d_1} = i^0_{d_2} = \varnothing, \quad \begin{array}{ccc} \succ^1_{h_1} & \succ^1_{h_2} & \succ^1_{h_3} \\ d_1 & d_3 & d_3 \\ d_2 & d_1 & d_1 \\ d_3 & d_2 & d_2 \end{array} \ .$$

Step 2   Suppose d_1 is picked; she makes an offer to h_1, which is accepted; thus

$$\mu_1 = \left\{ \begin{array}{ccccc} d_1 & d_2 & d_3 & \varnothing & \varnothing \\ h_1 & \varnothing & \varnothing & h_3 & h_2 \end{array} \right\} ,$$

$$i^0_{d_1} = i^0_{d_3} = 1, i^0_{d_2} = \varnothing, \quad \begin{array}{ccc} \succ^1_{h_1} & \succ^1_{h_2} & \succ^1_{h_3} \\ d_1 & d_1 & d_1 \\ d_2 & d_3 & d_3 \\ d_3 & d_2 & d_2 \end{array} \ .$$

Step 3   Suppose d_3 is picked; she makes an offer to h_3, which is accepted; thus

$$\mu_3 = \left\{ \begin{array}{cccc} d_1 & d_2 & d_3 & \varnothing \\ h_1 & \varnothing & h_3 & h_2 \end{array} \right\} ,$$

$$i^0_{d_1} = i^0_{d_3} = 1, i^0_{d_2} = \varnothing, \succ^3 = \succ^2 \ .$$

Step 4   Suppose d_2 is picked; she makes an offer to h_1, which is rejected; thus $\mu_4 = \mu_3, \succ^4 = \succ^3$.

Step 5 Suppose d_2 is picked; she makes an offer to h_3, which is rejected; thus $\mu_5 = \mu_4, \succ^5 = \succ^4$.

Step 6 Suppose d_3 is picked; she makes an offer to h_2, which is accepted; the outcome matching is

$$\mu_{\text{order2}} = \left\{ \begin{array}{ccc} d_1 & d_2 & d_3 \\ h_1 & h_2 & h_3 \end{array} \right\}.$$

$\square$

3.3 Concluding Remarks

Thus, given the primitives of the model, the outcome matching of the DAEP mechanism depends on the order in which doctors are picked to make offers. More interestingly in Example 1, μ_{order1} and μ_{order2} are the only two achievable matchings for any possible order. It happens that neither all doctors prefer μ_{order1} to μ_{order2}, nor the reverse. In this specific sense, the set of stable matching has no upper nor lower bounds for doctors. The existence of such matchings, thus, is not a perquisite for the DA to properly operate in such settings. It is also easy to see that the mechanism is manipulable for the side of the market that makes offers.

References

Alkan, A.: On the property of stable many- to- many matchings under responsive preferences. In: Alkan, A., Aliprantis, C.D., Yannelis, N.C. (eds.) Current Trends in Economics: Theory and Applications. Studies in Economic Theory, vol. 8. Springer, Berlin, Heidelberg, New York (1999)

Alkan, A.: On preferences over subsets and the lattice structure of stable matchings. Rev. Econ. Des. **6**, 99–111 (2001)

Alkan, A.: A class of multipartner matching markets with a strong lattice structure. Econ. Theory. **19**, 737–746 (2002)

Alkan A., Gale, D.: Stable schedule matching under revealed preference. J. Econ. Theory. **112**, 289–306 (2003)

Aygün, O., Sönmez, T.: Matching with contracts: comment. Am. Econ. Rev. **103**(5), 2050–2051 (2013)

Baiou, M., Balinski, M.L.: Many-to-many matchings: polyandrous polygamy (or polygamous polyandry). Disc. Appl. Math. **101**(1–3), 1–12 (2000)

Blair, C.: The lattice structure of the set of stable matchings with multiple partners. Math. Oper. Res. **13**, 619–629 (1988)

Cramton, P., Shoham, Y., Steinberg, R.: Combinatorial Auctions. MIT, Cambridge, MA (2006)

Echenique, F.: Contracts versus salaries in matching. Am. Econ. Rev. **102**(1), 594–601 (2012)

Echenique, F., Oviedo, J.: Core of many-to-one matchings by fixed point methods. J. Econ. Theory **115**(2), 358–376 (2004)

Echenique, F., Oviedo, J.: A theory of stability in many-to-many matching markets. Theor. Econ. **1**(2), 233–273 (2006)

Fleiner, T.: A fixed-point approach to stable matchings and some applications. Math. Oper. Res. **28**(1), 103–126 (2003)

Gale, D., Shapley, L.S.: College admissions and the stability of marriage. Am. Math. Mon. **69**, 9–14 (1962)

Hatfiel, J.M., Kojima, F.: Matching with contracts: comment. Am. Econ. Rev. **98**(3), 1189–1194 (2008)

Hatfield, J.M., Milgrom, P.: Matching with contracts. Am. Econ. Rev. **95**(4), 913–935 (2005)

Kelso, A.S. Jr., Crawford, V.P.: Job matching, coalition formation, and gross substitute. Econometrica **50**, 1483–1504 (1982)

Knuth, D.E.: Mariages stables. Les Presses de l'Université de Montréal, Montreal (1976)

Martínez, R., Massó, J., Neme, A., Oviedo, J.: On the lattice structure of the set of stable matchings for a many-to-one model. Optimization **50**, 439–457 (2001)

Milgrom, P.R.: Putting Auction Theory to Work. Cambridge University Press, Cambridge (2004)

Roth, A.E., Sotomayor, M.O.A.: Two-sidedmatching: A Study in Game Theoretical Modeling and Analysis. Econometric Society Monograph, vol. 18. Cambridge University Press, Cambridge (1990)

Chapter 4
Externalities, Optimal Subsidy and Growth

Enrique R. Casares and Horacio Sobarzo

Abstract The experiences of East Asian countries (China is an example of this) have brought the role of subsidies in promoting economic growth back into the public discussion. Hence, we study the relationship between subsidies and economic growth with a multi-sector endogenous growth model. The economy has two sectors, manufacturing and nonmanufacturing. The manufacturing (learning) sector is the only sector that generates domestic technological knowledge through learning by doing. This knowledge is used in the nonmanufacturing (non-learning) sector. We find the planner's solution in order to obtain the optimal subsidy for the market economy. We study how the economy responds, in the steady state, when the government establishes the optimal rate of investment subsidy in the manufacturing sector. Thus, the proportion of labor in the manufacturing sector, the relative price of the nonmanufacturing good and the ratio of consumption to nonmanufacturing capital are higher, and the ratio of nonmanufacturing to manufacturing capital is lower. Therefore, the market economy has a higher growth rate.

Keywords Two-sector model • Manufacturing sector • Learning by doing • Market economy • Command economy • Optimal subsidy • Endogenous growth

4.1 Introduction

There has been a long debate in the economic literature over whether governments can play an important role in helping the market to internalize externalities that may be important for developing a nascent industry (by subsidizing industries).

E.R. Casares (✉)
Departamento de Economia, Universidad Autonoma Metropolitana Unidad Azcapotzalco,
Av. San Pablo 180, Col. Reynosa Tamaulipas, Delegacion Azcapotzalco,
02200, Mexico City, Mexico
e-mail: ercg@correo.azc.uam.mx

H. Sobarzo
El Colegio de Mexico, Centro de Estudios Economicos, Camino al Ajusco 20, Tlalpan, Pedregal
de Santa Teresa, 10740, Mexico City, Mexico
e-mail: hsobarzo@colmex.mx

© Springer International Publishing Switzerland 2016
A.A. Pinto et al. (eds.), *Trends in Mathematical Economics*,
DOI 10.1007/978-3-319-32543-9_4

The experiences of East Asian countries (China is an outstanding example of this) have brought the role of subsidies in promoting economic growth back into the public discussion (see Haley and Haley 2013; Stiglitz and Greenwald 2014).[1] The theory of endogenous growth with learning externalities has pointed out the positive impact of subsidies on economic growth. Therefore, in this paper, we study the relationship between subsidies and economic growth with a multi-sector dynamic general equilibrium approach.

Consequently, we develop an endogenous growth model with two sectors, manufacturing (learning) and nonmanufacturing (non-learning), with two types of capital. The economy is closed.[2] We assume that the manufacturing (learning) sector is the only sector that generates domestic technological knowledge through learning by doing. The knowledge produced in the manufacturing sector is available to the nonmanufacturing (non-learning) sector. Thus, the model has two learning externalities. Therefore, the manufacturing sector drives the market economy to a sustained positive growth rate. We assume that the two goods are consumed and accumulated. The government taxes households with a lump-sum tax to finance an investment subsidy in the manufacturing sector. Households own both types of capital. The main objective of this paper is to obtain the optimal subsidy in the steady state. Our model is related to models with two types of physical capital and externalities. Thus, Brock and Turnovsky (1994) and Turnovsky (1996) develop models with two types of physical capital. In particular, Korinek and Serven (2010) develop an endogenous growth model where the tradable sector generates higher learning externalities than the non-tradable sector.

First, we present a market economy with zero subsidies. With the aim of identifying the optimal subsidy, we find the planner's solution where both externalities are internalized. Thus, with the optimal solution, we obtain the optimal rate of investment subsidy to the manufacturing sector in the market economy. Next, we study how the economy responds, in the steady state, when the government establishes the optimal rate of investment subsidy in the manufacturing sector. Thus, when the subsidy is increased, the manufacturing sector is encouraged, and the proportion of labor in the manufacturing sector increases initially. Likewise, investment in the manufacturing sector expands, and investment in the nonmanufacturing sector falls. Consequently, the ratio of nonmanufacturing to manufacturing

---

[1]The Chinese government, for instance, has been subsidizing the shipbuilding industry (and all the suppliers of this industry) and in this way generating large sources of domestic employment. These subsidies translate also to the transportation costs for the manufacturing goods that China exports. This could help to explain how a subsidy can be effective in both ways, first, by creating domestic jobs and, second, by helping to export goods to compete in international markets (see Haley and Haley 2013). Stiglitz and Greenwald (2014) recommend encouraging the industrial sector (as in East Asia).

[2]The economy also can be interpreted as open but without capital mobility (trade balance is zero at all time) where the manufacturing sector would correspond to the tradable (learning) sector and the nonmanufacturing sector to the non-tradable (non-learning) sector. In this interpretation, the relative price of the nonmanufacturing good is also understood as the real exchange rate.

capital decreases slowly. Given that the price of the nonmanufacturing good is determined by supply and demand, the relative price of the nonmanufacturing good decreases initially. This produces an additional initial increase in the proportion of labor in the manufacturing sector. However, the level of the relative price of the nonmanufacturing good is higher in the optimal steady state. Moreover, as total wealth increases, the ratio of consumption to nonmanufacturing capital increases, as well. We suggest that the optimal rate of investment subsidy increases in the transition.

In summary, in the optimal solution, the proportion of labor in the manufacturing sector is higher, the ratio of nonmanufacturing to manufacturing capital is lower, the relative price of the nonmanufacturing good is higher, and the ratio of consumption to nonmanufacturing capital is higher. Therefore, since the manufacturing sector is the leader in technological terms, the optimal subsidy produces a higher long-run growth rate in the market economy. We remember that in exogenous growth models (the long-run growth rate of the economy is determined by exogenous technical progress), an increase in the subsidy rate has a level effect, that is, the income level of the economy increases in the long run; in endogenous growth models (the long-run growth rate of the economy is determined internally by the model), a rise in the subsidy rate has a growth effect, that is, the growth rate of the economy increases in the long run.

Thus, we have generalized in an economy with two learning externalities, two capital goods, and endogenous growth, the basic conclusion of the learning-by-doing literature that the first best response of the government is to establish an investment subsidy in the learning sector. Thus, the optimal policy is to encourage the sources of the learning process, the manufacturing sector (see Clemhout and Wan 1970; Bardhan 1971; Succar 1987; Boldrin and Schienkman 1988; Young 1991; Rauch 1992; Aizenman and Lee 2010). Therefore, the results of the impact of the subsidy on the relative price and the allocation of labor between sectors and growth that we have obtained are not present in the literature and contribute to a better understanding of the relationship between subsidies and economic growth. However, whether subsidies are permitted or not, or whether governments have the ability to deal appropriately with externalities or not, it still remains to be discussed as to what extent these subsidy processes can be carried out in a democratic country, that is, how a government can justify subsidizing one particular sector. These questions belong to the arena of political economy.

In Sect. 4.2, we develop a model of a competitive market economy, and we find the steady-state solution. In Sect. 4.3, we discuss the planner's solution, and we conclude that the optimal growth rate is higher than that achieved in the market economy. In Sect. 4.4, we deduce the optimal rate of investment subsidy to the manufacturing sector. In Sect. 4.5, we present our conclusions.

4.2 The Competitive Market Economy

In this section, we develop a dynamic general equilibrium model of a competitive market economy.[3] There are two production sectors: the manufacturing (learning) and nonmanufacturing (non-learning) sectors. There are a large number of competitive manufacturing and nonmanufacturing firms with the same production function. The manufacturing good and the nonmanufacturing good are produced, accumulated, and consumed. The output in each sector is produced through physical capital, labor, and technological knowledge. The total labor supply is constant. Labor is freely mobile between the two sectors. The representative household maximizes the present value of a utility function. The consumption basket is formed by the manufacturing and nonmanufacturing goods. The government collects taxes and gives subsidies.

4.2.1 The Manufacturing Sector

We assume that the production function of the manufacturing (learning) firm i ($i = 1, \ldots, N$, where N is large) is Cobb–Douglas:

$$Y_{M_i} = A_{M_i} K_{M_i}^{\alpha} L_{M_i}^{1-\alpha} E_1$$

where Y_{M_i} is the output of the manufacturing firm i; A_{M_i} is a positive parameter of efficiency; K_{M_i} is the stock of physical capital accumulated of the manufacturing good in the manufacturing firm i; L_{M_i} is the labor employed in the manufacturing firm i; α and $1 - \alpha$ are the shares of K_{M_i} and L_{M_i}, respectively, with $0 < \alpha < 1$; and E_1 is a learning externality.

---

[3]We have used a general equilibrium approach in previous research. Thus, in a static general equilibrium model with scale economies and imperfect competition in the Mexican industry, Sobarzo (1994) evaluates the effects that an eventual free-trade agreement between Mexico, Canada, and the United States would have on the Mexican economy. Also, with a static model, Sobarzo (2000) shows the interaction between trade and tax reform in Mexico. The general conclusion is that changes in value-added tax and public pricing policy do not have strong effects on trade performance and, more generally, on reallocation of resources. Moreover, Casares (2004) develops an export sector-led endogenous growth model with two learning externalities. First, he shows theoretically that when the tariff rate is reduced, the labor factor flows to the export sector, the capital accumulation increases in this sector, and the growth rate of the economy rises. Second, using data of the Mexican manufacturing sector, 1988–2000 (before and after NAFTA), he concludes that the highly exporter manufacturing sector behaved as predicted by the model. Also, Casares (2007) develops an endogenous growth model with two sectors, manufacturing and nonmanufacturing. The manufacturing sector is the source of productivity growth. The main conclusion is that when productivity increases in the manufacturing sector, the fraction of labor employed in the manufacturing sector follows an inverted V curve as the documented pattern of development for the share of manufacturing employment.

Let K_M be the aggregate stock of physical capital accumulated of the manufacturing good. Domestic technological knowledge is created through learning by doing in the manufacturing sector, so knowledge is a by-product of investment (Arrow 1962). Since knowledge is a public good, there are spillover effects of knowledge across manufacturing firms. Therefore, E_1 is the external effect of K_M in the production function of the manufacturing firm i. In order to generate endogenous growth, we assume $E_1 = K_M^{1-\alpha}$ (Romer 1986, 1989).

Given that all the manufacturing firms make the same choice, we obtain the aggregate production function of the manufacturing sector:

$$Y_M = A_M K_M^\alpha L_M^{1-\alpha} \left[K_M^{1-\alpha} \right] \tag{4.1}$$

where Y_M is the aggregate output in the manufacturing sector, A_M is the aggregate positive parameter of efficiency, and L_M is the aggregate labor employed in the sector. We assume that K_M is used only in the manufacturing sector.

Considering that the rate of depreciation of K_M is zero and that the price of the manufacturing good is the numéraire, the rental price of K_M is $R_M = r$, where r is the interest rate. As we will see, the optimal government policy is to establish an investment subsidy in the manufacturing sector. Thus, we introduce a rate of investment subsidy, μ, where $0 < \mu < 1$. Taking the externality as given, the manufacturing firms maximize profit $\pi_M = A_M K_M^\alpha L_M^{1-\alpha} \left[K_M^{1-\alpha} \right] - w_M L_M - R_M (1 - \mu) K_M$, where w_M is the wage rate in the sector. The first-order conditions are:

$$w_M = A_M K_M^\alpha (1-\alpha) L_M^{-\alpha} \left[K_M^{1-\alpha} \right] = A_M K_M (1-\alpha) L_M^{-\alpha} \tag{4.2}$$

$$R_M (1-\mu) = r (1-\mu) = A_M \alpha K_M^{\alpha-1} L_M^{1-\alpha} \left[K_M^{1-\alpha} \right] = A_M \alpha L_M^{1-\alpha} \tag{4.3}$$

Equation (4.2) states that the wage rate is equal to the value of the marginal product of L_M. Equation (4.3) states that the interest rate, net of subsidy, is equal to the marginal product of K_M.

4.2.2 The Nonmanufacturing Sector

We assume that the production function of the nonmanufacturing (non-learning) firm i is Cobb–Douglas:

$$Y_{N_i} = A_{N_i} K_{N_i}^\beta L_{N_i}^{1-\beta} E_2$$

where Y_{N_i} is the output of the nonmanufacturing firm i; A_{N_i} is a positive parameter of efficiency; K_{N_i} is the stock of physical capital accumulated of the nonmanufacturing good in the nonmanufacturing firm i; L_{N_i} is labor employed in the nonmanufacturing firm i; β and $1 - \beta$ are the shares of K_{N_i} and L_{N_i}, respectively, with $0 < \beta < 1$; and E_2 is a learning externality.

Since there are spillover effects of knowledge between the sectors, E_2 is technological knowledge generated in the manufacturing sector, but used in the nonmanufacturing sector. We consider that these interindustry benefits of knowledge are purely external to the nonmanufacturing firm i. Thus, E_2 is the external effect of K_M in the production function of the nonmanufacturing firm i. We assume $E_2 = K_M^{1-\beta}$.

Given that all the nonmanufacturing firms make the same choice, we obtain the aggregate production function of the nonmanufacturing sector:

$$Y_N = A_N K_N^{\beta} L_N^{1-\beta} \left[K_M^{1-\beta} \right] \tag{4.4}$$

where Y_N is the aggregate output in the nonmanufacturing sector, A_N is the aggregate positive parameter of efficiency, K_N is the aggregate stock of physical capital accumulated of the nonmanufacturing good, and L_N is the total labor employed in the nonmanufacturing sector. We assume that K_N is used only in the nonmanufacturing sector.

We define p_N as the relative price of the nonmanufacturing to the manufacturing good. Considering that the rate of depreciation of K_N is zero, the rental price of K_N is $R_N = p_N(r - \dot{p}_N/p_N)$, where $\dot{p}_N/p_N$ is the growth rate of p_N (capital gains of K_N). Taking the externality as given, the nonmanufacturing firms maximize profit $\pi_N = p_N A_N K_N^{\beta} L_N^{1-\beta} \left[K_M^{1-\alpha} \right] - w_N L_N - R_N K_N$ where w_N is the wage rate in the sector. The first-order conditions are:

$$w_N = p_N A_N K_N^{\beta} (1 - \beta) L_N^{-\beta} \left[K_M^{1-\beta} \right] = p_N A_N K_N^{\beta} K_M^{1-\beta} (1 - \beta) L_N^{-\beta} \tag{4.5}$$

$$R_N = p_N(r - \dot{p}_N/p_N) = p_N A_N \beta K_N^{\beta-1} L_N^{1-\beta} \left[K_M^{1-\beta} \right]$$

$$= p_N A_N \beta K_N^{\beta-1} K_M^{1-\beta} L_N^{1-\beta} \tag{4.6}$$

Equation (4.5) states that the wage rate is equal to the value of the marginal product of L_N. Equation (4.6) states that the rental price of K_N is equal to the marginal product of K_N or the interest rate is equal to the marginal product of K_N plus capital gains.

4.2.3 The Government

The investment subsidy is financed through lump-sum taxes, T, to the households. The government has a balanced government budget constraint:

$$T = \mu R_M K_M \tag{4.7}$$

where $\mu R_M K_M$ is the amount of investment subsidy in the manufacturing sector.

4.2.4 The Representative Household

The household disposable income is the sum of labor income and interest on assets less lump-sum taxes. This disposable income is allocated to consumption or saving. Thus, the budget constraint of the representative household is:

$$w_M L_M + w_N L_N + R_M K_M + R_N K_N - T = C_M + p_N C_N + I_M + p_N I_N \qquad (4.8)$$

where $w_M L_M + w_N L_N$ is wage income, $R_M K_M + R_N K_N$ is capital income, C_M is consumption of the manufacturing good, C_N is consumption of the nonmanufacturing good, $I_M = \dot{K}_M$ is the net investment in K_M, and $I_N = \dot{K}_N$ is the net investment in K_N. Next, we can define C (aggregate consumption) as a homothetic index of C_M and C_N: $C = D C_M^\gamma C_N^{1-\gamma}$, where $D = \gamma^{-\gamma}(1-\gamma)^{-(1-\gamma)}$ is a parameter and γ and $1-\gamma$ are the shares of C_M and C_N in the total expenditure on consumption, respectively, with $0 < \gamma < 1$. The consumer price index, p_C, is defined as $p_C = p_N^{1-\gamma}$. Thus, the total expenditure on consumption is:

$$p_C C = C_M + p_N C_N \qquad (4.9)$$

Households can borrow and lend in the debt market (zero net loans in the aggregate). Also, we define $A = K_M + p_N K_N$, where A are assets, and $\dot{A} = \dot{K}_M + p_N \dot{K}_N + \dot{p}_N K_N$. Using the previous concepts, the budget constraint, Eq. (4.8), becomes:

$$w_M L_M + w_N L_N + rA - T = p_C C + \dot{A} \qquad (4.10)$$

The decision problem of the representative household is to choose a path of aggregate consumption that maximizes the present value of a utility function with a constant elasticity of intertemporal substitution, σ, and a constant subjective discount factor, ρ, with $\rho > 0$, so:

$$\max U(0) = \int_0^\infty \frac{C^{1-1/\sigma}}{1 - 1/\sigma} e^{-\rho t} dt$$

where $C = D C_M^\gamma C_N^{1-\gamma}$, subject to the budget constraint, Eq. (4.10), and to the solvency condition $\lim_{t \to \infty} A e^{-\int_0^t r_v dv} \geq 0$.

The first-order conditions are:

$$\frac{\dot{\lambda}_A}{\lambda_A} = \rho - r \qquad (4.11)$$

$$\lambda_A = \frac{C^{-1/\sigma}}{p_C} \qquad (4.12)$$

and $\lim_{t \to \infty} \lambda_A e^{-\rho t} A = 0$, where λ_A is the shadow price, as of time t, of A at time t. Next, considering that p_N varies with time, we take logarithms and time derivatives of the consumer price index and obtain:

$$\frac{\dot{p}_C}{p_C} = (1 - \gamma)\frac{\dot{p}_N}{p_N} \tag{4.13}$$

Also, we take logarithms and time derivatives of Eq. (4.12) and obtain:

$$\frac{\dot{\lambda}_A}{\lambda_A} = -\left(\frac{1}{\sigma}\right)\frac{\dot{C}}{C} - \frac{\dot{p}_C}{p_C} \tag{4.14}$$

Substituting Eqs. (4.11) and (4.13) in (4.14), we obtain the dynamic allocation condition for aggregate consumption:

$$\frac{\dot{C}}{C} = \sigma\left[r - (1 - \gamma)\frac{\dot{p}_N}{p_N} - \rho\right] \tag{4.15}$$

The optimal consumption basket of C_M and C_N results from static maximization of the utility function $DC_M^\gamma C_N^{1-\gamma}$ subject to the total expenditure on consumption, Eq. (4.9). The static first-order conditions are:

$$C_M = \gamma p_C C \tag{4.16}$$

$$C_N = (1 - \gamma)\frac{p_C C}{p_N} \tag{4.17}$$

4.2.5 Equilibrium in Goods and Labor Markets

We can now proceed to obtain the resource constraint of the economy. Substituting Eqs. (4.2), (4.3), (4.5)–(4.7) in the budget constraint of the representative household, Eq. (4.8), we obtain:

$$Y_M + p_N Y_N = C_M + p_N C_N + I_M + p_N I_N \tag{4.18}$$

Equation (4.18) is the aggregate equilibrium condition for the goods market, where the value of the total output, Y, is $Y = Y_M + p_N Y_N$. Next, we define the equilibrium condition for the nonmanufacturing good market. The relative price of the nonmanufacturing good is flexible, ensuring that the supply of the nonmanufacturing good is always equal to its demand:

$$Y_N = C_N + I_N \tag{4.19}$$

With the equilibrium condition for the nonmanufacturing good market, we can obtain the equilibrium condition for the manufacturing good market. Thus,

Eq. (4.18) becomes:

$$Y_M = C_M + I_M \tag{4.20}$$

The equilibrium condition in the labor market is:

$$L_M + L_N = L \tag{4.21}$$

where L is the total labor supply and we assume that it is constant.

4.2.6 The Model in Stationary Variables

Given that C, K_M, K_N, Y_M, Y_N, and Y are growing at all times, to solve the model we define the variables in terms of stationary variables. The characteristic of these variables is that they remain constant in the steady state (see Barro and Sala-i-Martin 2004). Thus, we define the variables $z = K_N/K_M$ and $v = C/K_N$ as stationary variables. As L is constant, it is normalized to one. Thus, the equilibrium condition in the labor market is $n + (1 - n) = 1$, where n is the fraction of labor employed in the manufacturing sector and $(1 - n)$ is the fraction of labor employed in the nonmanufacturing sector. Given that n is constant in the steady state, we can use it as another stationary variable. Therefore, we can rewrite the aggregate production functions as:

$$Y_M = A_M K_M n^{1-\alpha} \tag{4.22}$$

$$Y_N = A_N K_M z^\beta (1 - n)^{1-\beta} \tag{4.23}$$

We can rewrite the first-order conditions (4.2), (4.3), (4.5), and (4.6) as:

$$w_M = A_M K_M (1 - \alpha) n^{-\alpha} \tag{4.24}$$

$$r(1 - \mu) = A_M \alpha n^{1-\alpha} \tag{4.25}$$

$$w_N = p_N A_N K_M z^\beta (1 - \beta)(1 - n)^{-\beta} \tag{4.26}$$

$$r - \frac{\dot{p}_N}{p_N} = \frac{A_N \beta (1 - n)^{1-\beta}}{z^{1-\beta}} \tag{4.27}$$

Equating the value of the marginal product of labor in both sectors, Eq. (4.24) and (4.26), we find the static efficient allocation condition for labor between the sectors:

$$A_M (1 - \alpha) n^{-\alpha} = p_N A_N z^\beta (1 - \beta)(1 - n)^{-\beta} \tag{4.28}$$

With Eqs. (4.25) and (4.27), we obtain the dynamic arbitrage condition for the two capital goods:

$$\frac{A_M \alpha n^{1-\alpha}}{(1-\mu)} = \frac{A_N \beta (1-n)^{1-\beta}}{z^{1-\beta}} + \frac{\dot{p}_N}{p_N} \tag{4.29}$$

where the total private returns for both types of capital must be the same. Thus, Eq. (4.29) states that the private marginal product of K_M is equal to the private marginal product of K_N plus capital gains on K_N. We assume that $\alpha > \beta$, so the manufacturing sector is more capital intensive than the nonmanufacturing sector.

Using Eqs. (4.15) and (4.25), we can define the growth rate of aggregate consumption as:

$$\frac{\dot{C}}{C} = \sigma \left[\frac{A_M \alpha n^{1-\alpha}}{(1-\mu)} - (1-\gamma)\frac{\dot{p}_N}{p_N} - \rho \right] \tag{4.30}$$

where $\dot{C}/C = g_C$ is the growth rate of C. Alternatively, with Eqs. (4.15) and (4.27), we can obtain the growth rate of aggregate consumption as:

$$\frac{\dot{C}}{C} = \sigma \left[\frac{A_N \beta (1-n)^{1-\beta}}{z^{1-\beta}} + \gamma\frac{\dot{p}_N}{p_N} - \rho \right] \tag{4.31}$$

Finally, we can rewrite the equilibrium conditions (4.19) and (4.20) in terms of the stationary variables. Considering the production function of the manufacturing sector, Eq. (4.22); the definition of $v = C/K_N$, the level of C_M, Eq. (4.16); and the identity $I_M = \dot{K}_M$ and that $p_C = p_N^{1-\gamma}$, we can rewrite the equilibrium condition for the market of the manufacturing good, Eq. (4.20), as:

$$\frac{\dot{K}_M}{K_M} = A_M n^{1-\alpha} - \gamma p_N^{1-\gamma} vz \tag{4.32}$$

where $\dot{K}_M/K_M = g_{K_M}$ is the growth rate of K_M. Also, with the production function of the nonmanufacturing sector, Eq. (4.23); the level of C_N, Eq. (4.17); and the identity $I_N = \dot{K}_N$ and that $p_C = p_N^{1-\gamma}$, we can rewrite the equilibrium condition for the market of the nonmanufacturing good, Eq. (4.19), as:

$$\frac{\dot{K}_N}{K_N} = \frac{A_N (1-n)^{1-\beta}}{z^{1-\beta}} - \frac{(1-\gamma)\,v}{p_N^{\gamma}} \tag{4.33}$$

where $\dot{K}_N/K_N = g_{K_N}$ is the growth rate of K_N.

4.2.7 The Steady-State Solution in the Market Economy

We can obtain a system of three nonlinear equations in three variables, z, n, and v. First, using the definition of z, the growth rate of z is $\dot{z}/z = \dot{K}_N/K_N - \dot{K}_M/K_M$. Next, we obtain the growth rates of K_M and K_N in terms of z, n, and v. From the efficient allocation condition for labor market, Eq. (4.28), we obtain the level of p_N in terms of stationary variables:

$$p_N = \frac{A_M (1 - \alpha) (1 - n)^\beta}{A_N z^\beta (1 - \beta) n^\alpha} \tag{4.34}$$

Given that p_N depends on z, n, and its parameters, we have that p_N is constant in the steady state. Using Eq. (4.34), we can rewrite Eqs. (4.32) and (4.33) in the steady state:

$$g_{K_M}^* = A_M n^{*(1-\alpha)} - \gamma \left[\frac{A_M (1 - \alpha) (1 - n^*)^\beta}{A_N z^{*\beta} (1 - \beta) n^{*\alpha}} \right]^{1-\gamma} v^* z^* \tag{4.35}$$

$$g_{K_N}^* = \frac{A_N (1 - n^*)^{1-\beta}}{z^{*(1-\beta)}} - (1 - \gamma) \left[\frac{A_N z^{*\beta} (1 - \beta) n^{*\alpha}}{A_M (1 - \alpha) (1 - n^*)^\beta} \right]^\gamma v^* \tag{4.36}$$

where the steady state levels are denoted with $*$. In the steady state, the growth rate of z is zero, so $g_{K_M}^* = g_{K_N}^*$. Using Eqs. (4.35) and (4.36), we have:

$$A_M n^{*(1-\alpha)} - \gamma \left[\frac{A_M (1 - \alpha) (1 - n^*)^\beta}{A_N z^{*\beta} (1 - \beta) n^{*\alpha}} \right]^{1-\gamma} v^* z^*$$

$$= \frac{A_N (1 - n^*)^{1-\beta}}{z^{*(1-\beta)}} - (1 - \gamma) \left[\frac{A_N z^{*\beta} (1 - \beta) n^{*\alpha}}{A_M (1 - \alpha) (1 - n^*)^\beta} \right]^\gamma v^* \tag{4.37}$$

We know that the growth rate of v is $\dot{v}/v = \dot{C}/C - \dot{K}_N/K_N$. Given that $\dot{p}_N/p_N = 0$, the growth rate of C, Eq. (4.30), in the steady state is:

$$g_C^* = \sigma \left[\frac{A_M \alpha n^{*(1-\alpha)}}{(1 - \mu)} - \rho \right] \tag{4.38}$$

alternatively, the growth rate of C, Eq. (4.31), in the steady state is:

$$g_C^* = \sigma \left[\frac{A_N \beta (1 - n^*)^{1-\beta}}{z^{*(1-\beta)}} - \rho \right] \tag{4.39}$$

In the steady state, the growth rate of v is zero, so $g_C^* = g_{K_N}^*$. With Eqs. (4.38) and (4.36), we obtain:

$$\sigma \left[\frac{A_M \alpha n^{*(1-\alpha)}}{(1-\mu)} - \rho \right] = \frac{A_N(1-n^*)^{1-\beta}}{z^{*(1-\beta)}} - (1-\gamma) \left[\frac{A_N z^{*\beta}(1-\beta)n^{*\alpha}}{A_M(1-\alpha)(1-n^*)^\beta} \right]^\gamma v^* \tag{4.40}$$

alternatively, with Eqs. (4.39) and (4.36), we have:

$$\sigma \left[\frac{A_N \beta(1-n^*)^{1-\beta}}{z^{*(1-\beta)}} - \rho \right] = \frac{A_N(1-n^*)^{1-\beta}}{z^{*(1-\beta)}} - (1-\gamma) \left[\frac{A_N z^{*\beta}(1-\beta)n^{*\alpha}}{A_M(1-\alpha)(1-n^*)^\beta} \right]^\gamma v^* \tag{4.41}$$

Given that $\dot{p}_N/p_N = 0$, the dynamic arbitrage condition for the two capital goods, Eq. (4.29), is:

$$\frac{A_M \alpha n^{*(1-\alpha)}}{(1-\mu)} = \frac{A_N \beta(1-n^*)^{1-\beta}}{z^{*(1-\beta)}} \tag{4.42}$$

Therefore, we have obtained a system of three nonlinear equations, (4.37), (4.40) or (4.41), and (4.42), in three variables, z, n, and v. Finally, given that $\dot{p}_N/p_N = 0$ and $\dot{n} = 0$, we can show that the growth rate of Y, g_Y, is:

$$g_Y^* = \frac{Y_M}{Y} g_{Y_M}^* + \frac{P_N Y_N}{Y} g_{Y_N}^* \tag{4.43}$$

where $Y_M/Y = 1/\{1 + [p_N^* A_N z^{*\beta}(1-n^*)^{1-\beta}/A_M n^{*(1-\alpha)}]\}$ is the share of Y_M in the value of total output, $p_N Y_N/Y = 1/\{[A_M n^{*(1-\alpha)}/(p_N^* A_N z^{*\beta}(1-n^*)^{1-\beta})] + 1\}$ is the share of $p_N Y_N$ in the value of total output, p_N^* is given by Eq. (4.34) in the steady state, $g_{Y_M}^*$ is the growth rate of Y_M, and $g_{Y_N}^*$ is the growth rate of Y_N. With Eqs. (4.22) and (4.23), we obtain in the steady state:

$$g^* = g_Y^* = g_{Y_M}^* = g_{Y_N}^* = g_{K_M}^* = g_{K_N}^* = g_C^* \tag{4.44}$$

so Y, Y_M, and Y_N grow at the same rate as K_M, K_N, and C. Thus, in the steady state, the long-run growth rate is defined as g^*.

We solve numerically the system of equations, (4.37), (4.40) or (4.41), and (4.42), with fsolve/MATLAB. Roe et al. (2010) show numerical algorithms for the solution of some multi-sector growth models. We use the following parameter values: Valentinyi and Herrendorf (2008) show (US economy) that the tradable sector (agriculture, manufactured consumption, and equipment investment) is more capital intensive than the non-tradable sector (services and construction investment); thus, $\alpha = 0.37$ and $\beta = 0.32$. We use $\rho = 0.02$ as in Barro and Sala-i-Martin (2004). We set $\gamma = 0.4$ (see Rabanal and Tuesta 2013). We give $\sigma = 0.2$ (see Yogo 2004). As the magnitude of A_M and A_N depends on the unique characteristics of an economy, they are set only for explanatory purposes as $A_M = 0.4$ and $A_N = 0.4$. For the moment, we impose $\mu = 0$. We obtain that $z^* = 1.20$, $n^* = 0.383$, $v^* = 0.411$, $p_N^* = 1.06$, and $g^* = 0.012$. Thus, the steady-state growth rate is 1.2 % per annum. In the next section, we develop and solve the command economy.

4.3 The Command Economy

Since there are two externalities, the market economy is inefficient. To identify the optimal solution, we need to find the planner's solution, that is, we need to internalize the externalities. Given that in the command economy there are no markets and prices, the social coordinator maximizes the present value of a constant intertemporal elasticity of substitution utility function:

$$\max\ U(0) = \int_0^\infty \frac{\left(DC_M^\gamma C_N^{1-\gamma}\right)^{1-1/\sigma}}{1-1/\sigma} e^{-pt} dt$$

subject to $Y_M = C_M + \dot{K}_M$ and $Y_N = C_N + \dot{K}_N$ where $Y_M = A_M K_M n^{1-\alpha}$ and $Y_N = A_N K_N^\beta K_M^{1-\beta}(1-n)^{1-\beta}$, which explicitly take into account the externalities and the labor market equilibrium condition.

The Hamiltonian is:

$$H = \left\{ \frac{\left(DC_M^\gamma C_N^{1-\gamma}\right)^{1-1/\sigma}}{1-1/\sigma} + \lambda_M \left[A_M K_M n^{1-\alpha} - C_M\right] \right.$$
$$\left. + \lambda_N \left[A_N K_N^\beta K_M^{1-\beta}(1-n)^{1-\beta} - C_N\right] \right\} e^{-pt}$$

where λ_M and λ_N are the shadow prices as of time t and of an additional unit of K_M and K_N at time t, respectively. The first-order conditions with respect to C_M, C_N, and n are:

$$\left(DC_M^\gamma C_N^{1-\gamma}\right)^{-1/\sigma} D\gamma C_M^{\gamma-1} C_N^{1-\gamma} = \lambda_M \tag{4.45}$$

$$\left(DC_M^\gamma C_N^{1-\gamma}\right)^{-1/\sigma} DC_M^\gamma (1-\gamma) C_N^{-\gamma} = \lambda_N \tag{4.46}$$

$$A_M K_M (1-\alpha) n^{-\alpha} = \frac{\lambda_N}{\lambda_M} \left[A_N K_N^\beta K_M^{1-\beta}(1-\beta)(1-n)^{-\beta}\right] \tag{4.47}$$

The first-order conditions with respect to K_M and K_N are:

$$A_M n^{1-\alpha} + \frac{\lambda_N}{\lambda_M}\left[A_N K_N^\beta (1-\beta) K_M^{-\beta}(1-n)^{1-\beta}\right] + \frac{\dot{\lambda}_M}{\lambda_M} = \rho \qquad (4.48)$$

$$A_N \beta K_N^{\beta-1} K_M^{1-\beta}(1-n)^{1-\beta} + \frac{\dot{\lambda}_N}{\lambda_N} = \rho \qquad (4.49)$$

with $\lim_{t\to\infty} \lambda_M e^{-\rho t} K_M = 0$ and $\lim_{t\to\infty} \lambda_N e^{-\rho t} K_N = 0$.

Let $p_N = \lambda_N/\lambda_M$, then we can define aggregate conditions (see Barro and Sala-i-Martin 2004). Using $z = K_N/K_M$, Eq. (4.47) is the static efficient allocation condition for labor: $A_M(1-\alpha)n^{-\alpha} = p_N A_N z^\beta(1-\beta)(1-n)^{-\beta}$. We see the static efficient allocation condition for labor in the command economy is equal to Eq. (4.28) in the market economy. Next, substituting $\lambda_N = p_N \lambda_M$ in Eq. (4.46) and equating the result in Eq. (4.45), we obtain:

$$\frac{\gamma}{(1-\gamma)}\frac{C_N}{C_M} = \frac{1}{p_N} \qquad (4.50)$$

Equation (4.50) states that the marginal rate of substitution between C_M and C_N is equal to the relative price. With Eq. (4.50) and $p_C C = C_M + p_N C_N$, where $p_C = p_N^{1-\gamma} = (\lambda_N/\lambda_M)^{1-\gamma}$ and $\dot{p}_C/p_C = (1-\gamma)\dot{p}_N/p_N$, we obtain the levels of C_M and C_N: $C_M = \gamma p_C C$ and $C_N = (1-\gamma)p_C C/p_N$. Using Eq. (4.45), $C = D C_M^\gamma C_N^{1-\gamma}$ and $C_M = \gamma p_C C$, we obtain:

$$C^{-1/\sigma} = \lambda_M p_C \qquad (4.51)$$

With Eq. (4.46), $C = D C_M^\gamma C_N^{1-\gamma}$ and $C_N = (1-\gamma)p_C C/p_N$, we find:

$$C^{-1/\sigma} p_N^\gamma = \lambda_N \qquad (4.52)$$

Taking logarithms and time derivatives of Eq. (4.51), we obtain $\dot{\lambda}_M/\lambda_M = -(1/\sigma)\dot{C}/C - (1-\gamma)\dot{p}/p$. Using $z = K_N/K_M$ and equating $\dot{\lambda}_M/\lambda_M$ in Eq. (4.48), we have:

$$\frac{\dot{C}}{C} = \sigma\left[A_M n^{1-\alpha} + p_N A_N z^\beta(1-\beta)(1-n)^{1-\beta} - (1-\gamma)\frac{\dot{p}_N}{p_N} - \rho\right] \qquad (4.53)$$

alternatively, taking logarithms and time derivatives of Eq. (4.52), we have $\dot{\lambda}_N/\lambda_N = -(1/\sigma)\dot{C}/C + \gamma\dot{p}/p$. Using $z = K_N/K_M$ and equating $\dot{\lambda}_N/\lambda_N$ in Eq. (4.49), we obtain:

$$\frac{\dot{C}}{C} = \sigma\left[A_N \beta z^{\beta-1}(1-n)^{1-\beta} + \gamma\frac{\dot{p}_N}{p_N} - \rho\right] \qquad (4.54)$$

Equating Eqs. (4.53) and (4.54), we obtain the optimal dynamic arbitrage condition for the two capital goods:

$$A_M n^{1-\alpha} + p_N A_N z^\beta (1-\beta)(1-n)^{1-\beta} = A_N \beta z^{\beta-1}(1-n)^{1-\beta} + \frac{\dot{p}_N}{p_N} \quad (4.55)$$

indicating that the total social return of K_M is equal to the total social return of K_N. When the externalities are internalized, the total social return of K_M is formed by the social marginal product of K_M in the manufacturing sector plus the social marginal product of K_M in the nonmanufacturing sector, all expressed relative to the price of the manufacturing good. The total social return of K_N is equal to the social marginal product of K_N plus capital gains. When we compare Eq. (4.55) with Eq. (4.29) with zero subsidy and $0 < \alpha < 1$, we conclude that the private return of K_M, $A_M \alpha n^{1-\alpha}$ is lower than the total social return of K_M, $A_M n^{1-\alpha} + p_N A_N z^\beta (1-\beta)(1-n)^{1-\beta}$. Thus, the market economy is under-accumulating, implying that the market economy has a lower growth rate than the optimal growth rate.

4.3.1 The Steady-State Solution in the Command Economy

Now, we solve the command economy in the steady state. We need to form a system as we did for the case of the market economy. Using p_N, Eq. (4.34), and $Y_M = C_M + I_M$, the growth rate of K_M (following the procedure of Sect. 4.2) is given by Eq. (4.35). Using Eq. (4.34) and $Y_N = C_N + I_N$, the growth rate of K_N is given by Eq. (4.36). In the steady state, $g_{K_M}^* = g_{K_N}^*$, so we obtain:

$$A_M n^{*(1-\alpha)} - \gamma \left[\frac{A_M (1-\alpha)(1-n^*)^\beta}{A_N z^{*\beta}(1-\beta) n^{*\alpha}} \right]^{1-\gamma} v^* z^* \quad (4.56)$$

$$= \frac{A_N (1-n^*)^{1-\beta}}{z^{*(1-\beta)}} - (1-\gamma) \left[\frac{A_N z^{*\beta}(1-\beta) n^{*\alpha}}{A_M (1-\alpha)(1-n^*)^\beta} \right]^\gamma v^*$$

Next, we know that the growth rate of v is $\dot{v}/v = \dot{C}/C - \dot{K}_N/K_N$. Given that $\dot{p}_N/p_N = 0$, and using p_N, Eq. (4.34), the growth rate of C, Eq. (4.53), in the steady state is:

$$g_C^* = \sigma \left[A_M n^{*(1-\alpha)} + \frac{A_M (1-\alpha)(1-n^*)}{n^{*\alpha}} - \rho \right] \quad (4.57)$$

alternatively, with Eq. (4.54), we obtain:

$$g_C^* = \sigma \left[A_N \beta z^{*(\beta-1)}(1-n^*)^{1-\beta} - \rho \right] \quad (4.58)$$

In the steady state, $g_C^* = g_{K_N}^*$, from Eqs. (4.57) and (4.36), we have:

$$\sigma \left[A_M n^{*(1-\alpha)} + \frac{A_M (1-\alpha)(1-n^*)}{n^{*\alpha}} - \rho \right] \tag{4.59}$$

$$= \frac{A_N(1-n^*)^{1-\beta}}{z^{*(1-\beta)}} - (1-\gamma) \left[\frac{A_N z^{*\beta}(1-\beta) n^{*\alpha}}{A_M(1-\alpha)(1-n^*)^\beta} \right]^\gamma v^*$$

alternatively, with Eqs. (4.58) and (4.36):

$$\sigma \left[A_N \beta z^{*(\beta-1)}(1-n^*)^{1-\beta} - \rho \right]$$

$$= \frac{A_N(1-n^*)^{1-\beta}}{z^{*(1-\beta)}} - (1-\gamma) \left[\frac{A_N z^{*\beta}(1-\beta) n^{*\alpha}}{A_M(1-\alpha)(1-n^*)^\beta} \right]^\gamma v^* \tag{4.60}$$

Equating Eqs. (4.57) and (4.58), we obtain the dynamic arbitrage condition for the two capital goods in the steady state:

$$A_M n^{*(1-\alpha)} + \frac{A_M(1-\alpha)(1-n^*)}{n^{*\alpha}} = A_N \beta z^{*(\beta-1)}(1-n^*)^{1-\beta} \tag{4.61}$$

We obtain a system of three nonlinear equations, (4.56), (4.59) or (4.60), and (4.61), in three variables, z, n, and v, and parameters. Next, using the parameter values of Sect. 4.2, we solve the dynamic system for the planned economy in the steady state, obtaining $z^* = 0.091$, $n^* = 0.464$, $v^* = 2.83$, $p_N^* = 2.16$, and $g^* = 0.081$. We can see that the steady-state optimal growth rate is 8.1 % per annum. When we compare the optimal steady-state growth rate with the steady-state growth rate of the market economy, with $\mu = 0$, we deduce that there is opportunity for improving the steady-state growth rate in the market economy. Thus, the government can increase the steady-state growth rate. The correct policy to achieve the optimal steady-state growth rate is through an investment subsidy in the manufacturing sector.

4.4 The Optimal Investment Subsidy in the Market Economy

The objective of the government in a market economy is to maximize social welfare and to reach the optimal growth rate. The optimal government policy is to establish an investment subsidy in the manufacturing sector, stimulating the source of the learning process.

Using the optimal steady-state solution and Eq. (4.38), the optimal investment subsidy in the steady state is $\mu = 0.785$. Using this optimal investment subsidy, we solve the system for z, n, and v in the steady state, Eqs. (4.37), (4.40) or (4.41), and (4.42). We obtain $z^* = 0.091$, $n^* = 0.464$, $v^* = 2.83$, $p_N^* = 2.16$, and

$g^* = 0.081$. Thus, the steady-state growth rate is 8.1 % per annum. Note that all these levels correspond to the optimal solution.

Now, we are ready to analyze how the variables of the economy respond to an increase in the rate of investment subsidy. First, using Eqs. (4.28) and (4.42), we can obtain a useful relationship in the steady state:

$$n^* = \frac{1}{(1-\mu)^{\beta/(\alpha-\beta)}} \frac{1}{p_N^{*(1-\beta)/(\alpha-\beta)}} B \tag{4.62}$$

where $B = [A_M \alpha / A_N \beta]^{\beta/(\alpha-\beta)} [A_M (1-\alpha)/A_N (1-\beta)]^{(1-\beta)/(\alpha-\beta)}$.

Next, we show the response of the variables when the government establishes the optimal rate of investment subsidy. Considering that p_N^* is constant for the moment and that $\alpha > \beta$, we can see in Eq. (4.62) that when μ increases, the manufacturing sector is stimulated, and the proportion of labor in the manufacturing sector increases initially. Likewise, the incentive to invest (disinvest) in the manufacturing (nonmanufacturing) sector increases (decreases). Consequently, the level of z decreases slowly. Also, as the relative price of the nonmanufacturing good is flexible, we can see in Eq. (4.34) that when n increases, the relative price decreases initially. This movement of the relative price confirms the initial increase in n [see Eq. (4.62)]. However, in the optimal steady state, the level of the relative price of the nonmanufacturing good is higher. Moreover, given that total wealth increases, the level of v increases. We suggest that μ increases in the transition. Therefore, in the optimal steady state, the level of z^* decreases from 1.20 to 0.091, the proportion of labor in the manufacturing sector increases from 0.383 to 0.464, v^* increases from 0.411 to 2.83, and p_N^* increases from 1.06 to 2.16. Therefore, as the manufacturing sector is the leading sector in technological terms, the economy has a higher growth rate. The growth rate increases from 1.2 to 8.1 % per annum.

4.5   Conclusions

We have studied an economy with manufacturing and nonmanufacturing goods with two externalities. The relative price of the nonmanufacturing good is endogenously determined by supply and demand for the nonmanufacturing good. We have also shown that the optimal growth rate is achieved with an investment subsidy in the manufacturing sector.

We have studied how the economy responds when the government establishes the optimal investment subsidy. When the rate of subsidy is increased, the manufacturing sector is stimulated. Thus, the proportion of labor in the manufacturing sector increases, and the proportion of labor in the nonmanufacturing sector decreases. Likewise, investment in the manufacturing sector increases, and investment decreases in the nonmanufacturing sector. Thus, the ratio of nonmanufacturing to manufacturing capital decreases slowly. In addition, given that the relative price of the nonmanufacturing good is flexible, the relative price decreases initially. Also, this relative price adjustment confirms the initial increase in the proportion of labor in the manufacturing sector. Nevertheless, the relative price of the nonmanufacturing good is higher in the optimal steady state. Also, given that total wealth increases, the ratio of aggregate consumption to nonmanufacturing capital increases.

In summary, in the optimal solution, the proportion of labor in the manufacturing sector, the relative price of the nonmanufacturing good, and the ratio of consumption to nonmanufacturing capital are higher, and the ratio of nonmanufacturing to manufacturing capital is lower. Therefore, as the manufacturing sector is leader in technological terms, the market economy has a higher growth rate.

Thus, if the economy is technologically commanded by the manufacturing sector and there is strong intra- and inter-learning by doing among firms and sectors, the government should establish an optimal investment subsidy in the manufacturing sector. Thus, this paper has presented in an overall manner a general conclusion, concerning models with production externalities, two types of capital and endogenous growth: that the optimal policy is to stimulate the sources of the learning process (see Bardhan 1993). However, if subsidies are permitted or not, or if governments have the ability to manage an economy with externalities or not, there still remains another question that is not solved: how a government can justify subsidizing a particular sector in a democratic society, since these practices certainly have political costs.

References

Aizenman, J., Lee, J.: Real exchange rate, mercantilism and the learning by doing externality. Pac. Econ. Rev. **15**(3), 324–335 (2010)

Arrow, K.J.: The economic implication of learning by doing. Rev. Econ. Stud. **29**(3), 155–173 (1962)

Bardhan, P.K.: On optimum subsidy to a learning industry: an aspect of the theory of infant-industry protection. Int. Econ. Rev. **12**(1), 54–70 (1971)

Bardhan, P.: The new growth theory, trade and development. In: Hansson, G. (ed.) Trade, Growth and Development. Routledge, London (1993)

Barro, R.J., Sala-i-Martin, X.: Economic Growth, 2nd edn. MIT, Cambridge (2004)

Boldrin, M., Schienkman, J.: Learning-by-doing, international trade and growth: a note. In: Anderson, P., Arrow, K., Pines, D. (eds.) The Economy as an Evolving Complex System. Addison-Wesley, Reading, MA (1988)

Brock, P.L., Turnovsky, S.J.: The dependent-economy model with both traded and nontraded capital goods. Rev. Int. Econ. **2**, 306–325 (1994)

Casares, E.R.: Liberalización Comercial, Ajuste Sectorial y Crecimiento en Mexico. In: Casares, E.R., Sobarzo, H. (eds.) Diez Años del TLCAN en Mexico: Una Perspectiva Analitica. Lecturas de El Trimestre Económico 95. Fondo de Cultura Económica, Mexico (2004)

Casares, E.R.: Productivity, structural change in employment and economic growth. Estudios Económicos 22(2), 335–355 (2007)

Clemhout, S., Wan, H.Y.: Learning-by-doing and infant industry protection. Rev. Econ. Stud. **37**, 33–56 (1970)

Haley, U.C.V., Haley, G.T.: Subsidies to Chinese Industry: State Capitalism. Business Strategy and Trade Policy. Oxford University Press, Oxford (2013)

Korinek, A., Serven, L.: Undervaluation through foreign reserve accumulation. Static losses, dynamic gains. Policy Research Working Paper 5250, World Bank (2010)

Rabanal, P., Tuesta, V.: Nonmanufacturing goods and the real exchange rate. Open Econ. Rev. **24**(3), 495–535 (2013)

Rauch, J.E.: A note on the optimum subsidy to a learning industry. J. Dev. Econ. **38**(1), 233–243 (1992)

Roe, T.L., Smith, R.B.W., Saracoğlu, D.Ş.: Multisector Growth Models. Springer, Berlin (2010)

Romer, P.M.: Increasing returns and long-run growth. J. Polit. Econ. **94**(5), 1002–1037 (1986)

Romer, P.M.: Capital accumulation in the theory of long run growth. In: Barro, R. (ed.) Modern Business Cycle Theory. Blackwell, Oxford (1989)

Sobarzo, H.: The gains for mexico from a north american free trade agreement. In: Francois, J.F.,Shiells, C.R. (eds.) Modeling Trade Policy. Cambridge University Press, Cambridge (1994)

Stiglitz, J.E., Greenwald, B.C.: Creating a learning society: A New Approach to Growth, Development, and Social Progress. Columbia University Press, New York (2014)

Stiglitz, J.E., Greenwald, B.C.: Creating a learning society. In: A New Approach to Growth, Development, and Social Progress. Columbia University Press, New York (2014)

Succar, P.: The need for industrial policy in LDC's: a restatement of the infant industry argument. Int. Econ. Rev. **28**(2), 521–534 (1987)

Turnovsky, S.J.: Endogenous growth in a dependent economy with traded and nontraded capital. Rev. Int. Econ. **4**, 300–321 (1996)

Valentinyi, A., Herrendorf, B.: Measuring factor income shares at the sectoral level. Rev. Econ. Dyn. **11**, 820–835 (2008)

Yogo, M.: Estimating the elasticity of intertemporal substitution when instruments are weak. Rev. Econ. Stat. **86**(3), 797–810 (2004)

Young, A.: Learning by doing and the dynamic effects of international trade. Q. J. Econ. **106**(2), 369–405 (1991)

Chapter 5
The Fractal Nature of Bitcoin: Evidence from Wavelet Power Spectra

Rafael Delfin-Vidal and Guillermo Romero-Meléndez

Abstract In this study, a continuous wavelet transform is performed on bitcoin's historical returns. Despite the asset's novelty and high volatility, evidence from the wavelet power spectra shows clear dominance of specific investment horizons during periods of high volatility. Thanks to wavelet analysis, it is also possible to observe the presence of fractal dynamics in the asset's behavior. Wavelet analysis is a method to decompose a time series into several layers of time scales, making it possible to analyze how the local variance, or wavelet power, changes both in the frequency and time domain. Although relatively new to finance and economic, wavelet analysis represents a powerful tool that can be used to study how economic phenomena operate at simultaneous time horizons, as well as aggregated processes that are the result of several agents or variables with different term objectives.

Keywords Fractal market hypothesis • Bitcoin • Wavelet power spectrum • Wolfram Mathematica • Economics and finance • Cryptocurrencies • Wavelet analysis

5.1 Introduction

Bitcoin is a digital currency that relies on cryptographic technology to control its creation and distribution. Just like banknotes or coins, transactions in bitcoin can be performed directly between two individuals without the need of an intermediary. However, bitcoins are not issued by any government or other legal entities; they are produced by a large number of people running computers around the world, using software that solves mathematical problems. It is the first example of a growing category of money known as cryptocurrency.

Unlike fiat currencies, whose value is derived through regulation or law and underwritten by the state, bitcoin's technology has currency, platform, and equity properties that make it extremely difficult to assess its intrinsic value

R. Delfin-Vidal • G. Romero-Meléndez (✉)
Departamento de Actuaría, Física y Matemáticas, Universidad de las Américas Puebla, Ex Hacienda Sta. Catarina Mártir, San Andrés, Cholula, Puebla 72810, México
e-mail: raffadelfin@gmail.com; guillermoa.romero@udlap.mx

© Springer International Publishing Switzerland 2016
A.A. Pinto et al. (eds.), *Trends in Mathematical Economics*,
DOI 10.1007/978-3-319-32543-9_5

(Weisenthal 2013). As a consequence, most of bitcoin's value is based on a highly volatile demand—what people are willing to pay and receive for them at any given time. In April 2011, less than 1 year after the first transactions using bitcoins took place, a single bitcoin (currency ticker BTC) was worth about $0.80. Three years later, as of October 29, 2014, one bitcoin is now worth $348, having reached a historical maximum value of $1132 in December 2013.

It is widely known that the bitcoin economy has experienced a recurring volatility cycle over its short existence. As media coverage on the cryptocurrency increased, this attracted new waves of investors pushing bitcoin's price to unprecedented highs, leading to an eventual crash of the BTC/USD exchange rate. Before reaching its $1120 historical maximum in December 2013, bitcoin's price rose 40-fold from around $0.80 in April 2011 to more than $30 by June 2011 to then fall below $2 by November 2011 before stabilizing at around $5 in early 2012. After the initial boom and bust, bitcoin's price gradually stabilized between $4.30 and $5.48 during the first half of 2012. In the second half of 2012, BTC prices climbed from $5.15 in June to $13.59 by December 2012. This pattern repeated itself twice during 2013. From $13.50 at the start of the year, bitcoin's value soared to $237 in May and then crashed to $68 later that same month. After the first volatility cycle in 2013, BTC prices ranged between $68 and $130 until October 2013; then by the end of November, bitcoin prices reached $1120. Finally, during the first half of 2014, the USD/BTC exchange rate has steadily decreased to around $400–500.

The volatility pattern observed in BTC price behavior suggests three important features in the asset's price behavior. First, the uncharacteristically large price changes in the USD/BTC exchange rate suggest that the frequency distribution of BTC returns does not follow a normal distribution, i.e., extreme events that deviate from the mean by five or more standard deviations have a greater probability of occurrence than that predicted by the normal distribution.

Figure 5.1 shows the quantile–quantile plot for BTC historical returns and illustrates the evidence of long tails and over-dispersion in the series, represented by the blue thick line.

Second, clear clustering periods of high and low volatility in the BTC price data suggest that while asset returns may be random, its periods of volatility are not. This is illustrated in Fig. 5.2. The top graphic shows the autocorrelation of BTC returns, suggesting no sign of serial correlation between returns. The bottom graphic shows the correlation of BTC volatility, i.e., the second moment of the asset's returns.

The second graphic in Fig. 5.2 shows a clear positive trend in the autocorrelation of the asset's volatility, a clear sign of long memory, or persistent behavior. Finally, bitcoin price data exhibits evidence of scale invariance, or self-similar statistical structures, at different price levels. For example, BTC returns follow the same frequency distribution regardless of time scale, while bitcoin's price volatility cycles show the same behavior, independent of price level.

These features directly violate the fundamental assumptions of Gaussian distribution required by the established efficient market hypothesis (EMH), rendering most financial modeling approaches unsuitable to study bitcoin price behavior. Moreover, after decades of statistical analysis of price fluctuations across different markets,

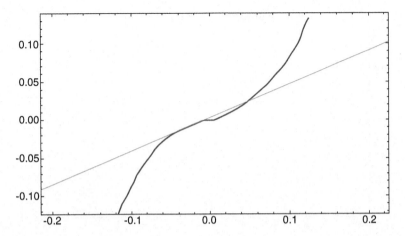

Fig. 5.1 BTC returns Q–Q plot

asset types, and time periods, there are a large number of studies documenting the failure of EMH to mirror or model the empirical evidence of financial time series (Mandelbrot 1963, 1997; Blackledge 2010). Despite the widespread use of the Brownian motion and Gaussian distribution paradigms in financial economics, a number of systematic statistical departures from the EMF have been identified and are now widely acknowledged as "stylized facts" of financial time series (Rama 2001, 2005; Borland et al. 2005; Ehrentreich 2008; Dermietzel 2008).

Notably, the main stylized facts standing out in the literature include the three prominent features of bitcoin's volatility cycle previously mentioned: heavy tails or non-normal distribution of returns, long memory effect in squared returns also known as volatility clustering, and presence of fractal dynamics. Therefore, given the strong deviations from the EMH framework readily observable in the BTC price data, an alternative analytical framework is used to study financial data with likely presence of non-normality, self-similarity, and persistent volatility.

The fractal market hypothesis (FMH) is a theoretical framework developed by Peters. He proposes a more realistic market structure that places no statistical requirements on the process; he explains why self-similar statistical structures exist and how risk is shared and distributed among investors (Peters 1991a,b, 1994). Under the FMH approach, market stability is maintained only when many investors participate and they can cover a large number of investment horizons, thus ensuring ample liquidity for trading (Peters 1994). Peters argues that after adjusting for scale of investment horizon, all investors must share same risk levels, which explains why the frequency of distribution of BTC returns exhibits self-similar behavior at different scales (Peters 1994). According to the FMH, a market becomes unstable when its self-similar structure breaks down, i.e., when investors with long-term horizons either stop participating in the market or become short-term investors themselves. When long-term fundamental information is no longer important or unreliable, markets become unstable and are characterized by extreme high levels

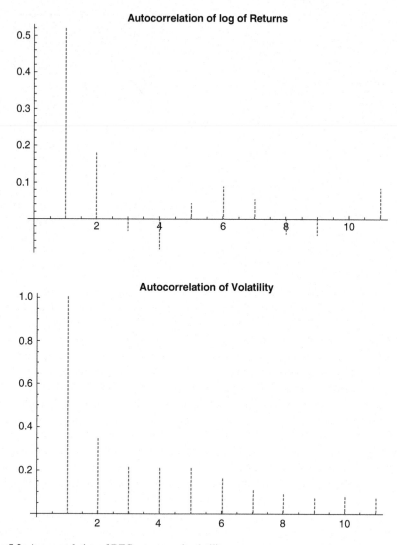

Fig. 5.2 Autocorrelation of BTC returns and volatility

of short-term volatility. This approach explains the presence of periods of clustering volatility in the BTC time series and the occurrence of extreme events that violate the normality of the frequency distribution of bitcoin returns.

The FMH suggests that stable markets are characterized by equal representation of all investment horizons in the market so that supply and demand are efficiently cleared. When investors at one horizon (or group of horizons) become dominant, the selling or buying signals of the investors at these horizons will not be met with a corresponding order of the remaining horizons, and periods of high volatility might occur. Thanks to time–frequency analysis, it is possible to investigate whether BTC returns follow the market dynamics established by the fractal market hypothesis and its focus on liquidity and investment horizons. According to Kristoufek, after performing a continuous wavelet transform to a time series and obtaining its wavelet power spectra, it is possible detect the dominance of specific investment horizons during periods of high volatility (Kristoufek 2013).

Wavelet analysis is a method to decompose a time series into time–frequency space; it uses mathematical expansions that transform data from the time domain into different layers of frequency levels. This makes possible to observe and analyze data at different scales. Although this approach is relatively new to economics, wavelets have been used in a wide range of fields, for example, for the analysis of oceanic and atmospheric flow phenomena in geophysics (Torrence and Compo 1998), image processing for computer and image compression (Grapps 1995), as well as in medicine for heart rate monitoring (Thurner et al. 1998) and for molecular dynamics simulation and energy transfer in physics (McCowan 2007), just to name a few. Among the most well-known applications of wavelet analysis are the FBI algorithm for fingerprint data compression and the JPEG algorithm for image compression (Grapps 1995; Li 2003).

5.1.1 Scope of This Manuscript

The main goal of this study is to provide empirical evidence supporting the fractal market hypothesis. To do so, the BTC returns time series is analyzed to determine the existence of dominance of short investment horizons during periods of high market turbulence. This objective is accomplished using a continuous wavelet transform analysis to obtain information about bitcoin's price volatility across time and different scales of investment horizons.

There are several reasons for the importance of this study. First, to date, this is the only study using wavelet analysis to detect dominance of investment horizons in BTC price returns. Second, the results of the continuous wavelet transform of the time series show supporting evidence in favor of the fractal market hypothesis. Third, the wavelet analysis performed suggests that while bitcoin's price has been characterized by high volatility, it follows the same market dynamics as other currencies and equity markets (e.g., government bonds, stocks, and commodities). Finally, the use of wavelet to analyze economic phenomena is relatively recent; this work will show original contributions to the applications of wavelet analysis in economics, finance, and cryptocurrencies.[1]

---

[1]This manuscript is based on the undergraduate thesis project of the first author (Delfin 2014) and supervised by the second author.

5.1.2   Organization of the Manuscript

The remainder of this work is organized as follows. The "Theoretical Framework" section presents an overview of the Bitcoin payment system and an introduction to wavelet analysis and its relation to FMH. The following section presents the continuous wavelet transform methodology, results, and discussion. Then, the final section concludes with a general discussion, future research subjects, and benefits of wavelet analysis to the study of economic phenomena.

5.2   Theoretical Framework

This background section should provide an introductory understanding to the topics presented in the following sections, although it is far from a complete examination of the concepts covered in this study. Although the intersection of these subjects has yet to gain wider recognition, studies on Bitcoin's public ledger technology along with wavelet analysis span several fields within economics and finance in general. It is highly recommended to consult the sources referenced in this section should the reader be interested in a more comprehensive understanding of the topics covered in this work.

5.2.1   The Bitcoin Protocol

Bitcoin is a peer-to-peer payment system introduced as open-source software in January 2009 by a computer programmer using the pseudonym Satoshi Nakamoto (Nakamoto 2009). It is referred to as a cryptocurrency because it relies on cryptographic principles to validate transaction in the system and ultimately control the production of the currency itself. Each transaction in the system is recorded in a public ledger, also known as the Bitcoin block chain, using the network's own unit of account, also called bitcoin.[2] The block chain ledger is a database where transactions are sequentially stored, and the file containing it is visible to all members on the network.

Bitcoin's block chain is a unique technology since it solves several problems at once: it avoids forgery or counterfeiting, it also avoids the need for a trusted intermediary, and it regulates the creation of new bitcoins in a controlled way

---

[2] According to the Bitcoin wiki website (https://en.bitcoin.it/wiki/Introduction#Capitalization_.2F_ Nomenclature), capitalization and nomenclature can be confusing since Bitcoin is both a currency and a protocol. Bitcoin, singular with an uppercase letter B, will be used to label the protocol, software, and community, and bitcoins, with a lowercase b, will be used to label units of the currency.

(Congresional Research Service 2013; Velde 2013). Since validation for each transaction is a computationally intensive task, the Bitcoin protocol solves these problems by rewarding those who devote computing power to validate transactions with the privilege to create new bitcoins in a controlled way.

According to (Barber et al. 2012), there are several reasons why Bitcoin, despite more than three decades of previous attempts at digital money by cryptography researchers (see, e.g., Chaum 1983; Chaum et al. 1990; Szabo 2008), has witnessed enormous success since its invention. Among the number of reasons are no central point of trust, economic incentives to participate, predictable money supply, divisibility and fungibility, transaction irreversibility, low transaction costs, and readily available implementation.

Contrary to earlier implementations of e-cash, Bitcoin is a decentralized network that lacks a central trusted entity. The network assumes that the majority of its nodes are honest, and as mentioned earlier, the task of validating transactions for dispute resolution and to avoid double spending is carried out by members on the network dedicating computing power for those purposes. The absence of a central point of trust guarantees that the currency cannot be subverted by any single entity—government, bank, or authority—for its own benefit, and while this feature can be used for illegal purposes, there are also numerous legitimate reasons for using this technology.

Regarding the economic incentives for participation in the Bitcoin network, Kroll et al. argue that if all parties act according to their incentives, the Bitcoin protocol can be stable, meaning the system will continue to operate (Kroll et al. 2013). Since the generation of new bitcoins is rewarded only to those individuals who devote computing power to validate transactions, also known as bitcoin mining, this reward ensures that users have clear economic incentives to invest unused computing power in the network. In addition to rewards from dedicating computational cycles to verify transactions, miners can charge small transaction fees for performing the said validation. Finally, Barber et al. argue that the open-source nature of the project also gives incentives for new applications within the protocol and the creation of a large ecosystem of new businesses (Barber et al. 2012), for example, new applications that add better anonymity measures or payment processing services that allow merchants to receive payments in bitcoin and send money internationally at significant low cost.

In addition to a predictable money supply, Barber et al. argue that the divisibility, fungibility, and transaction irreversibility of Bitcoin give it an advantage over other e-cash systems since the coins can be easily divided, up to eight decimal places, and recombined which allows to create a large number of denominations, while the irreversibility of transactions means that merchants concerned with credit card fraud and charge-backs can conduct business with customers in countries with high prevalence of credit card fraud. Moreover, thanks to its high divisibility, Bitcoin has great potential as a platform for enabling micropayments, payments much smaller than what the traditional financial system can handle (Barber et al. 2012).

After Nakamoto's publication of the Bitcoin protocol in January 2009, the homonymous currency remained a modest project undertaken by a small community of cryptographers during its first year. However, Nakamoto's creation soon spread beyond the initial community and took a life of his own. In October 2009, the first USD/BTC exchange rates were published by New Liberty Standard (2009); $1 was valued at 1309.03 BTC. In May 2010, Laszlo Hanyecz, a Florida programmer, conducted what is thought to be the first real-world bitcoin transaction, agreeing to pay 10,000 bitcoins for two pizzas from Papa John's worth around $25 at the time (Mack 2013). Two months later in July 2010, bitcoin's exchange value began a 10× increase over a 5-day period, from about $0.008/BTC to $0.08/BTC. By November of that year, bitcoin had reached a market capitalization of $1 million, while the exchange rate was $0.50 for 1 bitcoin (Bitcoin 2014). The next important milestone for the currency occurred in February 2011 when bitcoin reached parity with the US dollar at the now defunct Japanese exchange MtGox.

During the spring of 2011 after several stories on the new cryptocurrency by High-profile media outlets, one from Time (Brito 2011) and another one by Forbes reporter Timothy Lee (Lee 2011) and also from popular design and technology blog Gizmodo (Biddle 2011), the price of bitcoin skyrocketed from around 86 cents in early April to $9 at the end of May. Additionally, on June 1, media outlet Gawker published a story about the use of bitcoin in the online black market Silk Road to buy drugs, weapons, and stolen personal information, thanks to the currency's pseudo-anonymous features Chen (2011a,b). One week later, bitcoin's exchange rate increased threefold from $9/BTC to $31/BTC.

As the price of bitcoin rose and stories of return on investment in the order of thousands, mining became more popular. Now real-money stakes and the dramatic price rise had attracted people who saw bitcoin as a commodity in which to speculate. However, given the novelty of this asset and how its uncharacteristic behavior clearly violates the fundamental assumptions of most financial modeling approaches, an alternative analytical framework is used to study bitcoin price behavior.

5.2.2 Wavelet Analysis and the Fractal Market Hypothesis

As mentioned in the introductory section, the FMH suggests that stable markets are characterized by equal representation of all investment horizons, while market volatility occurs when the selling or buying signals of a dominant investment horizon are not met with a corresponding order from the remaining horizons. However, simultaneous operation at different time horizons is not only restricted to currency and equity markets. Aguiar-Conraria and Soares argue that many economic processes are the result of actions of several agents who have different time objectives, and therefore, many economic time series are an aggregation of components operating on different frequencies (spanning milliseconds in high-frequency trading to several decades for institutional investors) (Aguiar-Conraria and Soares 2011). Moreover, Ramsey and Lampart argue that economists have long acknowledged the importance of time scale, but only until recently, it had been difficult to decompose economic time series into time scale components (Ramsey and Lampart 1997a). Central banks, for example, have different objectives in the short and long run and therefore operate at different time scales.

The main advantage of using the continuous wavelet transform (referred as CWT from now on) in economic time series is its ability to analyze how the wavelet power of the underlying process changes in both the time and frequency domain. In terms of financial economics, the wavelet power spectrum (WPS) is defined by Rua as the contribution to the variance around each time and scale (Rua 2012). Formally, the WPS is defined as the squared absolute value of the wavelet coefficients resulting from the transform. According to FMH, a high-power spectrum is associated with dominant investment horizons, i.e., the selling or buying signals of investors at the dominant horizons are not being met with a corresponding order from the remaining horizons, and periods of short-term volatility might occur. Therefore, high-power spectrum values should be observed at low time scales (high frequencies) during periods of high volatility.

5.2.3   Origins of Wavelet Analysis

In order to talk about wavelet analysis, it is necessary to talk about Fourier analysis first since the former has various points of similarity and contrast with the later. The Fourier transform is based on using a sum of sine and cosine functions of different wavelengths to represent any other function. The Fourier transform of a time series $f(t)$ is a function $F(\omega)$ in the frequency domain, $F(\omega) = \int_{-\infty}^{\infty} f(t)e^{-i\omega t}dt$, where ω is the angular frequency and $e^{-i\omega t} = \cos(\omega t) - i\sin(\omega t)$ according to Euler's formula. However, the Fourier transform does not allow the frequency content of the signal to change over time, making it unsuitable for analyzing processes that have time-varying features. This means that if a single frequency is present in a process but it varies over time, the Fourier transform does not allow to identify when in time the frequency component changes (Rua 2012).

To illustrate the shortcomings of the Fourier transform when reproducing signals that have time-varying features, the following example is based on Wolfram's presentation on wavelet concepts (see Wolfram 2014a). Considering the stationary process, $s(t) = \cos([2\pi]20t) + \cos([2\pi]40t)$. This process is composed of two signals, one at 20 Hz and another at 40 Hz. When the Fourier transform of this data is performed, two frequencies are correctly identified, at two times the frequency in the x-axis, i.e., 40 Hz and 80 Hz, respectively (see Fig. 5.3). While the Fourier transform provides frequency information, it lacks time information about these frequencies, i.e., at what time did these frequencies occur and for how long? Considering now a nonstationary process with three frequency components defined by

$$s(t) = \begin{cases} \sin([2\pi]t) & 0 \le t \le 2 \\ \sin([2\pi]5t) & 2 \le t \le 4 \\ \sin([2\pi]10t) & 4 \le t \le 8 \end{cases}$$

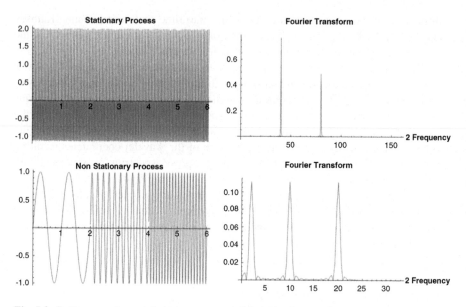

Fig. 5.3 Stationary and nonstationary processes and their Fourier transform

the Fourier transform correctly shows three peaks (Fig. 5.3) at the corresponding frequencies (1, 5, and 10 Hz); however, the transform does not provide information about the time-varying components of this process. According to McCowan, the Fourier transform gives optimal results only when a single frequency is present (McCowan 2007). When multiple frequencies are present in a process, the transform may have difficulties separating noise or assigning accurate relative amplitudes for each frequency.

A possible way to overcome the previous limitations is the short-time or windowed Fourier transform, a Fourier-related transform used to obtain frequency information of local sections of a signal as it changes over time. As its name suggests, the Fourier transform is performed for short periods of time, sliding a segment of length T across all the data. However, the windowed Fourier transform (WFT) imposes the use of constant-length windows.

This restriction makes the WFT an inaccurate method for time–frequency analysis since many high- and low-frequency components of the process or signal will not fall within the frequency range of the window. Relatively small windows will fail to detect frequencies whose wavelengths are larger than the size of the window, while relatively large windows will decrease the temporal resolution because larger intervals of signal are analyzed at once.

Torrence and Compo argue that for analyses where a predetermined scaling may not be appropriate because of a wide range of dominant frequencies are present in the process, a method of time–frequency localization that is scale independent, such as wavelet analysis, should be employed (Torrence and Compo 1998).

5.2.4   The Continuous Wavelet Transform

Just as the windowed Fourier transform, the aim of the continuous wavelet transform (CWT) is to detect the frequency, or spectral, content of a signal and describe how it changes over time. The CWT however uses a base function that can be stretched and translated with a flexible resolution in both frequency and time, making it possible to analyze nonstationary time series that contain many different frequencies. Moreover, the CWT intrinsically adjusts the time resolution to the frequency content. This means the analyzing window width will narrow when focusing on high frequencies (short time periods) and widen when assessing low frequencies (long time scales).

The CWT of a time series x(t) can be formally defined as $W_x(\tau, s) = \int_{-\infty}^{\infty} x(t)\psi_{\tau,s}^*(t)dt$, where $*$ denotes the complex conjugate. Starting with a mother wavelet $\psi(t)$, the CWT decomposes the time series $x(t)$ in terms of analyzing wavelets $\psi_{\tau,s}(t)$. The analyzing wavelets are obtained by scaling and translating $\psi(t)$, which is defined as $\psi_{\tau,s}(t) = (1/\sqrt{|s|})\psi((t - \tau)/s)$, where s is the scale and τ the translation parameters. The wavelets can be stretched (if $|s| > 1$) or compressed (if $|s| < 1$), while translating the wavelet means shifting their position in time. Thanks to the CWT flexible resolution in both frequency and time, rapidly changing feature can be captured at low scales, or wavelengths, whereas slow-changing, or higher time scales, components can be detected with dilated analyzing wavelets (Torrence and Compo 1998; Aguiar-Conraria and Soares 2011).

Mother wavelets must fulfill certain mathematical criteria in order to be considered analytical wavelets; in economics and finance, the Morlet wavelet is the most widely used mother wavelet (Torrence and Compo 1998; Aguiar-Conraria and Soares 2011; Rua 2012; Kristoufek 2013). The Morlet wavelet consists of a complex sine wave modulated by a Gaussian envelope, and it is formally defined as: $\psi(t) = \pi^{-1/4}e^{i\omega_0 t}e^{-t^2/2}$. The term ω_0 controls the nondimensional frequency, i.e., the number of oscillations within the Gaussian envelope, and is set equal to six to satisfy the admissibility criteria as analytic wavelet (see Lee and Yamamoto 1994 and Adisson 2002 for a detailed analysis of wavelet admissibility criteria).

Figure 5.4 shows the Morlet wavelet, which unlike sines and cosines, it is localized in both time and frequency.

5.2.5   Wavelet Power Spectrum and Other Definitions in the Wavelet Domain

Once the CWT has been defined, we offer two definitions from the wavelet domain to analyze an asset's volatility as well as its local covariance with other assets (see (Ranta 2010) for additional definitions regarding correlation and contagion in the time–frequency domain). First, the wavelet power spectrum (WPS) can be

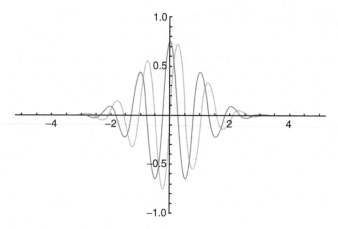

Fig. 5.4 Morlet wavelet

defined as $|W_x(\tau, s)|^2$, i.e., the square of the absolute value of each coefficient at each time and scale, and measures the local contribution to the variance of the series. Second, the cross-wavelet transform (XWT) of two time series $x(t)$ and $y(t)$, with continuous wavelet transforms $W_x(\tau, s)$ and $W_y(\tau, s)$, is defined as $W_{xy}(\tau, s) = W_x(\tau, s) \, W_y^*(\tau, s)$. The corresponding cross-wavelet spectrum is defined as $XWP_{xy} = |W_{xy}|$. According to Aguiar-Conraria and Soares, the cross-wavelet power of two time series can be defined as the local covariance between them in the time–frequency domain, giving the researcher a quantified indication of the similarity of volatility between the time series (Aguiar-Conraria and Soares 2011).

5.3 Methods and Results

In this section the CWT will be implemented on the BTC historical returns time series to provide evidence for the dominance of short investment horizons during periods of high volatility. Additionally, since all the analysis in this study was performed using the computational software Mathematica, the code used to perform the computations will be used to provide the reader new tools for wavelet analysis. The findings of this analysis will be discussed afterward.

5.3.1 Data

A time series for the price of bitcoin against the US dollar will be analyzed to find their respective wavelet power spectrum. The oldest available date for bitcoin prices is July 17, 2010. The time series cover the oldest available BTC price until October 29, 2014.

The data used for this study was obtained from the data platform Qandl's website, a search engine for numerical data with access to a large collection of financial, economic, and social datasets.

5.3.2   Method: Basic Wavelet Concepts

Performing a CWT in Mathematica can be done with very few commands. Before the main analysis of this study, three examples will be presented to overview basic wavelet transform concepts and their advantage over time or frequency analysis.

The first example is based on (Aguiar-Conraria and Soares 2011). Fifty years of monthly data are generated according to the process:

$$y(t) = \sin\left(\frac{2\pi}{p_1}t\right) + \sin\left(\frac{2\pi}{p_2}t\right) + \epsilon_t; \qquad t = \frac{1}{12}, \frac{2}{12}, \dots, 50$$

$p_1 = 10$ and $p_2 = 5$ for $20 \leq t \leq 30$, and $p_2 = 3$ otherwise. It can be seen that this process is the sum of two periodic components: a 10-year cycle and 3-year cycle that briefly change to a 5-year cycle during between the second and third decades. Although Fig. 5.5 shows the process $y(t)$ in the time domain, it is not possible to clearly observe any of the cyclic dynamics of the series.

Figure 5.6 shows a visualization, also called wavelet scalogram, of the wavelet power spectrum, $|W_x(\tau,s)|^2$, of the $y(t)$ process. The wavelet scalogram functionality in Mathematica plots the absolute value of the wavelet transform coefficients at each time and scale. In Fig. 5.6, the wavelet scalogram is able to capture the

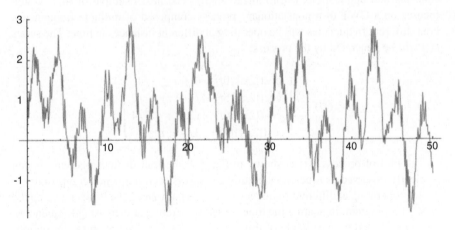

Fig. 5.5   Time series $y(t)$

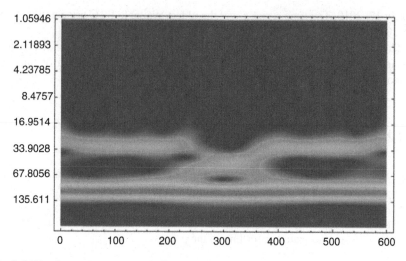

Fig. 5.6 Wavelet power spectrum of $y(t)$

three cyclic dynamics of the time series. The time dimension is represented in the horizontal axis, while the vertical axis represents the scales, or frequencies, analyzed. The wavelet power is represented by color, ranging from blue for low power to red for high wavelet power. The lower region in red from Fig. 5.6 shows the 10-year cycle of the time series, while the light-green regions in the middle section of the graphic show how the second component of $y(t)$ transitions from a 3-year cycle to a 5-year cycle between the second and third decades. Since the series is given in monthly data, the second and third decades fall within observations 240–360 on the horizontal axis.

The following two examples are based on Wolfram's presentations on wavelet concepts and applications (Wolfram 2014a,b). The next example of this section focuses on a CWT of a nonstationary process composed of multiple frequencies. Four different frequencies will be operating at different instances in time. The series $f(t)$ will be generated by the process:

$$
f(t) = \begin{cases}
\cos([2\pi]10t) & 0 \leq t \leq 1/4 \\
\cos([2\pi]25t) & 1/4 < t \leq 1/2 \\
\cos([2\pi]50t) & 1/2 < t \leq 3/4 \\
\cos([2\pi]100t) & 3/4 < t \leq 1
\end{cases}
$$

The plot of process $f(t)$ is shown in Fig. 5.7; the four distinct frequencies can be clearly observed as t increases. Figure 5.8 shows each frequency composing the process operating at different frequency bands; as t advances, the bands move up the wavelet scalogram, indicating the time series is operating at increasing frequencies.

The third and final example of this section is used to illustrate how discontinuities, or in economic terms structural changes and regime shifts, can be identified

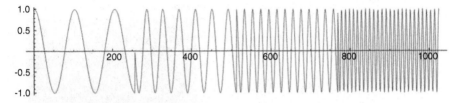

Fig. 5.7 Time series $f(t)$ with four frequency components

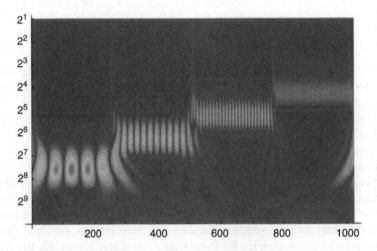

Fig. 5.8 Wavelet power spectrum of $f(t)$

using wavelet analysis, considering a process $h(t)$ determined by a simple cosine function. Overlapping the cosine function, an extreme event of small duration occurs at time t; hence, the process is defined as: $h(t) = \cos([2\pi]2t) + 2e^{-10^5([1/3]-t)^2}$. After performing a CWT on the series $h(t)$, the wavelet scalogram can provide a clear picture of the process' behavior in the time–frequency domain. Figure 5.9 shows the wavelet power of the series at various frequencies or scales. The left indexes on the vertical are associated with each scale, while the right indexes represent the voice per scale. At large scales (low frequencies), the wavelet scalogram is able to capture the signal from the cosine function, but as we move upward to lower scales (higher frequencies), the extreme and short-lived event can be localized in both frequency and time.

5.3.3   Wavelet Transform of Bitcoin's Returns

Once the basic concepts of wavelet analysis in the Mathematica platform have been established, the BTC returns time series will be decomposed using the CWT. The

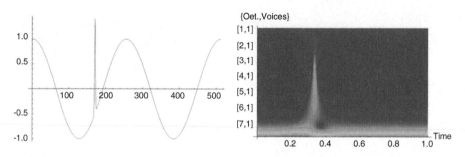

Fig. 5.9 WPS of a cosine signal overlapped by an extreme and short-lived anomaly

first steps are importing bitcoin historical prices to Mathematica, defining the time
series for bitcoin returns, and creating a list with the data. The following three lines
of code perform each step, respectively:

btcprice=Import["C: \ \ Users \ \ ... \ \ BTCAVERAGE-USD.xlsx",{"Data",1,
{All},2}];
returns[x_]:=Log[(btcprice[[x+1]]/btcprice[[x]])];
btcreturns=Array[returns,(Length[btcprice])-1];

After the BTC returns data is defined, a CWT can be applied with the following
command: cwt=ContinuousWaveletTransform[btcreturns,MorletWavelet[],{9,10}].
The CWT command gives the wavelet transform of btcreturns, using the complex
MorletWavelet[], and decomposes the data into nine octaves, or scales, and ten
subsequent voices, or samples for each scale. The scales chosen for the wavelet
transform are defined as fractional powers of two, $s_j = s_0\,2^{j/voc}$; $j = 0, 1, \ldots, J$;
and $J = \mathrm{oct}\cdot\mathrm{voc}$, where s_0 is the smallest resolvable scale and J determines the total
number of layers in which the signal will be decomposed, i.e., $J = \#\mathrm{octaves}\cdot\#\mathrm{voices}$.
Additionally, the smallest resolvable scale s_0 is computed automatically as the
inverse of Fourier wavelet length of the wavelet (Wolfram 2014c). For the CWT
of the BTC returns time series, the smallest resolvable scale computed is 0.86
days or about 20 h; therefore, the scales and samples per scale will be computed
as $s_j = 0.86 \cdot 2^{j/90}$; $j = 0, 1, \ldots, 90$.

By default, Mathematica computes the number of scales used in each transform
as $\mathrm{Log}_2(N/2)$, where N is the length of the time series, while the default value for
the number of voices per octave is four. Computing $\mathrm{Log}_2(N/2)$, where $N = 1547$
for the BTC returns time series, results in 9.59. Mathematica correctly computes the
number of scales to be used in the wavelet transform; however, the number of scales
was explicitly indicated in the CWT command in order to specify the number of
voices per scale as well. The more voices per scale are used in the CWT, the better
the time–frequency resolution; hence, it was increased to ten from the default value
of four.

Mathematica evaluates the CWT command and the output continuous wavelet
data object (CWDo) in the form $\{\{\mathrm{oct}_1, \mathrm{voc}_1\} \rightarrow coef_1, \ldots, coef_n\}$, with N wavelet
coefficients $coef_i$ corresponding to $\{\mathrm{oct}_i, \mathrm{voc}_i\}\}$. The CWDo also contains additional
information that can be later accessed and manipulated. For example, each octave

and voice pair is associated with a certain scale; these can be accessed using the property "Scales":

cwt["Scales"]

$\{\{1, 1\} \rightarrow 0.925992,$

$\{1, 2\} \rightarrow 0.992453,$

$\dots,$

$\{5, 1\} \rightarrow 14.8159,$

$\{5, 2\} \rightarrow 15.8793,$

$\dots,$

$\{9, 9\} \rightarrow 412.735,$

$\{9, 10\} \rightarrow 442.358\}$

5.3.4   Wavelet Power Spectrum of Bitcoin's Returns

As mentioned in previous sections, the scalogram is a visual method to represent the absolute value of each coefficient, or wavelet power. The wavelet scalogram displays three axes: the horizontal axis represents time, the vertical axis the time scales or frequencies, and the transform's coefficient values. The coefficient values are plotted as rows of colorized rectangles whose color corresponds to the magnitude of each coefficient. Figure 5.10 shows three graphics. The middle graphic shows the wavelet scalogram for the bitcoin returns CWT. The top graphic shows the historical BTC returns, and the bottom graphic is a plot of the historical observed volatility.

The regions with significant wavelet powers against the null hypothesis of a white noise (AR[1] process) are denoted by orange and yellow colors. According to Torrence and Compo and Aguiar-Conraria and Soares, the use of CWT for finite-length series will suffer from border distortions at the beginning and end of the wavelet power spectrum because the wavelet function will be defined beyond the limits of the time series (Torrence and Compo 1998; Aguiar-Conraria and Soares 2011). The cone of influence (COI) is the region in the time–frequency plane where border distortions become important and in Fig. 5.10 by the region above the white contour line. The COI can be defined as the set of all observations t included in the effective support of the wavelet at a given position and scale. This set is defined by $|t - \tau| \leq sB$ where τ is the translation parameter of the analyzing wavelets $\psi_{\tau,s}(t) = (1/\sqrt{|s|})\psi([t - \tau]/s)$, s is the scale parameter, and $[-B, B]$ is the effective support of the daughter wavelets, i.e., the initial and final values of the time series $[1, 1548]$.

As mentioned at the beginning of this section, the scalogram displays the wavelet transform in three dimensions: time, frequency, and wavelet power. Figure 5.11 shows a three-dimensional representation of the BTC returns power spectrum.

Several features can be observed in the previous two figures. First, the highest wavelet power regions (colored in red, orange, and yellow) are associated with periods of highest volatility. This can be confirmed with the top and bottom graphics of Fig. 5.10, where returns and volatility are, respectively, plotted. Second, for most

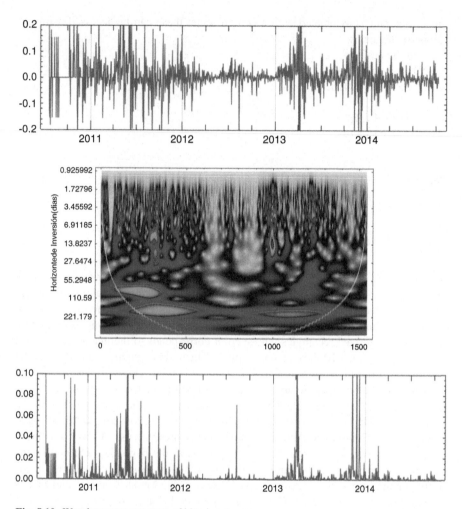

Fig. 5.10 Wavelet power spectrum of bitcoin returns

of the analyzed period, no investment horizon, or scale power, dominates the series. However, the wavelet scalogram correctly captures the biggest price movements in bitcoin: the 40-fold increase around mid-2011 from around $0.80 to more than $30, a low-variance period during 2012, and the two price bubbles from 2013 during May and late November. Third, during the periods of high volatility, the BTC power spectrum shows clear dominance of short investment horizons. Moreover, these dominant investment horizons are located within the 3.5- to 7-day band, and during the price increase in May 2013, dominant investment horizons can also be observed in the 7- to 14-day band. Larger investment horizons (lower frequencies) only show moderate wavelet power. However, since the cryptocurrency was created

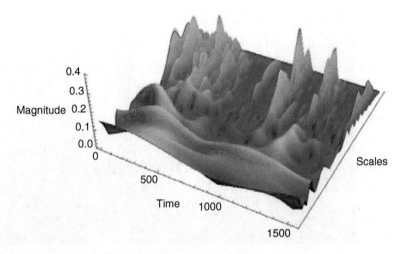

Fig. 5.11 Three-dimensional representation of Bitcoin's WPS

little less than 6 years ago, it is not possible to draw significant conclusions for large investment horizons.

Finally, the results presented in Fig. 5.10 support the thesis of Peters of dominant investment horizons during periods of turbulence and provide further evidence in favor of FMH's prediction of market stability only under equal representation of all investment horizons (Peters 1991a,b, 1994). According to the FMH, a market becomes unstable when its self-similar structure breaks down. This can happen for a number of reasons, if investors with long-term horizons stop participating in the market, become short-term investors themselves, or when long-term fundamental information is no longer important or unreliable. Given the novel nature of Bitcoin, the large price swings and low liquidity of cryptocurrencies in general might make clearer the fractal dynamics of these markets.

Indeed, after closer examination of the wavelet scalogram, it is possible to magnify certain regions of the time–frequency plane to observe the presence of fractal dynamics in the series. The following section will present evidence of bitcoin's self-similar behavior in the time–frequency plane.

5.3.5   Self-Similarity in Bitcoin Returns

Contrary to their mathematical counterparts, real-life fractal processes exhibit self-similar behavior over a finite range of scales. Bitcoin returns time series however exhibit fractal properties over a sufficiently large range of scales to allow wavelet transform analysis to examine the process. Since a process with fractal behavior displays self-similar structures regardless of scale, wavelet analysis is adequately suited to detect these properties. The basic principle for studying fractal processes

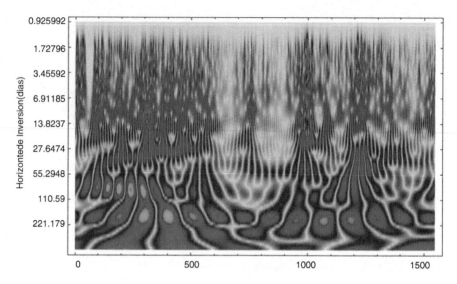

Fig. 5.12 Self-similar behavior of Bitcoin price returns

with wavelet transform is that since the signal is self-similar at any scale, the wavelet coefficients of the transform too will be self-similar, and this can be observed plotting the power spectrum of the signal or series.

In order to show the self-similarity of BTC returns in the time–frequency domain, first only the real values of each wavelet coefficient are taken:

$Rcwt = ReplacePart[cwt, 1 \rightarrow Re[cwt[[1]]]]$

Once the real part of the wavelet transform is defined, the wavelet scalogram is plotted and shown in Fig. 5.12. Self-similar curves in the time–frequency plane are visible at first glance. The fractal pattern is present throughout the series, irrespective of scale or wavelet power.

Figures 5.13a, b, c, d are magnifications of Fig. 5.12 at varying scales. The top left chart depicts scales 1–4 and the top right figure scales 4–7, while the bottom left chart shows scales 5–8 and the bottom right chart scales 7–10. Specific scales can be plotted as follows:

c1=WaveletScalogram[Rcwt,{1 | 2 | 3 | 4,_}];
c2=WaveletScalogram[Rcwt,{4 | 5 | 6 | 7,_}];
c3=WaveletScalogram[Rcwt,{5 | 6 | 7 | 8,_}];
c4=WaveletScalogram[Rcwt,{7 | 8 | 9 | 10,_}];
Grid[{{c1,c2},{c3,c4}}]

5.4 Conclusion

Concluding remarks are presented in this section. The contribution of wavelet analysis to the fractal market hypothesis and the economic sciences in general are discussed, as well as the future possible areas of research using time–frequency analysis.

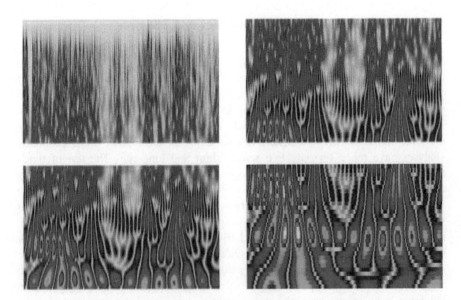

Fig. 5.13 Self-similar behavior of Bitcoin price returns at scales: 1-4 (top left), 4-7 (top right), 5-8 (bottom left) and 7-10 (bottom right)

5.4.1   Fractal Market Hypothesis: Evidence from Wavelet Power Spectrum

In spite of the novelty of the Bitcoin protocol and the uncharacteristically high volatility of the homonymous currency, the predictions made by the fractal market hypothesis correctly capture the asset's behavior. Thanks to the ability of wavelet analysis to decompose a time series into different scales, it is possible to observe the dominance of short investment horizons during periods of price volatility. The theoretical framework developed by Peters takes into account heterogeneous agents who operate at simultaneous time horizons and react to market information with respect to their investment horizon; therefore, it is possible to account for the statistical departures to the efficient market hypothesis observed in the cryptocurrency's returns (Peters 1991a,b, 1994).

Additionally, the use of wavelet analysis allowed to observe the presence of self-similar dynamics in the time series through the wavelet power spectrum. This methodology has also been used to detect fractal properties in a wide range of natural phenomena, from fluid turbulence to DNA sequences and breathing rate variability (Adisson 2002).

5.4.2 Wavelet Analysis in Economics and Finance

Many authors argue the importance of multiscale relationships in economics and finance (In and Kim 2012; Aguiar-Conraria and Soares 2011; Ramsey and Lampart 1997a). In financial terms, market dynamics are affected by investment horizons that range from high-frequency trading to individual stockbrokers, hedge funds, multinational corporations, pension funds, and government debt. However, despite the wide range of investment horizons operating in the market, most economic analyses have relied on only two scales, short and long run. The use of time–frequency analysis is being rapidly adopted in economics to study how a process operates across a wide range of time scales. Wavelet analysis has been used to investigate the multiscale relationship between the stock and futures markets over various time horizons, the interest rate swap market in the time–frequency domain, long memory in rates and volatility of LIBOR, and the relationship between stock returns and risk factors at various time scales (In and Kim 2012).

Ramsey and Lampart list many benefits of incorporating wavelet analysis to the discipline, for example, estimators for novel situations, greater estimation efficiency, robustness of modeling error, reduction of biased estimations, and, most importantly, discovering new insights into the properties of economic phenomena. For example, they mention previous studies using wavelet analysis, or time scale decomposition, to study the term structure of interest rates, the distinction between permanent and transitory shocks, or the relationship by time scale of money and income, and expenditure and income (Ramsey 2002; Ramsey and Lampart 1997b, 1998a,b).

In their studies of money, income, and expenditure in the time–frequency domain, Ramsey and Lampart found evidence of complex behavior in the relationships between these variables (Ramsey and Lampart 1998a,b). The authors found that the delays observed between variables are a function of time and scale, contrary to the accepted assumption that delays between variables are fixed. This provides opportunities for future research examining the underlying mechanisms of the time-varying delays. Ramsey speculates that the "timing" of actions by economics agents can explain time-varying delays and provides as an example a 2001 push by auto-manufacturers to lower the purchase price on cars (Ramsey 2002). The author argues the automakers' decision had two effects. First, undoubtedly, increased quantity was demanded in reaction to an implicit price decline, but it also shortened the delay between income and expenditure.

Wavelet analysis can also be used to study structural change and regime shifts, for example, to model the impact of minimum wage and tax legislation and innovation. Aggregate time series can be decomposed into long-term structural components, medium-term seasonal components, and short-term random components. This approach can allow to characterize a robust system at high scales that also permits fluctuations that are not entirely random and shorter time scales. As it was shown in Fig. 5.6, wavelet analysis allows for the study of transitional changes that were previously impossible to observe thanks to wavelet's ability to capture hidden dynamics.

Ramsey also points to the analysis of term structure of interest rates as a field where wavelet applications should provide extensive and deep insights since the role of the horizon of the decision-maker on market outcomes is so clearly indicated (Ramsey 2002). Kiermeier, for example, analyzes the risk factors of the European term structure of interest rates and finds good forecasting results, with up to 1 month of significant forecasts even during times of financial market distress (Kiermeier 2014).

Time–frequency analysis also allows for new developments on forecasting. By decomposing a time series into its global and local aspects, specific forecasting techniques can be applied to each scale of the time series.

5.4.3   Future Research

The results presented in this study indicate new areas for both empirical and theoretical research, for example, how agents operate at several scales simultaneously, both at the individual and aggregate level, or what are the long- and short-term structural components underlying cryptocurrencies and their relationship to other economic variables in the time–frequency domain. The application of wavelet analysis to economics and finance is still in its infancy when compared with other fields. However, the wavelet literature in economics is rapidly growing and expanding to other areas in the fields. While most of wavelet analysis has fallen into three broad categories (macroeconomics, volatility and asset pricing, and forecasting and spectral analysis), this new approach can provide not only novel techniques but also new insights in many fields of economics.

References

Adisson, P.S.: The Illustrated Wavelet Transform Handbook: Introductory Theory and Applications in Science, Engineering, Medicine and Finance. CRC Press, Boca Raton, FL (2002)

Aguiar-Conraria, L., Soares, M.J.: The continuous wavelet transform: a primer. Núcleo de Investigação em Políticas Económicas [Research Unit in Economic Policy] (2011). Accessed on May 10, 2014. Available online at http://www.nipe.eeg.uminho.pt/Uploads/WP_2011/NIPE_WP_16_2011.pdf

Barber, S., Boyen, X., Shi, E., Uzun, E.: Bitter to better—how to make bitcoin a better currency. Finance Crypt. Data Secur. **7379**, 399–414 (2012). Accessed on May 10, 2014. Available online at http://crypto.stanford.edu/~xb/fc12/bitcoin.pdf

Biddle, S.: What is Bitcoin? Gizmodo (2011). Accessed on May 10, 2014. Available online at http://gizmodo.com/5803124/what-is-bitcoin

Bitcoin: Important Milestones of the Bitcoin Project. Bitcoin (2014). Accessed on May 10, 2014. Available online at https://en.bitcoin.it/wiki/History

Blackledge, J.M.: The Fractal Market Hypothesis: Applications to Financial Forecasting. Dublin Institute of Technology, Dublin (2010). Accessed on May 10, 2014. Available online at http://eleceng.dit.ie/papers/182.pdf

Borland, L., Bouchaud, J.P., Muzy, J.F., Zumbach, G.: The dynamics of financial markets: Mandelbrot's multifractal cascades, and beyond. Wilmott Magazine (2005). Accessed on May 10, 2014. Available online at http://arxiv.org/pdf/cond-mat/0501292.pdf

Brito, J.: Online cash bitcoin could challenge governments, Banks. Time. (2011). Accessed on May 10, 2014. Available online at http://techland.time.com/2011/04/16/online-cash-bitcoin-could-challenge-governments/

Chaum, D.: Blind signatures for untraceable payments. Adv. Cryptol. **82**(3), 199–203 (1983). Accessed on May 10, 2014. Available online at http://blog.koehntopp.de/uploads/Chaum. BlindSigForPayment.1982.PDF

Chaum, D., Fiat A., Naor, M.: Untraceable electronic cash. Lect. Notes Comput. Sci. **403**, 319–327 (1990). Accessed on May 10, 2014. Available online at http://blog.koehntopp.de/uploads/ chaum_fiat_naor_ecash.pdf

Chen, A.: The underground website where you can buy any drug imaginable. Gawker (2011a). Accessed on May 10, 2014. Available online at http://gawker.com/the-underground-website-where-you-can-buy-any-drug-imag-30818160

Chen, A.: Everyone wants bitcoins after learning they can buy drugs with them. Gawker (2011b). Accessed on May 10, 2014. Available online at http://gawker.com/5808314/everyone-wants-bitcoins-after-learning-they-can-buy-drugs-with-them

Congresional Research Service: Bitcoin: questions, answers, and analysis of legal issues. Federation of American Scientists (2013). Accessed on May 10, 2014. Available online at https:// www.fas.org/sgp/crs/misc/R43339.pdf

Delfin, R.: The fractal nature of bitcoin: evidence from wavelet power spectra. Universidad de las Américas Puebla (2014). Accessed on May 10, 2014. Available online at http://goo.gl/qCGmfn

Dermietzel, J.: The heterogeneous agents approach to financial markets - development and milestones. In: Seese, D., Weinhardt, C., Schlottmann, F. (eds.) Handbook on Information Technology in Finance: International Handbooks Information System, pp. 443–464. Springer, Berlin, Heidelberg, Alemania (2008). Accessed on May 10, 2014. Available online at http:// link.springer.com/chapter/10.1007%2F978-3-540-49487-4_19

Ehrentreich, N.: Replicating the stylized facts of financial markets. In: Ehrentreich, N. (ed.) Agent-Based Modeling The Santa Fe Institute Artificial Stock Market Model Revisited, pp. 51–88. Springer, Berlin, Heidelberg (2008). Accessed on May 10, 2014. Available online at http://link. springer.com/chapter/10.1007/978-3-540-73879-4_5

Grapps, A.: Introduction to wavelets. IEEE Comput. Sci. Eng. **2**(2), 50–61 (1995). Accessed on November 10, 2014. Available online at http://cs.haifa.ac.il/~nimrod/Compression/Wavelets/ Wavelets_Graps.pdf

In, F., Kim, S.: An Introduction to Wavelet Theory in Finance: A Wavelet Multiscale Approach. World Scientific, Singapore (2012)

Kiermeier, M.M.: Essay on wavelet analysis and the European term structure of interest rates. Bus. Econ. Horiz. **9**(4), 18–26 (2014). https://ideas.repec.org/a/pdc/jrnbeh/v9y2014i4p18-26.html

Kristoufek, L.: Fractal market hypothesis and the global financial crisis: wavelet power evidence. Scientific Reports (2013). Accessed on May 10, 2014. Available online at http://www.nature. com/srep/2013/131004/srep02857/full/srep02857.html

Kroll, J.A., Davey, I.C., Felten, E.W.: The economics of bitcoin mining, or bitcoin in the presence of adversaries. Workshop on the Economics of Information Security (2013). Accessed on May 10, 2014. Available online at http://weis2013.econinfosec.org/papers/ KrollDaveyFeltenWEIS2013.pdf

Lee, T.B.: The Bitcoin bubble. Bottom-up (2011). Accessed on May 10, 2014. Available online at http://timothyblee.com/2011/04/18/the-bitcoin-bubble/

Lee, D.T.L., Yamamoto, A.: Wavelet analysis: theory and applications. Hewlett-Packard J. (1994) Accessed on May 10, 2014. Available online at http://www.hpl.hp.com/hpjournal/94dec/ dec94a6.pdf

Li, J.: Image compression: the mathematics of JPEG 2000. Mod. Signal Process. **46**, 185–221 (2003). Accessed on May 10, 2014. Available online at http://library.msri.org/books/Book46/ files/08li.pdf

Mack, E.: The bitcoin pizza purchase that's worth $7 million today. Forbes (2013). Accessed on May 10, 2014. Available online at http://www.forbes.com/sites/ericmack/2013/12/23/the-bitcoin-pizza-purchase-thats-worth-7-million-today/

Mandelbrot, B.: The variation of certain speculative prices. J. Bus. 36(4), 394–419 (1963). Accessed on May 10, 2014. Available online at http://ideas.repec.org/p/cwl/cwldpp/1164.html

Mandelbrot, B.: Fractals and Scaling in Finance: Discontinuity, Concentration, Risk. Springer, Berlin (1997)

McCowan, D.: Spectral estimation with wavelets. The University of Chicago (2007). Accessed on November 10, 2014. Available online at http://home.uchicago.edu/~mccowan/research/wavelets/mccowan_htc_thesis.pdf

Nakamoto, S.: Bitcoin: a peer-to-peer electronic cash system. Bitcoin (2009). Accessed on May 10, 2014. Available online at https://bitcoin.org/bitcoin.pdf

Peters, E.: Chaos and Order in the Capital Markets: A New View of Cycles, Prices, and Market Volatility. Wiley, New York (1991a)

Peters, E.: A chaotic attractor for the S&P 500. Finance Anal. J. 47(2), 55–62 (1991b). Accessed on May 10, 2014. Available online at http://harpgroup.org/muthuswamy/talks/cs498Spring2013/AChaoticAttractorForTheSAndP500.pdf

Peters, E.: Fractal Market Analysis: Applying Chaos Theory to Investment and Economics. Wiley, New York (1994)

Rama, C.: Empirical properties of asset returns: stylized facts and statistical issues. Quant. Finance 1(2), 223–236 (2001). Accessed on May 10, 2014. Available online at http://www-stat.wharton.upenn.edu/~steele/Resources/FTSResources/StylizedFacts/Cont2001.pdf

Rama, C.: Volatility clustering in financial markets: empirical facts and agent-based models. Social Science Research Network (2005). Accessed on May 10, 2014. Available online at https://www.newton.ac.uk/preprints/NI05015.pdf

Ramsey, J.B., Lampart, C.: The Decomposition of Economic Relationships by Time Scale Using Wavelets. New York University, New York (1997a). Accessed on May 10, 2014. Available online at http://econ.as.nyu.edu/docs/IO/9382/RR97-08.PDF

Ramsey, J.B., Lampart, C.: The analysis of foreign exchange rates using waveform dictionaries. J. Empir. Finance (1997b). Accessed on May 10, 2014. Available online at http://citeseerx.ist.psu.edu/viewdoc/summary?doi=10.1.1.52.9744

Ramsey, J.B., Lampart, C.: The decomposition of economic relationships by time scale using wavelets: money and income. Macroecon. Dyn. 2, 49–71 (1998a). Accessed on May 10, 2014. Available online at http://journals.cambridge.org/article_S1365100598006038

Ramsey, J.B., Lampart, C.: The decomposition of economic relationships by time scale using wavelets: expenditure and income. Stud. Nonlinear Dyn. Econ. 3(4), 49–71 (1998b). Accessed on May 10, 2014. Available online at https://ideas.repec.org/a/bpj/sndecm/v3y1998i1n2.html

Ramsey, J.B.: Wavelets in economics and finance: past and future. Stud. Nonlinear Dyn. Econ. 6(1), 1–27 (2002). Accessed on November 10, 2014. Available online at http://plaza.ufl.edu/yiz21cn/refer/wavelet%20in%20economics%20and%20finance.pdf

Ranta, M.: Wavelet multiresolution analysis of financial time series. University of Vaasa (2010). Accessed on May 10, 2014. Available online at http://www.uva.fi/materiaali/pdf/isbn_978-952-476-303-5.pdf

Rua, A.: Wavelets in economics. Banco de Portugal (2012). Accessed on May 10, 2014. Available online at https://www.bportugal.pt/en-US/BdP%20Publications%20Research/AB201208_e.pdf

Szabo, N.: Bit gold. Unenumerated (2008). Accessed on May 10, 2014. Available online at http://unenumerated.blogspot.mx/2005/12/bit-gold.html

Thurner, S., Feurstein, M.C., Teich, M.C.: Multiresolution wavelet analysis of heartbeat intervals discriminates healthy patiens from those with cardia pathology. Phys. Rev. Lett. 80(7), 1544–1547 (1998). Accessed on May 10, 2014. Available online at http://sws.bu.edu/teich/pdfs/PRL-80-1544-1998.pdf

Torrence, C., Compo, G.P.: A practical guide to wavelet analysis. Program in Atmospheric and Oceanic Sciences (1998). Accessed on November 10, 2014. Available online at http://paos.colorado.edu/research/wavelets/bams_79_01_0061.pdf

Velde, F.R.: Chicago fed letter - bitcoin: a primer. The Federal Reserve Bank of Chicago (2013). Accessed on May 10, 2014. Available online at http://www.chicagofed.org/webpages/publications/chicago_fed_letter/2013/december_317.cfm

Weisenthal, J.: Here's the answer to Paul Krugman's difficult question about bitcoin. Business Insider (2013). Accessed on May 10, 2014. Available online at http://www.businessinsider.com/why-bitcoin-has-value-2013-12

Wolfram, S.: Wavelet analysis: concepts. Wolfram (2014a). Accessed on May 10, 2014. Available online at http://www.wolfram.com/training/videos/ENG811/

Wolfram, S.: Wavelet analysis: applications. Wolfram (2014b). Accessed on May 10, 2014. Available online at http://www.wolfram.com/training/videos/ENG812/

Wolfram, S.: WaveletScale. Wolfram (2014c). Accessed on May 10, 2014. Available online at http://reference.wolfram.com/language/ref/WaveletScale.html

Chapter 6
Computing Greeks for Lévy Models: The Fourier Transform Approach

Federico De Olivera and Ernesto Mordecki

Abstract We review the computation of Greeks for exponential Lévy models extending Lewis formula for the option value. This gives accurate approximations using fast Fourier transform. We present an exhaustive development of Greeks for call options. We provide error estimation in for all Greeks in the Black–Scholes model (where Greeks can be exactly computed) and consider other models used in the literature, such as the Merton and variance gamma models.

Keywords Exponential Lévy models • Option pricing • Lévy processes • Derivatives • Greeks • Lewis pricing formula • Black Scholes • Merton model • Variance gamma model

6.1 Introduction

We consider a Lévy process $X = \{X_t\}_{t \geq 0}$ defined on a probability space $(\Omega, \mathscr{F}, \mathbf{Q})$; a financial market model with two assets; a deterministic savings account $B = \{B_t\}_{t \geq 0}$, given by

$$B_t = B_0 e^{rt},$$

with $r \geq 0$ and $B_0 > 0$; and a stock $S = \{S_t\}_{t \geq 0}$, given by

$$S_t = S_0 e^{rt + X_t}, \tag{6.1}$$

F. De Olivera
Mathematics Center, School of Sciences, Universidad de la República, Montevideo, Uruguay

Departamento de Matemática, Federico Garcia Lorca entre Pastori y Goya, CeRP del Sur, Atlántida, 15200, Uruguay
e-mail: fededeo@gmail.com

E. Mordecki (✉)
Mathematics Center, School of Sciences, Universidad de la República, Montevideo, Uruguay
e-mail: mordecki@cmat.edu.uy

© Springer International Publishing Switzerland 2016
A.A. Pinto et al. (eds.), *Trends in Mathematical Economics*,
DOI 10.1007/978-3-319-32543-9_6

with $S_0 > 0$, where $X = \{X_t\}_{t\geq 0}$ is a Lévy process. When the process X has continuous paths, we obtain the classical Black–Scholes model (Merton 1973). For general reference on the subject, we refer to Kyprianou (2006) or Cont and Tankov (2004).

The aim of this paper is the computation of the price partial derivatives of a European option with general payoff with respect to any parameter of interest. These derivatives are usually named as "Greeks," and consequently, we use the term *Greek* to refer to any price partial derivative of the option (of any order and with respect to any parameter). Our approach departs from the subtle observation by Cont and Tankov (see Cont and Tankov 2004, p. 365):

Contrary to the classical Black–Scholes case, in exponential-Lévy models there are no explicit formulae for call option prices, because the probability density of a Lévy process is typically not known in closed form. However, the characteristic function of this density can be expressed in terms of elemen- elementary functions for the majority of Lévy processes discussed in the literature. This has led to the development of Fourier-based option pricing methods for exponential-Lévy models. In these methods, one needs to evaluate one Fourier transform numerically but since they simultaneously give option prices for a range of strikes and the Fourier transform can be efficiently computed using the FFT algorithm, the overall complexity of the algorithm per option price is comparable to that of evaluating the Black–Scholes formula.

In other words, in the need of computation of a range of option prices, from a practical point of view, the Lewis formula works as a closed formula, as it can be implemented and computed with approximately the same precision and in the same time as the Black–Scholes formula.

Some papers have addressed this problem. Eberlein et al. Eberlein et al. (2009) obtained a formula similar to the Lewis one and derived delta (Δ) and gamma (Γ), the price partial derivatives with respect to the initial value S_t of first and second order, for a European payoff function. The assumptions are similar to the ones we require. Takahashi and Yamazaki (2008) also obtain these Greeks in the case of call options, based on the Carr and Madan approach. The advantage of the Lewis formula is that it gives option prices for general European payoffs, while the Carr–Madan only price European vanilla options or some other type of strike-dependent options. It must be said that the Carr–Madan formula is applicable for general processes, not necessarily Lévy, but, as we are interested in the Lévy process, we heavily rely on Lewis' approach, being the main difference with the analysis in Lee (2004). Other works deal with the problem of Greek computation for more general payoff functions, including path-dependent options [e.g., see Boyarchenko and Levendorskií (2009), Chen and Glasserman (2007), Glasserman and Liu (2007), Glasserman and Liu (2008), Jeannin and Pistorius (2010) and Kienitz (2008)]. These works are based on different techniques, such as simulation or finite differences introducing a method error that has to be analyzed, whereas our approach does not.

In the present paper, we obtain closed formulas for Greeks based on the Lewis formula that computes efficiently and with arbitrary precision [as exposed in Cont and Tankov (2004)], for arbitrary payoff European options in the Lévy models with respect to any parameter and arbitrary order. As an example, we analyze the case of call options.

6.2 Greeks for General European Options in Exponential Lévy Models

Denote by $\mathbf{Q}$ the risk-neutral pricing measure, i.e.,

$$\mathbf{E}\,e^{X_1} = 1, \tag{6.2}$$

where $\mathbf{E}$ denotes expectation under $\mathbf{Q}$. Furthermore, by the Lévy–Khintchine theorem, we obtain that $\mathbf{E}\,e^{izX_t} = e^{t\Psi(z)}$, where the characteristic exponent is

$$\Psi(z) = -iz(1 - iz)\frac{\sigma^2}{2} + \int_{\mathbb{R}} \left(e^{izy} - 1 - iz(e^y - 1)\right)\nu(dy), \tag{6.3}$$

with $\sigma \geq 0$ as the standard deviation of the Gaussian part of the Lévy process and ν its jump measure.

Regarding the payoff, following Lewis (2001), denote $s = \ln S_T$ and consider a payoff $w(s)$ and its Fourier transform $\hat{w}(z) = \int_{-\infty}^{\infty} \exp(isx)w(x)dx$. For instance, if K is a strike price, the call option payoff and its respective transform are

$$w(s) = (e^s - K)^+, \qquad \hat{w} = \frac{-K^{iz+1}}{z^2 - iz} \quad (\Im(z) > 1). \tag{6.4}$$

Then, the Lewis formula (Lewis 2001) for the European options, valued at time t, and denoting $\tau = T - t$ the time to maturity, is:

$$V_t = \frac{e^{-r\tau}}{2\pi} \int_{iv+\mathbb{R}} e^{-iz(\ln(S_t)+r\tau)} e^{\tau\Psi(-z)} \hat{w}(z)dz, \tag{6.5}$$

where $z \in S_V = \{u + iv : u \in \mathbb{R}\}$ and v must be chosen depending on the payoff function (Lewis 2001). In this context, it is simple to obtain some general formulas for the Greeks.

In order to differentiate under the integral sign, we present the following classical result.

Lemma 1. *Let $\Theta \subset \mathbb{R}$ an interval and $\mathscr{I} = iv + \mathbb{R}$. Let $h : \mathscr{I} \times \Theta \to \mathbb{C}$ and $g : \mathscr{I} \to \mathbb{C}$ such that*

- $h(\cdot, \theta)g(\cdot)$ *is integrable for all $\theta \in \Theta$ and g is integrable.*
- $h(z, \cdot)$ *is differentiable in Θ for all $z \in \mathscr{I}$ and $\frac{\partial h}{\partial \theta}$ is bounded.*

Then, $\int_{\mathscr{I}} h(x,\theta)g(x)dx$ is differentiable and

$$\frac{\partial}{\partial\theta}\int_{\mathscr{I}} h(x,\theta)g(x)dx = \int_{\mathscr{I}} \frac{\partial h(x,\theta)}{\partial\theta}g(x)dx \qquad \forall\theta\in\Theta.$$

Proof. We observe that $\left|\frac{\partial h(z,\theta)g(z)}{\partial\theta}\right| \leq C|g(z)|$ for all $z\in\mathscr{I},\theta\in\Theta$. The result is obtained from Theorem 2.27 in Folland (1999).

In consequence, in the rest of the paper, we will always assume that the conditions in Lemma 1 are satisfied for the real part of the integrand because the price imaginary part integrate is zero.

6.2.1 First-Order Greeks

We introduce the auxiliary function

$$\vartheta(z) = e^{-iz(\ln(S_t)+r\tau)}e^{\tau\Psi(-z)}\hat{w}(z).$$

Departing from (6.5) and (6.3), by differentiation under the integral sign, we obtain

$$\Delta_t = \frac{\partial V_t}{\partial S_t} = -\frac{1}{S_t}\frac{e^{-r\tau}}{2\pi}\int_{iv+\mathbb{R}} iz\vartheta(z)dz,$$

$$\rho_t = \frac{\partial V_t}{\partial r} = -\tau\frac{e^{-r\tau}}{2\pi}\int_{iv+\mathbb{R}} (1+iz)\vartheta(z)dz,$$

$$\mathcal{V}_t = \frac{\partial V_t}{\partial\sigma} = \tau\sigma\frac{e^{-r\tau}}{2\pi}\int_{iv+\mathbb{R}} iz(1+iz)\vartheta(z)dz,$$

$$\Theta_t = \frac{\partial V_t}{\partial\tau} = \frac{e^{-r\tau}}{2\pi}\int_{iv+\mathbb{R}} (\Psi(-z) - (1+iz)r)\vartheta(z)dz.$$

Usually, the Lévy models used in the literature depend on a set of parameters that specify the jump measure. Therefore, we denote $v(dy) = v_\theta(dy)$ and $\Psi(z) = \Psi_\theta(z)$; then

$$\frac{\partial V_t}{\partial\theta} = \tau\frac{e^{-r\tau}}{2\pi}\int_{iv+\mathbb{R}} \frac{\partial\Psi_\theta(-z)}{\partial\theta}\vartheta(z)dz.$$

6.2.2 Second-Order Greeks

Similarly, we obtain

$$\Gamma_t = \frac{\partial^2 V_t}{\partial S_t^2} = \frac{1}{S_t^2}\frac{e^{-r\tau}}{2\pi}\int_{iv+\mathbb{R}} iz(1+iz)\vartheta(z)dz,$$

$$\text{Vanna}_t = \frac{\partial^2 V_t}{\partial \sigma \, \partial S_t} = \tau \sigma \frac{1}{S_t} \frac{e^{-r\tau}}{2\pi} \int_{iv+\mathbb{R}} z^2 (1+iz) \vartheta(z) dz,$$

$$\text{Vomma}_t = \frac{\partial^2 V_t}{\partial \sigma^2} = \tau \frac{e^{-r\tau}}{2\pi} \int_{iv+\mathbb{R}} \left(1 + \tau \sigma^2 iz(1+iz)\right) iz(iz+1) \vartheta(z) dz,$$

$$\text{Charm}_t = \frac{\partial^2 V_t}{\partial S_t \partial \tau} = -\frac{1}{S_t} \frac{e^{-r\tau}}{2\pi} \int_{iv+\mathbb{R}} iz \left(\Psi(-z) - (1+iz)r\right) \vartheta(z) dz,$$

$$\text{Veta}_t = \frac{\partial^2 V_t}{\partial \sigma \, \partial \tau} = \sigma \frac{e^{-r\tau}}{2\pi} \int_{iv+\mathbb{R}} iz(1+iz) \left(\tau \Psi(-z) - (iz+1)r\tau + 1\right) \vartheta(z) dz,$$

$$\text{Vera}_t = \frac{\partial^2 V_t}{\partial \sigma \, \partial r} = -\tau^2 \sigma \frac{e^{-r\tau}}{2\pi} \int_{iv+\mathbb{R}} iz(iz+1)^2 \vartheta(z) dz.$$

Other derivatives can be obtained analogously. In the next section, we will focus in the case of call options. This allows to obtain more explicit formulas.

6.3 Greeks for Call Options in Exponential Lévy Models

In order to exploit the particular payoff function, we exhaustively develop the Greeks for call options. The put option corresponding formulas can be obtained immediately via put-call parity. For other payoffs, the procedure to obtain the Greeks is analogous.

When the strike K is fixed, $x = \ln(K/S_t) - r\tau$ is variable in terms of S_t, r, and τ. Then, we must consider this for the computation of Greeks Δ, Γ, ρ, and others.

Lemma 2. *Let X_τ be a Lévy process with triplet (γ, σ, v) and characteristic exponent $\Psi(z)$ such that $\Psi(-i) = 0$ and $\int_{|y|>1} e^{vy} v(dy) < \infty$ with $v \geq 0$. Then, if $z \in iv + \mathbb{R}$,*

$$|\Psi_J(-z)| \leq (|z|^2 + |z|) \frac{e^v}{2} \int_{|y| \leq 1} y^2 v(dy) + 2 \int_{|y|>1} (e^{vy} + 1) v(dy)$$

and

$$|\Psi(-z)| \leq (|z|^2 + |z|) \left(\frac{e^v}{2} \int_{|y| \leq 1} y^2 v(dy) + \frac{\sigma^2}{2}\right) + 2 \int_{|y|>1} (e^{vy} + 1) v(dy),$$

$$(6.6)$$

where $\Psi_J(z) = \int_{\mathbb{R}} \left(e^{izy} - 1 - iz(e^y - 1)\right) v(dy)$.

Proof. Let $I(z) = \int_{\mathbb{R}} \left(e^{izy} - 1 - izy \mathbf{1}_{\{|y| \leq 1\}}\right) v(dy)$. Applying Taylor's expansion with Lagrange error form at point $y = 0$, there exists θ_y with $|\theta_y| \leq |y|$ such that

$$e^{izy} - 1 - izy\mathbf{1}_{\{|y|\le 1\}} = izy\mathbf{1}_{\{|y|>1\}} - z^2 y^2 \frac{e^{iz\theta_y}}{2}$$

$$= -z^2 y^2 \frac{e^{iz\theta_y}}{2} \mathbf{1}_{\{|y|\le 1\}} + \left(izy - z^2 y^2 \frac{e^{iz\theta_y}}{2} \right) \mathbf{1}_{\{|y|>1\}}$$

$$= -z^2 y^2 \frac{e^{iz\theta_y}}{2} \mathbf{1}_{\{|y|\le 1\}} + (e^{izy} - 1)\mathbf{1}_{\{|y|>1\}}.$$

Then

$$|I(-z)| \le \int_{\mathbb{R}} \left| -z^2 y^2 \frac{e^{-iz\theta_y}}{2} \mathbf{1}_{\{|y|\le 1\}} + (e^{-izy} - 1)\mathbf{1}_{\{|y|>1\}} \right| \nu(dy)$$

$$\le |z|^2 \int_{|y|\le 1} \frac{e^{v\theta_y}}{2} y^2 \nu(dy) + \int_{|y|>1} (e^{vy} + 1)\nu(dy)$$

$$\le |z|^2 \frac{e^v}{2} \int_{|y|\le 1} y^2 \nu(dy) + \int_{|y|>1} (e^{vy} + 1)\nu(dy). \tag{6.7}$$

Using (6.7), we have

$$|\Psi_J(-z)| = |I(-z) + izI(-i)|$$

$$\le (|z|^2 + |z|) \frac{e^v}{2} \int_{|y|\le 1} y^2 \nu(dy) + 2 \int_{|y|>1} (e^{vy} + 1)\nu(dy).$$

For the continuous part, let $\Psi_C(-z) = (iz - z^2)\frac{\sigma^2}{2}$; thus,

$$|\Psi(-z)| \le |\Psi_C(-z)| + |\Psi_J(-z)|$$

$$\le (|z|^2 + |z|) \left(\frac{e^v}{2} \int_{|y|\le 1} y^2 \nu(dy) + \frac{\sigma^2}{2} \right) + 2 \int_{|y|>1} (e^{vy} + 1)\nu(dy).$$

Lemma 3. *Let $\{X_\tau\}_{\tau\ge 0}$ be a Lévy process with triplet (γ, σ, ν) and characteristic exponent $\Psi(z)$, such that $\Psi(-i) = 0$ and $\mathbf{E}[e^{vX_\tau}] < \infty$ with $v > 0$.*

1. If $\int_{iv+\mathbb{R}} |z|^{-1} |e^{\tau\Psi(-z)}| dz < \infty$, then

$$\mathbf{P}(X_\tau > x) = -\frac{1}{2\pi} \int_{iv+\mathbb{R}} \frac{e^{izx}}{iz} e^{\tau\Psi(-z)} dz, \tag{6.8}$$

$$\mathbf{E}(e^{X_\tau} \mathbf{1}_{\{X_\tau > x\}}) = -\frac{1}{2\pi} \int_{iv+\mathbb{R}} \frac{e^{(1+iz)x}}{1+iz} e^{\tau\Psi(-z)} dz. \tag{6.9}$$

2. If

$$\int_{iv+\mathbb{R}} |z^n e^{\tau\Psi(-z)}| dz < \infty, \quad \text{for some } n = 0, 1, \ldots, \tag{6.10}$$

then X_τ has a density of class C^n and

$$\frac{\partial^n f(x)}{\partial x^n} = \frac{1}{2\pi} \int_{iv+\mathbb{R}} (iz)^n e^{izx} e^{\tau\Psi(-z)} dz.$$

Proof. *For a call option, the Fourier transform of the payoff function is $\hat{w}(z) = \frac{e^{iz(\ln(S_t)+r\tau+x)}}{iz(1+iz)}$. Then from the option value (6.5), we have, with $x = \log(K/S_t) - r\tau$,*

$$C_t(x) = S_t \frac{e^x}{2\pi} \int_{iv+\mathbb{R}} \frac{e^{izx} e^{\tau\Psi(-z)}}{iz(iz+1)} dz. \tag{6.11}$$

Then, being $x \in [\alpha, \beta]$ and $C_1 = \max_{\alpha \le x \le \beta} e^{(1-v)x}$,

$$\left| \frac{\partial e^{(1+iz)x} e^{\tau\Psi(-z)} [iz(iz+1)]^{-1}}{\partial x} \right| \le C_1 |z^{-1}| |e^{\tau\Psi(-z)}| \in L^1(iv+\mathbb{R}),$$

and by Theorem 2.27 in Folland (1999), we can differentiate under the integral sign. Therefore, with $S_t = 1$,

$$\mathbf{P}(X_\tau > x) = -e^{-x} \frac{\partial}{\partial x} \int_x^\infty (e^s - e^x) F(ds)$$

$$= -e^{-x} \frac{\partial C_t(x)}{\partial x} = -\frac{1}{2\pi} \int_{iv+\mathbb{R}} \frac{e^{izx}}{iz} e^{\tau\Psi(-z)} dz.$$

On the other hand, with $S_t = 1$,

$$\mathbf{E}(e^{X_\tau} \mathbf{1}_{\{X_\tau > x\}}) = \int_x^\infty e^s F(ds) = -e^x \frac{\partial}{\partial x} \int_x^\infty (e^{s-x} - 1) F(ds)$$

$$= -e^x \frac{\partial e^{-x} C_t(x)}{\partial x} = C_t(x) - \frac{\partial C_t(x)}{\partial x}$$

$$= \frac{1}{2\pi} \int_{iv+\mathbb{R}} \frac{e^{(1+iz)x}}{iz(1+iz)} e^{\tau\Psi(-z)} dz - \frac{1}{2\pi} \int_{iv+\mathbb{R}} \frac{e^{(1+iz)x}}{iz} e^{\tau\Psi(-z)} dz$$

$$= -\frac{1}{2\pi} \int_{iv+\mathbb{R}} \frac{e^{(1+iz)x}}{1+iz} e^{\tau\Psi(-z)} dz.$$

For the second part, observe in (6.8) that if $x \in [\alpha, \beta]$ and $C_2 = \max_{\alpha \le x \le \beta} e^{-vx}$,

$$\left| \frac{\partial^{n+1} \frac{e^{izx}}{iz} e^{\tau\Psi(-z)}}{\partial x^{n+1}} \right| \le C_2 |z|^n |e^{\tau\Psi(-z)}| \in L^1(iv+\mathbb{R}).$$

The result is obtained from Theorem 2.27 in Folland (1999).

We conclude the preliminaries of this section with some notations. As (6.2) holds, the relation $d\tilde{\mathbf{Q}} = e^{X_T} d\mathbf{Q}$ defines a probability measure $\tilde{\mathbf{Q}}$, known as the *dual martingale measure*. A call option price with payoff (6.4), and log-forward moneyness[1] $x = \ln(K/S_0) - r\tau$, can be computed by

$$C_t(x) = S_t \left(\tilde{\mathbf{Q}}(X_\tau > x) - e^x \mathbf{Q}(X_\tau > x) \right).$$

Furthermore, if X_t has a density $f_t(x)$ under $\mathbf{Q}$, i.e., if condition (6.10) holds with $n = 0$, its density under $\tilde{\mathbf{Q}}$ is $\tilde{f}_t(s) = e^s f_t(s)$. In order to obtain Greeks in terms of the risk-neutral measure, we replace $\mathbf{P}$ by $\mathbf{Q}$ in (6.8), and consequently (6.9), (6.10), and (6.11) are related to the probability measure $\mathbf{Q}$.

6.3.1   First-Order Greeks for Call Options

In this section, we do not assume general requirements. We specify the requirements in each case.

6.3.1.1   Delta

Assume that $\int_{iv+\mathbb{R}} |z|^{-1} |e^{\tau\Psi(-z)}| dz < \infty$ and $S_t \in [A, B]$. From (6.11), we obtain

$$
\begin{aligned}
\Delta_t^L &= \frac{\partial C_t(x(S_t))}{\partial S_t} \\
&= \frac{\partial}{\partial S_t} S_t \frac{1}{2\pi} \int_{iv+\mathbb{R}} \frac{e^{(1+iz)x(S_t)}}{iz(1+iz)} e^{\tau\Psi(-z)} dz \\
&= \frac{1}{2\pi} \int_{iv+\mathbb{R}} \frac{e^{(1+iz)x}}{iz(1+iz)} e^{\tau\Psi(-z)} dz - \frac{1}{2\pi} \int_{iv+\mathbb{R}} \frac{e^{(1+iz)x}}{iz} e^{\tau\Psi(-z)} dz \\
&= -\frac{1}{2\pi} \int_{iv+\mathbb{R}} \frac{e^{(1+iz)x}}{1+iz} e^{\tau\Psi(-z)} dz = \tilde{\mathbf{Q}}(X_\tau > x).
\end{aligned}
$$

6.3.1.2   Rho

Denote now $x = \ln(K/S_t) - r\tau$, that depends on the interest rate r. Assume that $\int_{iv+\mathbb{R}} |z|^{-1} |e^{\tau\Psi(-z)}| dz < \infty$ and $r \in [R_1, R_2]$. Then

---

[1]This seems to be the standard definition, although in Cont and Tankov (2004) it is defined as the opposite quantity.

$$\rho_t^L = \frac{\partial C_t(x(r))}{\partial r} = S_t \frac{1}{2\pi} \int_{iv+\mathbb{R}} -\tau \frac{e^{(1+iz)x}}{iz} e^{\tau\Psi(-z)} dz = \tau S_t e^x \, \mathbf{Q}(X_\tau > x).$$

6.3.1.3 Vega

In Black–Scholes, vega shows the change in variance of the log price. In exp-Lévy models, the derivative of $C_t(x)$ w.r.t. σ does not give exactly the same information. We assume that X_τ has density f [see (6.10)] and $z \in iv + \mathbb{R}$. Let

$$h(z,\sigma) = e^{\tau iz(1+iz)\frac{\sigma^2}{2}}, \qquad g(z) = \frac{e^{izx+\tau\int_{\mathbb{R}}(e^{izy}-1-iz(e^y-1))v(dy)}}{iz(1+iz)}.$$

Thus, $\frac{\partial}{\partial\sigma}h(z,\sigma)$ is bounded. On the other hand, $\int_{iv+\mathbb{R}} |g(z)|dz < \infty$ because $|\mathbf{E}(e^{-i(iv+s)J_\tau})| \le \mathbf{E}(e^{vJ_\tau}) < \infty$, where J_τ is the jump part of X_τ. By Lemma 1, we can differentiate under the integral sign. Then,

$$\mathcal{V}_t^L = \frac{\partial C_t(x)}{\partial\sigma} = S_t \frac{e^x}{2\pi} \int_{iv+\mathbb{R}} \frac{e^{izx} e^{\tau\Psi_\sigma(-z)}}{iz(iz+1)} \tau\sigma iz(1+iz)dz$$

$$= S_t \tau\sigma e^x f_\tau(x).$$

In order to complete the information provided by vega, we can calculate the derivative with respect to the jump intensity, when this intensity is finite, i.e., when $v(dy) = \lambda G(dy)$ for a probability distribution G, and assume the existence of a density (6.10). Denote

$$h(z,\lambda) = \frac{e^{\tau\lambda\int_{\mathbb{R}}(e^{-izy}-1+iz(e^y-1))G(dy)}}{\tau\int_{\mathbb{R}}(e^{-izy}-1+iz(e^y-1))\,G(dy)},$$

$$g(z) = \frac{e^{(iz+1)x}e^{\tau iz(1+iz)\frac{\sigma^2}{2}}}{iz(iz+1)} \tau\int_{\mathbb{R}}(e^{-izy}-1+iz(e^y-1))\,G(dy),$$

where $\frac{\partial h(z,\lambda)}{\partial\lambda}$ is bounded and from Lemma 2 $\int_{iv+\mathbb{R}}|g(z)|dz < \infty$; then by Lemma 1, we obtain

$$\frac{\partial C_t(x)}{\partial\lambda} = \frac{\partial}{\partial\lambda} S_t \frac{1}{2\pi} \int_{iv+\mathbb{R}} \frac{e^{(iz+1)x} e^{\tau\Psi(-z)}}{iz(iz+1)} dz$$

$$= \tau S_t \frac{1}{2\pi} \int_{iv+\mathbb{R}} \frac{e^{(iz+1)x} e^{\tau\Psi(-z)}}{iz(iz+1)} \overline{\Psi}_J(-z)dz,$$

with $\overline{\Psi}_J(-z) = \int_{\mathbb{R}} \left(e^{-izy} - 1 + iz(e^y - 1) \right) G(dy)$. Using Fubini's theorem, we obtain

$$\frac{\partial_\tau C_t(x)}{\partial \lambda} = \frac{\partial}{\partial \lambda} S_t \frac{1}{2\pi} \int_{iv+\mathbb{R}} \frac{e^{(iz+1)x} e^{\tau \Psi(-z)}}{iz(iz+1)} dz$$

$$= \tau S_t \frac{1}{2\pi} \int_{iv+\mathbb{R}} \frac{e^{(iz+1)x} e^{\tau \Psi(-z)}}{iz(iz+1)} \int_{\mathbb{R}} \left(e^{-izy} - 1 + iz(e^y - 1) \right) G(dy) dz$$

$$= \tau S_t \left[\int_{\mathbb{R}} \left(e^y \frac{e^{x-y}}{2\pi} \int_{iv+\mathbb{R}} e^{iz(x-y)} \frac{e^{\tau \Psi(-z)}}{iz(1+iz)} dz \right. \right.$$

$$- \frac{e^x}{2\pi} \int_{iv+\mathbb{R}} e^{izx} \frac{e^{\tau \Psi(-z)}}{iz(1+iz)} dz$$

$$\left. \left. + (e^y - 1) \frac{e^x}{2\pi} \int_{iv+\mathbb{R}} e^{izx} \frac{e^{\tau \Psi(-z)}}{1+iz} dz \right) G(dy) \right]$$

$$= \tau \left[\int_{\mathbb{R}} \left(e^y C_t(x-y) - C_t(x) - S_t(e^y - 1) \tilde{Q}(X_\tau > x) \right) G(dy) \right].$$

The use of Fubini's theorem is justified by (6.7) and the additional hypothesis $\int_{iv+\mathbb{R}} |e^{\tau \Psi(-z)}| dz < \infty$.

6.3.1.4 Theta

Assume condition (6.10) holds with $n = 2$, and let

$$h(z, \tau) = e^{(iz+1)x_\tau} e^{\tau \Psi(-z)}, \qquad g(z) = \frac{1}{iz(1+iz)}.$$

Then, $\int_{iv+\mathbb{R}} |g(z)| dz < \infty$; moreover, from (6.6) and $\int_{iv+\mathbb{R}} |z^2 e^{\tau \Psi(-z)}| dz < \infty$,

$$\frac{\partial h(z, \tau)}{\partial \tau} = e^{(iz+1)x_\tau} e^{\tau \Psi(-z)} \left(-r(1+iz) + \Psi(-z) \right)$$

is bounded, and by Lemma 1,

$$\Theta_t^L = \frac{\partial_\tau C_t(x_\tau)}{\partial \tau} = \frac{\partial}{\partial \tau} S_t \frac{1}{2\pi} \int_{iv+\mathbb{R}} \frac{e^{(iz+1)x_\tau} e^{\tau \Psi(-z)}}{iz(iz+1)} dz$$

$$= S_t \frac{1}{2\pi} \int_{iv+\mathbb{R}} \frac{e^{(iz+1)x_\tau} e^{\tau \Psi(-z)}}{iz(1+iz)} \left(\Psi(-z) - r(1+iz) \right) dz.$$

Using Fubini's theorem, we obtain

$$\Theta_t^L = S_t \frac{1}{2\pi} \int_{iv+\mathbb{R}} \frac{e^{(iz+1)x_\tau} e^{\tau \Psi(-z)}}{iz(1+iz)} \Big(\Psi(-z) - r(1+iz) \Big) dz$$

$$= S_t \Bigg[-\frac{r}{2\pi} \int_{iv+\mathbb{R}} \frac{e^{(iz+1)x_\tau} e^{\tau \Psi(-z)}}{iz} dz$$

$$+ \frac{1}{2\pi} \int_{iv+\mathbb{R}} \frac{e^{(iz+1)x_\tau} e^{\tau \Psi(-z)}}{iz(iz+1)} \Big(iz(1+iz)\frac{\sigma^2}{2}$$

$$+ \int_{\mathbb{R}} \big(e^{-izy} - 1 + iz(e^y - 1) \big) v(dy) \Big) dz \Bigg]$$

$$= S_t \Bigg[re^{x_\tau} \mathbf{Q}(X_\tau > x_\tau) + \frac{\sigma^2}{2} e^{x_\tau} f_\tau(x_\tau)$$

$$+ \int_{\mathbb{R}} \Big(e^y \frac{e^{x-y}}{2\pi} \int_{iv+\mathbb{R}} e^{iz(x-y)} \frac{e^{\tau \Psi(-z)}}{iz(1+iz)} dz - \frac{e^x}{2\pi} \int_{iv+\mathbb{R}} e^{izx} \frac{e^{\tau \Psi(-z)}}{iz(1+iz)} dz$$

$$+ (e^y - 1)\frac{e^x}{2\pi} \int_{iv+\mathbb{R}} e^{izx} \frac{e^{\tau \Psi(-z)}}{1+iz} dz \Big) v(dy) \Bigg]$$

$$= S_t \Bigg[re^{x_\tau} \mathbf{Q}(X_\tau > x_\tau) + \frac{\sigma^2}{2} e^{x_\tau} f_\tau(x_\tau) \Bigg]$$

$$+ \int_{\mathbb{R}} \Big(e^y C_t(x_\tau - y) - C_t(x_\tau) - S_t(e^y - 1)\tilde{\mathbf{Q}}(X_\tau > x_\tau) \Big) v(dy).$$

The use of Fubini's theorem is justified by (6.7) and the additional hypothesis $\int_{iv+\mathbb{R}} |z^2 e^{\tau \Psi(-z)}| dz < \infty$.

6.3.2 Second-Order Greeks for Call Options

6.3.2.1 Gamma

Once delta is obtained, we must only differentiate again with respect to S_t, to obtain gamma. We assume that X_τ has density f and $S_t \in [A, B]$, then

$$\Gamma_t^L = \frac{\partial^2 C_t(x(S_t))}{\partial S_t^2} = \frac{\partial \tilde{\mathbf{Q}}(X_\tau > x(S_t))}{\partial S_t} = \frac{1}{S_t} \tilde{f}(x) = \frac{e^x}{S_t} f(x).$$

6.3.2.2 Vanna

We assume that $\int_{iv+\mathbb{R}} |ze^{\tau \Psi(-z)}| dz < \infty$ and $0 < \Sigma_1 \leq \sigma \leq \Sigma_2$, then

$$\frac{\partial^2 C_t(x)}{\partial \sigma \partial S_t} = \frac{\partial \mathcal{V}_t^L}{\partial S_t} = \tau \sigma e^{x(S_t)} f_t(x(S_t)) - \tau \sigma e^{x(S_t)} \Big(f_t(x(S_t)) + f_t'(x(S_t)) \Big)$$

$$= -\tau \sigma e^x f_\tau'(x).$$

6.3.2.3 Vomma

We assume that $\int_{iv+\mathbb{R}} |z^2 e^{\tau \Psi(-z)}| dz < \infty$ and $0 < \Sigma_1 \leq \sigma \leq \Sigma_2$, let $z \in iv + \mathbb{R}$, and denote

$$h(z, \sigma) = z^2 e^{\tau i z(1+iz)\frac{\sigma^2}{2}}, \quad g(z) = \frac{e^{izx+\tau \int_{\mathbb{R}}(e^{izy}-1-iz(e^y-1))\nu(dy)}}{z^2}.$$

Thus, $\frac{\partial h(z,\sigma)}{\partial \sigma}$ is bounded, and $\int_{iv+\mathbb{R}} |g(z)| dz < \infty$, because $|\mathbf{E}(e^{-izJ_\tau})| \leq \mathbf{E}(e^{vJ_\tau}) < \infty$, where J_τ is the jump part of X_τ. By Lemma 1, we can differentiate under the integral sign

$$\frac{\partial^2 C_t(x)}{\partial \sigma^2} = \frac{\partial \mathcal{V}_t^L}{\partial \sigma} = S_t \tau e^x f_\tau(x) + S_t \tau \sigma \frac{e^x}{2\pi} \int_{iv+\mathbb{R}} e^{izx} e^{\tau \Psi(-z)} \tau \sigma (iz - z^2) dz$$

$$= S_t \tau e^x \left(f_\tau(x) + \tau \sigma^2 \left(f_\tau'(x) + f_\tau''(x) \right) \right).$$

6.3.2.4 Charm

Assume that (6.10) holds for $n = 3$, take $z \in iv + \mathbb{R}$, and let

$$h(z, \tau) = z e^{(iz+1)x_\tau} e^{\tau \Psi(-z)}, \quad g(z) = \frac{1}{z(1+iz)}.$$

Then, $\int_{iv+\mathbb{R}} |g(x)| dx < \infty$, and by Lemma 2,

$$\frac{\partial h(z, \tau)}{\partial \tau} = z e^{(iz+1)x_\tau} e^{\tau \Psi(-z)} \left(-r(1+iz) + \Psi(-z) \right)$$

is bounded. By Lemma 1,

$$\frac{\partial^2 C_t(x)}{\partial \tau \partial S_t} = \frac{\partial \tilde{\mathbf{Q}}(X_\tau > x_\tau)}{\partial \tau} = \frac{\partial}{\partial \tau} \frac{-1}{2\pi} \int_{iv+\mathbb{R}} e^{(iz+1)x_\tau} \frac{e^{\tau \Psi(-z)}}{1+iz} dz$$

$$= \frac{1}{2\pi} \int_{iv+\mathbb{R}} e^{(iz+1)x_\tau} \frac{e^{\tau \Psi(-z)}}{1+iz} \left(r(1+iz) - \Psi(-z) \right) dz. \qquad (6.12)$$

Using Fubini's theorem, we obtain

$$\frac{\partial^2 C_t(x)}{\partial \tau \partial S_t} = \frac{1}{2\pi} \int_{iv+\mathbb{R}} e^{(iz+1)x_\tau} \frac{e^{\tau \Psi(-z)}}{1+iz} \left(r(1+iz) - \Psi(-z) \right) dz$$

$$= r e^{x_\tau} f_\tau(x_\tau) - \frac{\sigma^2}{2} e^{x_\tau} f_\tau'(x_\tau)$$

$$-\int_{\mathbb{R}}\Big[e^y\frac{e^{x_\tau-y}}{2\pi}\int_{iv+\mathbb{R}}e^{iz(x_\tau-y)}\frac{e^{\tau\Psi(-z)}}{1+iz}dz-\frac{e^{x_\tau}}{2\pi}\int_{iv+\mathbb{R}}e^{izx_\tau}\frac{e^{\tau\Psi(-z)}}{1+iz}dz$$

$$+(e^y-1)\Big\{\frac{e^{x_\tau}}{2\pi}\int_{iv+\mathbb{R}}e^{izx_\tau}e^{\tau\Psi(-z)}dz$$

$$-\frac{e^{x_\tau}}{2\pi}\int_{iv+\mathbb{R}}e^{izx_\tau}\frac{e^{\tau\Psi(-z)}}{1+iz}dz\Big\}\Big]v(dy)$$

$$=-re^{x_\tau}f_\tau(x_\tau)+\frac{\sigma^2}{2}e^{x_\tau}f'_\tau(x_\tau)$$

$$+\int_{\mathbb{R}}\Big[-e^y\tilde{Q}(X_\tau>x_\tau-y)+\tilde{Q}(X_\tau>x_\tau)$$

$$+(e^x-1)\{e^xf_\tau(x)+\tilde{Q}(X_\tau>x_\tau)\}\Big]v(dy)$$

$$=re^{x_\tau}f_\tau(x_\tau)-\frac{\sigma^2}{2}e^{x_\tau}f'_\tau(x_\tau)$$

$$-\int_{\mathbb{R}}\Big[e^y\Big(\tilde{Q}(X_\tau>x_\tau)-\tilde{Q}(X_\tau>x_\tau-y)\Big)$$

$$+(e^y-1)e^{x_\tau}f_\tau(x_\tau)\Big]v(dy). \tag{6.13}$$

The use of Fubini's theorem is justified by (6.7) and the additional hypothesis $\int_{iv+\mathbb{R}}|z^3e^{\tau\Psi(-z)}|dz<\infty$.

6.3.2.5   Veta

We assume that $\int_{iv+\mathbb{R}}|z^4e^{\tau\Psi(-z)}|dz<\infty$. Similar to Charm$_t$, we assume that $\tau\in[\mathcal{T}_1,\mathcal{T}_2]$ and $z\in iv+\mathbb{R}$ and denote

$$h(z,\tau)=z^2e^{(iz+1)x_\tau}e^{\tau\Psi(-z)},\quad g(z)=\frac{1}{z^2}.$$

Then, $\int_{iv+\mathbb{R}}|g(z)|dz<\infty$, and by Lemma 2,

$$\frac{\partial h(z,\tau)}{\partial\tau}=z^2e^{(iz+1)x_\tau}e^{\tau\Psi(-z)}\Big(-r(1+iz)+\Psi(-z)\Big)$$

is bounded. By Lemma 1, we can differentiate under the integral sign,

$$\frac{\partial^2C_t(x_r)}{\partial\sigma\partial\tau}=\frac{\partial\mathcal{V}_t^L}{\partial\tau}=\frac{\partial S_t\tau\sigma e^{x_\tau}f_\tau(x_\tau)}{\partial\tau}$$

$$=S_t\sigma\Big[e^{x_\tau}f_\tau(x_\tau)-r\tau e^{x_\tau}f_\tau(x_\tau)+\frac{\tau e^{x_\tau}}{2\pi}\int_{iv+\mathbb{R}}\frac{\partial}{\partial\tau}e^{izx_\tau}e^{\tau\Psi(-z)}dz\Big]$$

$$=S_t\sigma\Big[e^{x_\tau}f_\tau(x_\tau)-r\tau e^{x_\tau}f_\tau(x_\tau)$$

$$+\frac{\tau e^{x_\tau}}{2\pi}\int_{iv+\mathbb{R}}e^{izx_\tau}e^{\tau\Psi(-z)}\Big(\Psi(-z)-riz\Big)dz\Big].\qquad(6.14)$$

Using Fubini's theorem, we obtain

$$\frac{\partial^2 C_t(x_r)}{\partial\sigma\,\partial\tau}=S_t\sigma\Big[e^{x_\tau}f_\tau(x_\tau)-r\tau e^{x_\tau}f_\tau(x_\tau)$$

$$+\frac{\tau e^{x_\tau}}{2\pi}\int_{iv+\mathbb{R}}e^{izx_\tau}e^{\tau\Psi(-z)}\Big(\Psi(-z)-riz\Big)dz$$

$$=S_t\sigma e^{x_\tau}\Big[f_\tau(x_\tau)-r\tau\big(f_\tau(x_\tau)+f_\tau'(x_\tau)\big)$$

$$+\frac{\tau}{2\pi}\int_{iv+\mathbb{R}}e^{izx_\tau}e^{\tau\Psi(-z)}\{\frac{\sigma^2}{2}(iz-z^2)$$

$$+\int_{\mathbb{R}}(e^{-izy}-1+iz(e^y-1))v(dy)\}dz\Big]$$

$$=S_t\sigma e^{x_\tau}\Big[f_\tau(x_\tau)-r\tau\big(f_\tau(x_\tau)+f_\tau'(x_\tau)\big)+\tau\frac{\sigma^2}{2}\big(f_\tau'(x_\tau)+f_\tau''(\tau)\big)$$

$$+\tau\int_{\mathbb{R}}\Big(f_\tau(x_\tau-y)-f_\tau(x_\tau)+(e^y-1)f_\tau'(x_\tau)\Big)v(dy)\Big].\qquad(6.15)$$

The use of Fubini's theorem is justified by (6.7) and the additional hypothesis $\int_{iv+\mathbb{R}}|z^4 e^{\tau\Psi(-z)}|dz<\infty$.

6.3.2.6 Vera

Assuming that $\int_{iv+\mathbb{R}}|ze^{\tau\Psi(-z)}|dz<\infty$ and $0<\Sigma_1\le\sigma\le\Sigma_2$,

$$\frac{\partial^2 C_t(x_r)}{\partial\sigma\,\partial r}=\frac{\partial\mathcal{V}_t^L}{\partial r}=S_t\tau\sigma e^{x_r}\Big(-\tau f_\tau(x_r)-\tau f_\tau'(x_r)\Big)$$

$$=-S_t\tau^2\sigma e^{x_r}\Big(f_\tau(x_r)+f_\tau'(x_r)\Big).$$

6.3.3 Third-Order Greeks for Call Options

6.3.3.1 Color

We assume that $\int_{iv+\mathbb{R}} |z^4 e^{\tau \Psi(-z)}| dz < \infty$, $\tau \in [\mathscr{T}_1, \mathscr{T}_2]$. Let $z \in iv + \mathbb{R}$ and

$$h(z, \tau) = z^2 e^{(iz+1)x_\tau} e^{\tau \Psi(-z)}$$

$$g(z) = \frac{1}{z^2}.$$

Then, $\int_{iv+\mathbb{R}} |g(x)| dx < \infty$, and by Lemma 2,

$$\frac{\partial h(z, \tau)}{\partial \tau} = z^2 e^{(iz+1)x_\tau} e^{\tau \Psi(-z)} \left(-r(1 + iz) + \Psi(-z) \right)$$

is bounded. By Lemma 1, we can differentiate under the integral sign.
Thus,

$$\frac{\partial^3 C_t(x)}{\partial S_t^2 \partial \tau} = \frac{\partial \Gamma_t^L}{\partial \tau} = \frac{1}{S_t 2\pi} \int_{iv+\mathbb{R}} e^{(iz+1)x_\tau} e^{\tau \Psi(-z)} \left(-r(iz + 1) + \Psi(-z) \right) dz.$$
(6.16)

Using Fubini's theorem, we obtain

$$\begin{aligned}
\frac{\partial^3 C_t(x)}{\partial S_t^2 \partial \tau} &= \frac{1}{S_t 2\pi} \int_{iv+\mathbb{R}} e^{(iz+1)x_\tau} e^{\tau \Psi(-z)} \left(-r(iz + 1) + \Psi(-z) \right) dz \\
&= \frac{e^x}{S_t} \Bigg[-r(f(x) + f'(x)) + \frac{\sigma^2}{2} (f'(x) + f''(x)) \\
&\quad + \frac{1}{2\pi} \int_{iv+\mathbb{R}} e^{izx} e^{\tau \Psi(-z)} \int_{\mathbb{R}} e^{-izy} - 1 + iz(e^y - 1)v(dy) dz \Bigg] \\
&= -\frac{e^x}{S_t} \Bigg[r\big(f_\tau(x) + f'_\tau(x)\big) - \frac{\sigma^2}{2} \big(f'_\tau(x) + f''_\tau(x)\big) \\
&\quad + \int_{\mathbb{R}} \big(f_\tau(x) - f_\tau(x - y) - (e^y - 1)f'_\tau(x)\big) v(dy) \Bigg].
\end{aligned}$$
(6.17)

Fubini is justified by (6.7) and the hypothesis $\int_{iv+\mathbb{R}} |z^4 e^{\tau \Psi(-z)}| dz < \infty$.

6.3.3.2 Speed

Assuming that $\int_{iv+\mathbb{R}} |z e^{\tau \Psi(-z)}| dz < \infty$,

$$\frac{\partial^3 C_t(x_r)}{\partial S_t^3} = \frac{\partial \Gamma_t^L}{\partial S_t} = \frac{e^{x(S_t)}\left(-\frac{1}{S_t}f_\tau(x(S_t)) - \frac{1}{S_t}f_\tau'(x(S_t))\right)S_t - e^{x(S_t)}f_\tau(x(S_t))}{S_t^2}$$

$$= -\frac{e^x}{S_t^2}\left(2f_\tau(x) + f_\tau'(x)\right).$$

6.3.3.3 Ultima

We assume that $\int_{iv+\mathbb{R}} |z^6 e^{\tau \Psi(-z)}| dz < \infty$. First, we calculate $\frac{\partial f_\tau^{(n)}(x)}{\partial \sigma}$ for $n = 0, 1, 2$. For $0 < \Sigma_1 \leq \sigma \leq \Sigma_2$ and $z \in iv + \mathbb{R}$, we denote

$$h_n(z, \sigma) = (iz)^{n+2} e^{\tau \Psi(-z)}, \quad g(z) = -\frac{e^{izx}}{z^2}.$$

Thus, $\int_{iv+\mathbb{R}} |g(z)| dz < \infty$ and $\frac{\partial h_n(z,\sigma)}{\partial \sigma}$ is bounded for $n = 0, 1, 2$. By Lemma 1, we can differentiate under the integral sign. Then,

$$\frac{\partial^n f_\tau(x)}{\partial \sigma} = \frac{\partial}{\partial \sigma} \frac{1}{2\pi} \int_{iv+\mathbb{R}} (iz)^n e^{izx} e^{\tau \Psi(-z)} dz$$

$$= \tau \sigma \frac{1}{2\pi} \int_{iv+\mathbb{R}} [(iz)^{n+1} - (iz)^{n+2}] e^{izx} e^{\tau \Psi(-z)} dz$$

$$= \tau \sigma \left(f_\tau^{(n+1)}(x) + f_\tau^{(n+2)}(x)\right). \tag{6.18}$$

Now, we have

$$\frac{\partial^3 C_t(x)}{\partial \sigma^3} = \frac{\partial S_t \tau e^x \left(f_\tau(x) + \tau \sigma^2 \left(f_\tau'(x) + f_\tau''(x)\right)\right)}{\partial \sigma}$$

$$= S_t \tau^2 \sigma e^x \left(3\left(f_\tau'(x) + f_\tau''(x)\right) + \tau \sigma^2 \left(f_\tau''(x) + 2f_\tau'''(x) + f_\tau^{(iv)}(x)\right)\right).$$

6.3.3.4 Zomma

We assume that $\int_{iv+\mathbb{R}} |z^2 e^{\tau \Psi(-z)}| dz < \infty$ and $0 < \Sigma_1 \leq \sigma \leq \Sigma_2$. Then,

$$\frac{\partial^3 C_t(x_r)}{\partial S_t^2 \partial \sigma} = \frac{\partial \text{Vanna}_t^L}{\partial S_t} = \frac{\partial \tau \sigma e^{x(S_t)} f_\tau'(x(S_t))}{\partial S_t}$$

$$= -\frac{\tau \sigma e^x}{S_t}\left(f_\tau'(x) + f_\tau''(x)\right).$$

6.4 Examples

6.4.1 The Black–Scholes Model

If we assume that the Gaussian distribution and density are exactly computed in R software, we can compare the Greeks for Black–Scholes model using Lewis representation. To approximate the Fourier transform, we cut the integral between $-A/2$ and $A/2$ and take a uniform partition of $[-A/2, A/2]$ of size N:

$$\int_{\mathbb{R}} e^{izx} g(z)dz \approx \int_{-A/2}^{A/2} e^{izx} g(z)dz \approx \frac{A}{N} \sum_{k=0}^{N-1} w_k e^{iz_k x} g(z_k),$$

where $z_k = -\frac{A}{2} + k\frac{A}{N-1}$ and w_k are weights that correspond to the integration numerical rule.

Table 6.1 shows the ℓ_∞-errors in the Black–Scholes model via Lewis representation and fast Fourier transform using $S_t = 1, r = 0.05, T = 1, \sigma = 0.1, A = 300$, and $N = 2^{22}$. Denoting by GL the value for the Greek given by our Lévy formula, and by G the direct computation, the ℓ_∞-errors are

$$\ell_\infty\text{-error}(GL) = \max_{x \in [-0.7, 0.7]} |GL - G|, \tag{6.19}$$

for $x = \ln(K/S_t) - r\tau$.

Table 6.1 ℓ_∞-errors in the Black–Scholes model via Lewis representation and fast Fourier transform using $S_t = 1, r = 0.05, T = 1, \sigma = 0.1, A = 300$, and $N = 2^{22}$

Greek	Expression	ℓ_∞-error
Call	$C = S\,\mathbf{E}(e^{X_\tau} - e^x)^+$	1.2e-07
Delta	$\partial_S C(x)$	2.4e-07
Rho	$\partial_r C_t(x)$	1.9e-07
Vega	$\partial_\sigma C(x)$	9.5e-08
Theta	$\partial_\tau C(x)$	1.2e-08
Gamma	$\partial_{SS}^2 C(x)$	9.5e-07
Vanna	$\partial_{\sigma S}^2 C(x)$	6.3e-07
Vomma	$\partial_{\sigma\sigma}^2 C(x)$	7.5e-07
Charm	$\partial_{S\tau}^2 C(x)$	6.8e-08
Veta	$\partial_{\sigma\tau}^2 C(x)$	8.9e-08
Vera	$\partial_{\sigma r}^2 C(x)$	5.8e-07
Color	$\partial_{SS\tau}^3 C(x)$	5.6e-07
Speed	$\partial_{SSS}^3 C(x)$	6.3e-06
Ultima	$\partial_{\sigma\sigma\sigma}^3 C(x)$	1.2e-05
Zomma	$\partial_{SS\sigma}^3 C(x)$	9.5e-06

6.4.2   The Merton Model

In this section, we show some results for the Merton model. The Merton model has four parameters $(\sigma, \mu_J, \sigma_J, \lambda)$ where σ is the diffusion parameter, λ is the jump intensity, and μ_J and σ_J are the mean and standard deviation of the jump which are Gaussianly distributed. The characteristic function for the Merton model is

$$\mathbf{E}(e^{izX_T}) = \exp\left\{iz\left[\frac{\sigma^2}{2} - \lambda\left(e^{\mu_J+\frac{\sigma_J^2}{2}} - 1\right)\right] + z^2\frac{\sigma^2}{2} + \lambda\left(e^{iz\mu_J - z^2\frac{\sigma_J^2}{2}} - 1\right)\right\}.$$
(6.20)

All Greeks for *At The Money* $(K = S_0 e^{-rT})$ are shown in Table 6.2 following Sect. 6.3. To evaluate errors, we compute first a reasonable value GL and then increase the precision with the corresponding time cost, to obtain G, that we assume is the true value, and apply (6.19). Hence, we took $A = 500, N = 2^{20}$ and $A = 500$, $N = 2^{22}$; the error is in all Greeks lower than 10^{-5}. In Fig. 6.1, the curves are shown in terms of $x = \ln(K/S_0) - rT$ for all Greeks with the comparison of the Black–Scholes model with volatility equal to implied volatility *At The Money*.

The characteristic function in this case is (6.20). To compute sensitivities w.r.t. μ_J, σ_J, and λ, we only need to differentiate the characteristic exponent with respect to these parameters:

$$\frac{\partial\Psi(-z)}{\partial\mu_j} = \lambda iz\left[e^{\mu_J+\sigma_J^2/2} - e^{-iz\mu_J - z^2\sigma_J^2/2}\right],$$

Table 6.2 Greeks in Merton model with: $S_0 = 1, r = 0.05, x = 0$, $T = 1, \sigma = 0.1, \mu_J = -0.005, \sigma_J = 0.1, \lambda = 1$

		$A = 500, N = 2^{20}$	$A = 500, N = 2^{21}$	Error
Call	C	0.0547129	0.0547129	2.6e-08
Delta	$\partial_S C$	0.5273560	0.5273562	2.5e-07
Rho	$\partial_r C$	0.4726431	0.4726433	2.2e-07
Vega	$\partial_\sigma C$	0.3077754	0.3077755	1.5e-07
Theta	$\partial_\tau C$	0.0524286	0.0524286	2.5e-08
Gamma	$\partial_{SS}^2 C$	3.0777536	3.0777550	1.5e-06
Vanna	$\partial_{\sigma S}^2 C$	0.1538877	0.1538878	7.3e-08
Vomma	$\partial_{\sigma\sigma}^2 C$	0.9091776	0.9091780	4.3e-07
Charm	$\partial_{S\tau}^2 C$	0.1682859	0.1682860	8.1e-08
Veta	$\partial_{\sigma\tau}^2 C$	0.1222075	0.1222076	5.8e-08
Vera	$\partial_{\sigma r}^2 C$	−0.1538877	−0.1538878	7.3e-08
Color	$\partial_{SS\tau}^3 C$	1.8556786	1.8556795	8.8e-07
Speed	$\partial_{SSS}^3 C$	−4.6166303	−4.6166325	2.2e-06
Ultima	$\partial_{\sigma\sigma\sigma}^3 C$	−11.5390901	−11.5390956	5.5e-06
Zomma	$\partial_{SS\sigma}^3 C$	−21.6857596	−21.6857699	1.0e-05

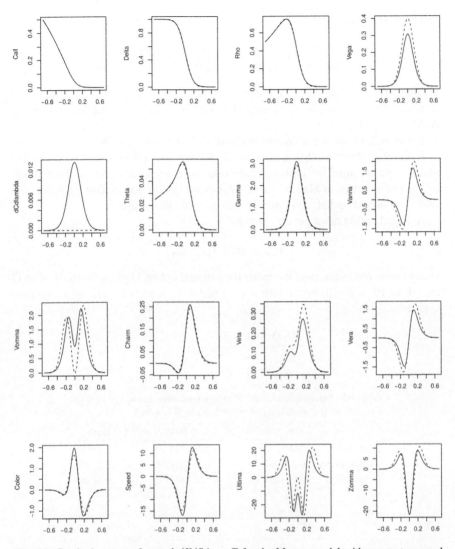

Fig. 6.1 Greeks in terms of $x = \ln(K/S_0) - rT$ for the Merton model with parameters equal to Table 6.2 (*continuous line*). *Discontinuous line*: Black–Scholes model with volatility equal to implied volatility in $x = 0$ ($\sigma_{\mathrm{imp}}(0) \approx 0.137$)

$$\frac{\partial \Psi(-z)}{\partial \sigma_j} = \lambda \sigma_J \left[i z e^{\mu_J + \sigma_j^2/2} - z^2 e^{-iz\mu_J - z^2\sigma_j^2/2} \right],$$

$$\frac{\partial \Psi(-z)}{\partial \lambda_j} = iz \left[e^{\mu_J + \sigma_j^2/2} - 1 \right] + e^{-iz\mu_J - z^2\sigma_j^2/2} - 1,$$

and for $\theta = \mu_J, \sigma_J, \lambda$,

$$\frac{\partial C_\theta(x)}{\partial \theta} = \tau S_t \frac{e^x}{2\pi} \int_{iv+\mathbb{R}} e^{izx} \frac{e^{\tau \Psi_\theta(-z)}}{iz(1+iz)} \frac{\partial \Psi_\theta(-z)}{\partial \theta} dz.$$

The differentiation under the integral sign is justified as above.

Using the same parameters presented in Table 6.2, we obtain the sensitivities for ATM given in Table 6.3.

In Fig. 6.2, we show the Greeks in terms of $x = \ln(K/S_0) - rT$.

In Kienitz (2008) are shown some results for a digital option in the Merton model, which were obtained by applying finite difference approximations to the formula for the option prices in Madan et al. (1998). Now we will deduce delta, gamma, and vega for a digital option, and thus, we will compare the results.

A digital option has a payoff given by

$$\mathbf{1}_{\{S_\tau - K > 0\}} = \mathbf{1}_{\{X_\tau - x > 0\}}.$$

Using Lewis representation, the value for a digital option is given by (6.8) (with $\mathbf{Q}$ instead of $\mathbf{P}$). If (6.10) holds with $n = 0$ and $n = 1$, respectively, differentiation leads to:

$$\frac{\partial D(x)}{\partial S_t} = \frac{1}{S_\tau} f_\tau(x), \tag{6.21}$$

Table 6.3 Sensitivities for the Merton model with: $S_0 = 1$, $r = 0.05$, $x = 0$, $T = 1$, $\sigma = 0.1$, $\mu_J = -0.005$, $\sigma_J = 0.1$, $\lambda = 1$

	$A = 500, N = 2^{20}$	$A = 1000, N = 2^{22}$	Error
μ_J-Sensitivity	0.006703850	0.006703855	4.7e-09
σ_J-Sensitivity	0.239001059	0.239001230	1.7e-07
λ-Sensitivity	0.013407701	0.013407711	9.6e-09

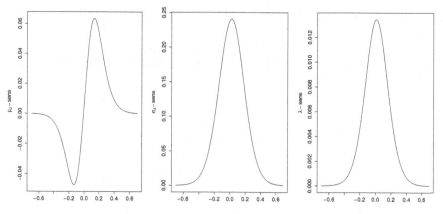

Fig. 6.2 Sensitivities in terms of $x = \ln(K/S_0) - rT$ for the Merton model with parameters equal to Table 6.3

Table 6.4 Digital option and greeks in the Merton model with: $\delta = 0.01$, $S_0 = 100$, $K = 100$, $T = 1$, $r = 0.07$, $\sigma = 0.2$, $\mu_J = 0.05$, $\sigma_J = 0.15$, and $\lambda = 0.5$

	D	D-Delta	D-Gamma	D-Vega
$N = 2^{20}$	0.531269863	0.016610445	-0.000280032	-0.560064360
$N = 2^{22}$	0.531270245	0.016610457	-0.000280032	-0.560064763
Error	3.8e-07	1.2e-08	2.0e-10	4.0e-07

Table 6.5 Greeks and ρ-sensitivity for variance gamma model with $(\rho, v, \theta) = (0.2, 1, -0.15)$, $r = 0.05$, $T = 1$, $S_0 = K = 100$ $(x = -0.05)$

	Call	Delta	Gamma	$\frac{\partial Call}{\partial \rho}$
$N = 2^{20}, \delta = 0.01$	11.26689113	0.72818427	0.01427437	23.04334371
$N = 2^{22}, \delta = 0.01$	11.26689919	0.72818479	0.01427438	23.04336021
err<	8.1e-06	5.2e-07	1.0e-08	1.6e-05

$$\frac{\partial^2 D(x)}{\partial S_t^2} = -\frac{1}{S_\tau^2}\left(f_\tau(x) + f'_\tau(x)\right), \tag{6.22}$$

$$\frac{\partial D(x)}{\partial \sigma} = -\tau\sigma\left(f_\tau(x) + f'_\tau(x)\right). \tag{6.23}$$

In (6.23), differentiation under the integral sign is similar to (6.18).

Then, our results via FFT are shown in Table 6.4. In Kienitz (2008), these values are (by finite difference): D= 0.531270, D-Delta= 0.016610, D-Gamma= -2.800324×10^{-4}, D-Vega= -0.560070. To obtain a given strike, we define $\delta = 2\pi\frac{N-1}{NA}$.

6.4.3 The Variance Gamma Model

In this section, we will compare some results from the literature. As an example, in Glasserman and Liu (2007), some results are shown for the variance gamma model with parameters (ρ, v, θ) where the characteristic function is:

$$\mathbf{E}[e^{izX_T}] = \exp\left\{\frac{T}{v}\left[iz\ln\left(1 - \theta v - \frac{\rho^2 v}{2}\right) - \ln\left(1 - iz\theta v + \frac{z^2\rho^2 v}{2}\right)\right]\right\}.$$

To obtain a given strike, we define $\delta = 2\pi\frac{N-1}{NA}$. Thus, in Table 6.5, we present two results for $N = 2^{20}$ and $N = 2^{22}$ with $\delta = 0.01$. The error shows the convergence of the complex integral. In Glasserman and Liu (2007), these results are obtained by applying finite difference approximations to the formula for the option prices in Madan et al. (1998), call= 11.2669, delta= 0.7282, and ρ-derivative $= 23.0434$, and in general with the LRM method, the error is worse than 10^{-2}.

6.5 Conclusions

Greeks are an important input for market makers in risk management. A lot of options are path dependent and they do not have an explicit formula. However, for the European options in the exponential Lévy models, we have the Lewis formula, which allows us to obtain closed formulas for Greeks, many of which are only density dependent; others require integration. In general, all Greeks can be approximated with high accuracy because they are a simple integral, similar to the Black–Scholes model.

A large number of papers are dedicated to obtain Greeks for more complex payoff functions. However, in order to estimate the accuracy of their methods, Greek approximations are computed through the finite difference technique.

For a fix strike K, we consider $x = \ln(K/S_t) - r\tau$, with $\tau = T - \tau$ the time to maturity. Thus, the Greeks for call options can be calculated through Table 6.6.

We observe that if the density of X_τ is known, then many of the Greeks can be exactly obtained. Some examples of these are normal inverse Gaussian, variance gamma, generalized hyperbolic, Meixner, and others.

Table 6.6 Greeks in exponential Lévy models in terms of $x = \ln(K/S) - r\tau$

First order	
Delta	$\partial C_S(x) = \tilde{\mathbf{Q}}(X_\tau > x)$
Rho	$\partial C_r(x) = \tau S e^x \, \mathbf{Q}(X_\tau > x)$
Vega	$\partial C_\sigma(x) = S \tau \sigma e^x f_\tau(x)$
If $\nu = \lambda \bar{\nu}$	$\partial C_\lambda(x) = \tau\left[\int_{\mathbb{R}} \left(e^y C(x-y) - C(x) \right) \right.$ $\left. - S(e^y - 1)\tilde{\mathbf{Q}}(X_\tau > x) \right) G(dy) \right]$
Theta	$\partial C_\tau(x) = S\left[re^x \, \mathbf{Q}(X_\tau > x) + \dfrac{\sigma^2}{2} e^x f_\tau(x) \right] + \dfrac{\lambda}{\tau} \partial C_\lambda(x)$
Second order	
Gamma	$\partial^2 C_{SS}(x) = S^{-1} e^x f_\tau(x)$
Vanna	$\partial^2 C_{\sigma S}(x) = -\tau \sigma e^x f'_\tau(x)$
Vomma	$\partial^2 C_{\sigma\sigma}(x) = S \tau e^x \left(f_\tau(x) + \tau \sigma^2 \left(f'_\tau(x) + f''_\tau(x) \right) \right)$
Charm	$\partial^2 C_{S\tau}(x) = $ see (6.12) and (6.13)
Veta	$\partial^2 C_{\sigma\tau}(x) = $ see (6.14) and (6.15)
Vera	$\partial^2 C_{\sigma r}(x) = -S \tau^2 \sigma e^x \left(f_\tau(x) + f'_\tau(x) \right)$
Third order	
Color	$\partial^3 C_{SS\tau}(x) = $ see (6.16) and (6.17)
Speed	$\partial^3 C_{SSS}(x) = -S^{-2} e^x \left(2f_\tau(x) + f'_\tau(x) \right)$
Ultima	$\partial^3 C_{\sigma\sigma\sigma}(x) = S\tau^2 \sigma e^x \left(3(f'_\tau(x) + f''_\tau(x)) \right.$ $\left. + \tau\sigma^2 \left(f''_\tau(x) + 2f'''_\tau(x) + f^{iv}_\tau(x) \right) \right)$
Zomma	$\partial^3 C_{SS\sigma}(x) = -\tau\sigma S^{-1} e^x \left(f'_\tau(x) + f''_\tau(x) \right)$

References

Boyarchenko, M., Levendorskií, S.: Prices and sensitivities of barrier and first-touch digital options in Lévy-driven models. Int. J. Theoret. Appl. Finance **12**(08), 1125–1170 (2009). doi:10.1142/S0219024909005610. http://www.worldscientific.com/doi/abs/10.1142/S0219024909005610

Chen, N., Glasserman, P.: Malliavin greeks without Malliavin calculus. Stoch. Process. Appl., **117**(11), 1689–1723 (2007). http://www.sciencedirect.com/science/article/B6V1B-4NWWWK6-2/1/01d688f30fb8cb8094354db2ad4b9495

Cont, R., Tankov, P.: Financial Modelling with Jump Processes. CRC Financial Mathematics Series. Chapman & Hall, London (2004)

Eberlein, E., Glau, K., Papapantoleon, A.: Analysis of Fourier transform valuation formulas and applications. Appl. Math. Finance 2010 **17**(3), 211–240 (2009). doi:10.1080/13504860903326669. http://arxiv.org/abs/0809.3405v4;http://arxiv.org/pdf/0809.3405v4

Folland, G.: Real Analysis: Modern Techniques and Their Applications. Pure and Applied Mathematics. Wiley, New York (1999). http://books.google.com.uy/books?id=uPkYAQAAIAAJ

Glasserman, P., Liu, Z.: Sensitivity estimates from characteristic functions. In: Proceedings of the 39th Conference on Winter Simulation: 40 Years! The Best is Yet to Come WSC '07, pp. 932–940. IEEE, Piscataway, NJ (2007). http://dl.acm.org/citation.cfm?id=1351542.1351707

Glasserman, P., Liu, Z.: Estimating greeks in simulating Lévy-driven models (2008). J. Comput. Finance **14**(2), 3–56 (2010/2011)

Jeannin, M., Pistorius, M.: A transform approach to compute prices and Greeks of barrier options driven by a class of Lévy processes. Quant. Finance **10**(6), 629–644 (2010)

Kienitz, J.: A note on Monte Carlo greeks for jump diffusions and other Lévy models (2008). Working paper

Kyprianou, A.E.: Introductory Lectures on Fluctuation of Lévy Processes with Applications, 1st edn. Springer, Berlin (2006)

Lee, R.: Option pricing by transform methods: extensions, unification, and error control. J. Comput. Finance **7**, 51–86 (2004)

Lewis, A.L.: A simple option formula for general jump-diffusion and other exponential Lévy processes (2001). Working paper. Envision Financial Systems and OptionCity.net Newport Beach, California. Available at http://www.optioncity.net

Madan, D., Carr, P., Chang, E.: The variance gamma process and option pricing. Eur. Finance Rev. **2**, 79–105 (1998)

Merton, R.: Theory of rational option pricing. Bell J. Econ. Manag. Sci. **4**, 141–183 (1973)

Takahashi, A., Yamazaki, A.: Efficient static replication of European options under exponential Lévy models (2008). CARF Working Paper

Chapter 7
Marginal Pricing and Marginal Cost Pricing Equilibria in Economies with Externalities and Infinitely Many Commodities

Matías Fuentes

Abstract This paper considers a general equilibrium model of an economy in which some firms may exhibit various types of non-convexities in production, there are external effects among agents and the commodity space is infinite dimensional. The consumption sets, the preferences of the consumers and the production possibilities are represented by correspondences in order to take into account the external effects. The firms are instructed to follow the marginal pricing rule from which we obtain an existence theorem. Then, the existence of a marginal cost pricing equilibrium is proved by adding additional assumptions. The simultaneous presence of externalities and infinitely many commodities are sources of technical difficulties when attempting to generalize previous existence results in the literature.

Keywords General equilibrium • Marginal pricing rules • Externalities • Increasing returns • Infinitely many commodities • Correspondences

JEL Classification: D50.

7.1 Introduction

It is well known that the presence of increasing returns in production constitutes a particular case of market failure that leads us to use an alternative criteria for producer behaviour rather than profit maximization. From the outset, beginning with Hotelling (1938), it has been argued that when the firms exhibit increasing returns to scale, prices should be proportional to marginal costs. This is the so-called *marginal cost pricing rule*. Hotelling also paid attention to the fact that in some cases, a firm

M. Fuentes (✉)
Escuela de Economía y Negocios, Centro de Investigación en Economía Teórica y Matemática Aplicada, Universidad Nacional de San Martín, Caseros 2241 San Martin,
Buenos Aires, Argentina
e-mail: mfuentes@unsam.edu.ar

© Springer International Publishing Switzerland 2016
A.A. Pinto et al. (eds.), *Trends in Mathematical Economics*,
DOI 10.1007/978-3-319-32543-9_7

or even an industry which adopts marginal cost pricing will run at a loss if there are high fixed costs. The deficit must then be financed from income taxes. Indeed, Hotelling argued that general government revenues should *be applied to cover fixed costs of electric power plants, waterworks, railroad, and other industries in which the fixed costs are large, so as to reduce to the level of marginal cost the prices charged for the services and products of these industries.*

There also exists another notion, that of *marginal pricing rule*. When the production set is smooth, this mechanism means that prices should be proportional to the gradient of the transformation functional, i.e. the producer fulfils the first-order necessary condition for profit maximization. Both the marginal pricing and the marginal cost pricing rules are closely related in such a way that they are treated as equivalents in most papers in the literature. However, as pointed out by Guesnerie (1990), they are often not equivalents at all and can often be very misleading. Bonnisseau and Cornet (1990a,b) investigated the link between both notions of equilibrium. To do this, the authors needed to introduce both the cost function and the iso-output set which required them to distinguish a priori between inputs and outputs and to propose additional assumptions. Accordingly, robust results are obtained relating both notions of equilibria.

Despite many criticisms, the marginal (cost) pricing doctrine is in force today. A rigorous and general proof of such doctrine was first offered by Guesnerie (1975) but only for economies with certain kinds of non-convex technologies. Indeed, Guesnerie considered the polar of the cone of interior displacements of the mathematicians A. Dubovickii and A. Miljurin to formalize the notion of marginal cost pricing when the production sets are non-convex. The problem with this approach comes when the production set has "inward kinks" since in this case, the normal to such a cone is only the null vector. If there is only one firm in the economy, then this problem will never arise if we simply assume that the boundary of the production set is smooth as in Mantel (1979) and Beato (1982). However, in a model with many firms, even if we assume that each firm has a smooth technology, the aggregate production set may exhibit inward kinks as Beato and Mas-Colell (1985) have shown. To avoid this difficulty, Cornet (1990) introduced in the economic literature the use of the Clarke tangent and normal cones of the mathematician F. Clarke to represent (through Clarke normal cone) the marginal (cost) pricing rule. This cone is always convex and coincides with the profit maximization behaviour (and with the cone of interior displacements) when the technologies are convex.

For economies with finitely many commodities, there are quite robust results concerning existence of marginal pricing equilibria (for a survey, we refer to Brown 1991). In contrast, for economies with infinitely many commodities, although there is a large literature on competitive equilibria (for a survey, we refer to Mas-Colell and Zame 1991), there are few results concerning marginal pricing or marginal cost pricing equilibria. Shannon (1996) stated the first proof of marginal cost pricing equilibrium in an infinite dimensional setting. She considered a private ownership economy with a finite number of consumers and only one firm. The production possibility frontier was assumed to be smooth. In the existence proof, she used the Leray-Schauder degree theory. Later, Bonnisseau (2002) generalized the results of

Shannon to the case of many firms with non-smooth production sets. He had to introduce a new and larger normal cone since the Clarke's cone does not have sufficient continuity properties in infinite-dimensional spaces. So far, there are no new results concerning marginal (cost) pricing equilibria with an infinite quantity of goods.

Furthermore, externalities constitute another basic market failure in the sense that when external effects are present, competitive equilibria are not Pareto optimal. Although it has been shown that if a competitive market exists for the externality, then optimality results (Villar 1997), this is not always the case. Take, for example, the case of an external effect produced by one individual on another. Here, price-taking behaviour is unrealistic. Moreover, by definition, the presence of external effects requires incorporating into the model the actions of other agents.

There is a large and growing literature on general equilibrium models with externalities. Laffont (1976, 1977), Laffont and Laroque (1972) and Bonnisseau (1997), among others, consider the very general case in which the action of any agent may affect the decisions on consumption and production, as well as the preferences, of the rest of the agents. In all cases, it is assumed that consumers have a non-cooperative behaviour in the sense that they maximize their preferences under their budget constraints taking the prices and the environment as given. More recently, this approach has been objected on the grounds that *price-taking assumptions inherent in the notion of competitive equilibrium are incompatible with the presence of agents who have market power-as all agents typically do when the total number of agents is finite* (Noguchi and Zame 2006). Consequently, there is also an important literature on competitive equilibria in exchange economies with externalities and a continuum of consumers (see also, Balder 2008 and Cornet and Topuzu 2005).

Another aspect of the external effects is that sometimes the presence of externalities leads to non-convexities in the underlying production processes (Mas-Colell et al. 1995, p. 375). Hence, models were proposed for combining both externalities and increasing returns. Given what was stated above on marginal pricing rule, we choose between these models, the one of Bonnisseau and Médecin (2001) where the authors develop a new marginal pricing rule with external factors. This is so because the pricing rule defined by means of Clarke's normal cone to the production set for a fixed environment does not have sufficient continuity properties. As a consequence, the pricing rule thus obtained is less precise since the new cone is larger than the former.

The purpose of this article is to provide an existence theorem with an arbitrary number of non-convex producers and externalities in an infinite dimensional setting. Infinite-dimensional commodity spaces arise naturally when we consider economic activity over an infinite time horizon, or with uncertainty about the states of the world, or when there are an infinite variety of commodity differentiation. For the sake of technical simplicity, we assume that every production set has a smooth boundary. Consequently, apart from this assumption, our existence result encompasses all the other existence results of marginal pricing equilibria in the literature.

As in Bonnisseau and Cornet (1990a,b), we show the relation between marginal pricing and marginal cost pricing equilibria. The model is not a direct extension of that of Bonnisseau and Cornet (1990a) since the presence of externalities does not allow us to claim that if a production plan belongs to a production set, the one with positive outputs also belongs to this set. We can say the same about consumers: if a consumption stream belongs to a consumption set restricted by an externality, we cannot claim that the same consumption stream belongs to a consumption set when the externality has changed by including non-negative outputs. Another important difference is that in the proof of marginal cost pricing equilibrium, they construct an argument which relies on a property of the gradient of the cost function that does not work in functional gradients. These drawbacks lead us to consider production vectors with non-negative outputs. An additional assumption on prices (which is weaker than what can generally be seen in the literature) allows us to obtain equilibrium production vectors with this property. So it is shown that a marginal pricing equilibrium is a marginal cost pricing equilibrium.

In the proof of the theorems, we roughly follow the method developed by Bewley (1972). The majority of the papers on general equilibrium with infinitely many commodities rely crucially on the First Welfare Theorem, which fails for marginal pricing and marginal cost pricing equilibria (see Guesnerie 1975). In addition to the two major drawbacks cited above, there are other technical difficulties such as those in Fuentes (2011). We take care of these problems in Sects. 7.4.2 and 7.6.1 in the same way we did in that paper.

The pricing rule in Fuentes (2011) encompasses general pricing rules. Nevertheless, we remove both bounded losses and continuity on pricing rule assumptions together with strong lower hemi-continuity in the truncated production correspondence.

Since we are interested in the relationship between non-convexities, marginal pricing and externalities in an infinite-dimensional setting, we do not follow the "continuum agents approach". It is well known that when there is an atomless measure space of agents, there are convexifying effects on preferences and technologies (Aumann 1966; Rustichini and Yannelis 1991), so we do not consider this possibility.

The paper proceeds as follows. Section 7.2 presents the model and the notation to deal with externalities, increasing returns and marginal pricing equilibrium with infinitely many commodities. Section 7.3 is devoted to the basic assumptions. In Sect. 7.4, we first define the finite-dimensional auxiliary economies, and we posit additional assumptions in order to deal with problems arising in the model. In Sect. 7.5, we state the marginal pricing equilibrium theorem. Section 7.6 is devoted to the proof of the existence result. In Sect. 7.7, we state the marginal cost pricing equilibrium theorem and give additional assumptions and definitions. Lengthy or tedious proofs are contained in the appendix.

7.2 The Model

We consider an economy with m consumers labelled by subscript $i = 1, \ldots, m$ and n producers, labelled by subscript $j = 1, \ldots, n$. The (infinite-dimensional) commodity space is represented by the space of essentially bounded, real-valued, measurable functions on a σ-finite positive measure space $(M, \mathcal{M}, \mu)$. In the following, we denote by L the space $\mathscr{L}_\infty (M, \mathcal{M}, \mu)$.[1] Each element $z = \left((x_i)_{i=1}^m, (y_j)_{j=1}^n \right)$ is an environment or externality.

Each consumer i has a *consumption set* and a *preference relation* which depends upon the actions of the other economic agents. Formally, the consumption set is represented by a correspondence X_i from L^{m+n} to L_+. For the environment $z \in L^{m+n}$, $X_i(z) \subset L_+$ is the set of possible consumption plans of the i-th consumer. We denote by $\succsim_{i,z}$ the (complete, reflexive, transitive and binary) preference relation which is influenced by the actions of all economic agents.

The *production set* of the j-th producer is defined by a correspondence Y_j from L^{m+n} to L. $Y_j(z)$ is the set of all feasible production plans for the j-th firm when the actions of the economic agents are given by z.

A *price system* is a continuous linear mapping on L. If L is endowed with the norm topology, the set of prices is $L^* = ba\,(M, \mathcal{M}, \mu)$, the space of bounded additive set functions on $(M, \mathcal{M})$ absolutely continuous with respect to μ. Thus, the value of a *commodity bundle* $x \in L_\infty$ is $\int_M x\,d\pi$ (Dunford and Schwarz 1958). If some price vector p belongs to $\mathscr{L}_1 (M, \mathcal{M}, \mu)$[2] $\subset ba\,(M, \mathcal{M}, \mu)$, then it is economically meaningful since for every $x \in L$, $p(x) = \int_{m \in M} p(m)\,x(m)\,d\mu(m)$ which is the natural generalization of the value of a commodity bundle concept in finite-dimensional spaces. The equilibrium prices can be chosen in the simplex $S = \{\pi \in ba_+(M, \mathcal{M}, \mu) : \pi(\chi_M) = 1\}$, where χ_M is the function equal to 1 for every m in M.

The weak-star topology $\sigma (\mathscr{L}_\infty, \mathscr{L}_1) = \sigma^\infty$ is the weakest topology for which the topological dual of L is $\mathscr{L}_1$. We denote by $\prod_{L^s} \sigma^\infty$ the product topology on the product space L^s. $\sigma(L, ba)$ and $\sigma(ba, L) = \sigma^{ba}$ are the weak and the weak-star topologies, respectively, on L and ba. Let $A : L^s \mapsto L$ be a correspondence. We say that A is $(\prod_{L^s} \sigma^\infty, \sigma^\infty)$-closed if it has a closed graph for the product of weak-star topologies. Let $\mathscr{S}$ be any topology on L^s. The net $(u^\alpha) \in L^s$ is said to $\mathscr{S}$-converge

---

[1] $\mathscr{L}_\infty (M, \mathcal{M}, \mu)$ is the set of equivalence classes of all μ-essentially bounded, $\mathcal{M}$-measurable functions on M. Let x be an element of $\mathscr{L}_\infty (M, \mathcal{M}, \mu)$, then $x \geq 0$ if $x(m) \geq 0$ μ-a.e. (almost everywhere); $x > 0$ if $x \geq 0$ and $x \neq 0$, and $x \gg 0$ if $x(m) > 0$ μ-a.e. Hence, if $x, \acute{x} \in \mathscr{L}_\infty (M, \mathcal{M}, \mu)$, then $x \geq \acute{x}$ (respectively, $x > \acute{x}$, $x \gg \acute{x}$) if $x - \acute{x} \geq 0$ (respectively, $x - \acute{x} > 0$, $x - \acute{x} \gg 0$). $L_+ = \{x \in L : x \geq 0\}$ is the positive cone of L, and $L_{++} = \{x \in L : x > 0\}$ is the strict positive cone or the quasi-interior of L. Let A and B be subsets of L. The difference of A and B is defined by $A \setminus B = \{x : x \in A \text{ and } x \notin B\}$. The open ball of centre x and radius ε is $B(x, \varepsilon) = \{\acute{x} \in L : \|\acute{x} - x\|_\infty < \varepsilon\}$, while the closed ball of centre x and radius ε is $\bar{B}(x, \varepsilon) = \{\acute{x} \in L : \|\acute{x} - x\|_\infty \leq \varepsilon\}$.

[2] $\mathscr{L}_1 (M, \mathcal{M}, \mu)$ is classes of all $\mathcal{M}$-measurable functions f on M such that $\int_{m \in M} |f(m)|\,d\mu(m) < \infty$.

to u if (u^α) converges for the topology $\mathscr{S}$. We denote by $\mathscr{T}$ the norm topology on L. The correspondence A is said to be $\left(\prod_{L^s} \sigma^\infty, \mathscr{T}\right)$-lower hemi-continuous (for short l.h.c.) if for every net (z^α) in L^s which $\prod_{L^s} \sigma^\infty$-converges to z and $a \in A(z)$, there is a net (a^α) such that $a^\alpha \in A(z^\alpha)$ for all α and a^α $\mathscr{T}$-converges to a. Let $\omega_i \in L_+$ be the *initial endowment* of the i-th agent and $\omega = \sum_{i=1}^m \omega_i$ the *total initial endowment* of the economy. Let $r_i : R^{1+n} \mapsto R$ be the *wealth function* of the i-th consumer. $r_i\left(\pi(\omega_i), \left(\pi(y_j)\right)_{j=1}^n\right)$ is his wealth whenever $\pi \in S$ and $(y_j)_{j=1}^n \in \prod_{j=1}^n Y_j(z)$. A special case of this structure is $r_i\left(\pi(\omega_i), \left(\pi(y_j)\right)_{j=1}^n\right) = \pi(\omega_i) + \sum_{j=1}^n \theta_{ij} y_j$ for $\theta_{ij} \geq 0$ and $\sum_{i=1}^m \theta_{ij} = 1$, which holds for a private ownership economy.

We now assume that the graph of every production correspondence is smooth.

Assumption P (Smoothness). For all j

(i) For every $z \in L^{m+n}$, $Y_j(z) = \{y \in L : f_j(y,z) \leq 0\}^3$ and $\partial_\infty Y_j(z) = \{y \in L : f_j(y,z) = 0\}$ where f_j is a transformation functional from $L \times L^{m+n}$ into R.

(ii) f_j is $\sigma^\infty \times \prod_{L^{m+n}} \sigma^\infty$-continuous on $L \times L^{m+n}$

(iii) For every $z \in L^{m+n}$, $f_j(\cdot, z)$ is Fréchet Differentiable, and if $f_j(y, z) \leq 0$ and $y' \leq y, f_j(y', z) \leq 0$ (free disposal)

(iv) $\nabla_1 f_j(y, z)^4 \in \mathscr{L}_1^+(M, \mathscr{M}, \mu) \setminus \{0\}$ if $f_j(y, z) = 0$ and $f_j(0, z) = 0$

(v) $\nabla_1 f_j$ is $\left(\sigma^\infty \times \prod_{L^{m+n}} \sigma^\infty\right)$-continuous on $L \times L^{m+n}$, that is, for all $y \in \partial_\infty Y_j(z)$, for all $\varepsilon > 0$, there exists a weak* open neighbourhood of (y, z), $U(y, z)$, in $L \times L^{m+n}$ such that $\nabla_1 f_j(y', z') \in B\left(\nabla_1 f_j(y, z), \varepsilon\right)$ for all $(y', z') \in U(y, z)$

Note that while non-convexities are allowed on the firms, they must be smooth ones (Assumptions P(i), P(ii) and P(iii)). However, no smoothness assumption is made in the aggregate production set $Y(z) = \sum_{j=1}^n Y_j(z)$, which would be far from being innocuous as Beato and Mas-Colell (1985) have shown. Assumption P(iii) also incorporates the free disposal condition. As for Assumption P(iv), we point out that $N_{Y_j(z)}(y_j) \subset ba_+(M, \mathscr{M}, \mu)$ for all $y_j \in \partial_\infty Y_j(z)$. Indeed, let $x \in L_+$. For all $t \in (0, \varepsilon), f_j(y + tx, z) \geq 0$ by Assumption P(i) and P(iii). Consequently, $\nabla_1 f_j(y_j, z)(x) = \lim_{t \downarrow 0} \frac{f_j(y_j + tx, z)}{t} \geq 0$. Thus, Assumption P(iv) only requires that prices be economically meaningful. Assumption P(v) says that f_j is continuously (Fréchet) differentiable on $L \times L^{m+n}$. This is a technical requirement for getting nice continuity properties in prices.

---

[3] We say that a production vector y is weakly efficient if and only if $y \in \partial_\infty Y(z)$. This is equivalent to say that $(\{y\} + \text{int}L_+) \cap Y(z) = \emptyset$. A stronger concept is that of efficiency. We say that a production vector y is efficient if and only if $(\{y\} + L_+) \cap Y(z) = \emptyset$.

[4] $\nabla_1 f_j(y, z)$ denotes the gradient vector of f_j with respect to y in the sense of Fréchet, that is, $\nabla_1 f_j(y_j, z)(x) = \lim_{t \to 0} \frac{f_j(y_j + tx, z) - f_j(y_j, z)}{t}$ for all $x \in L$, and the convergence is uniform with respect to x in bounded sets.

Remark 1. We point out that Assumptions P(i) and P(ii) imply that if $\left(y_j^\alpha\right) \in \partial_\infty Y_j (z^\alpha)$ for all α and $\left(y_j^\alpha, z^\alpha\right) \sigma^\infty \times \prod_{L^s} \sigma^\infty$—converges to $(\bar{y}_j, \bar{z})$, then $\bar{y}_j \in \partial_\infty Y_j (\bar{z})$.

Proposition 1. *Suppose that Assumption P holds. Then,* $Y_j : L^{m+n} \mapsto L$ *is a* $\left(\prod_{L^{m+n}} \sigma^\infty, \sigma^\infty\right)$-*closed and a* $\left(\prod_{L^{m+n}} \sigma^\infty, \mathscr{T}\right)$-*l.h.c correspondence.*

Proof. See Appendix

The smoothness assumptions allow us to introduce the marginal pricing rule for the j-th producer at $y \in \partial_\infty Y_j (z)$, as the closed half-line of outward normal vectors to $Y_j (z)$ at y_j, which also are in S, that is, $N_{Y_j(z)} (y_j) \cap S = \{\lambda \nabla_1 f_j (y, z) : \lambda \geq 0\} \cap S$. Indeed, for a given $z \in Z$, $N_{Y_j(z)} (y_j) = \{\lambda \nabla_1 f_j (y, z) : \lambda \geq 0\}$ since f is continuously differentiable on $L \times L^{m+n}$, $\nabla_1 f (y, z) \in \mathscr{L}_1^+ \setminus \{0\}$ and $f (y, z) = 0$ for all $y \in \partial_\infty Y_j (z)$ (Clarke 1983, Theorem 2.4.7, Corollary 2). Note that for all j and all $y_j \in \partial_\infty Y_j (z)$, $N_{Y_j(z)} (y_j) \cap S \neq \emptyset$, since $N_{Y_j(z)} (y_j) \subset \mathscr{L}_+^1 \setminus \{0\}$.

We characterize the economy by $\mathscr{E} = \left((X_i, \succsim_{i,z}, r_i)_{i=1}^m, (Y_j)_{j=1}^n, (\omega_i)_{i=1}^m\right)$. Before giving the definition of equilibrium, we need to introduce some useful definitions at first. The set of *weakly efficient allocations* is

$$Z = \{z \in L^{m+n} : \forall i \ x_i \in X_i (z), \forall j \ y_j \in \partial_\infty Y_j (z)\}.$$

We also define the set of *weakly efficient attainable allocations* corresponding to a given total initial endowment $\omega \in L$

$$A(\omega) = \{z \in Z : \textstyle\sum_{i=1}^m x_i \leq \sum_{j=1}^n y_j + \omega\}.$$

Finally, the set of *production equilibria* is

$$PE = \left\{(\pi, z) \in S \times Z : \pi \in \textstyle\bigcap_{j=1}^n N_{Y_j(z)} (y_j) \cap S\right\}.$$

We now formally define our notion of equilibrium.

Definition 1. A marginal pricing equilibrium of the economy $\mathscr{E}$ is an element $(\bar{z}, \bar{\pi}) = \left(\left((\bar{x}_i)_{i=1}^m, (\bar{y}_j)_{j=1}^n\right), \bar{\pi}\right)$ in $Z \times S$ such that:

a. For all i, $\bar{x}_i$ is $\succsim_{i,\bar{z}}$-maximal in $\left\{x_i \in X_i (\bar{z}) : \bar{\pi} (x_i) \leq r_i \left(\bar{\pi} (\omega_i), (\bar{\pi} (\bar{y}_j))_{j=1}^n\right)\right\}$

b. For all j, $\bar{\pi} \in N_{Y_j(\bar{z})} (\bar{y}_j) \cap S$ and $\bar{y}_j \in \partial_\infty Y_j (\bar{z})$

c. $\sum_{i=1}^m \bar{x}_i = \sum_{j=1}^n \bar{y}_j + \omega$

Condition *a.* says that for a given price $\bar{\pi}$, and a given externality $\bar{z}$, each consumer maximizes his preference relation under his budget constraint. Condition *b.* says that for a given externality $\bar{z}$ and for the same price vector $\bar{\pi}$, every producer satisfies his first-order necessary condition for profit maximization. Condition *c.* says that the demand is equal to the supply.

If we replace in the above definition, condition c. by condition c': $\sum_{i=1}^{m} \bar{x}_i \leq \sum_{j=1}^{n} \bar{y}_j + \omega$ and $\bar{\pi} \left(\sum_{i=1}^{m} \bar{x}_i \right) = \bar{\pi} \left(\sum_{j=1}^{n} \bar{y}_j + \omega \right)$, then we have the definition of *WA-equilibrium*.[5]

Remark 2. If Y_j is a convex-valued correspondence which satisfies Assumption P, then $N_{Y_j(z)} (y_j) \cap S = \{\pi \in S : \pi (y_j) \geq \pi (y), \forall y \in Y_j (z)\}$. Consequently, for a private ownership economy with convex-valued correspondences, the marginal pricing equilibria are equivalent to the notion of walrasian equilibria (see Clarke 1983, Proposition 2.4.4).

We end this section with the following proposition:

Proposition 2. *Let $(\Gamma, \leq)$ be a directed set. Let $(z^\alpha, \pi^\alpha)_{(\Gamma, \leq)}$ be a net of $Z \times S$, such that*

$$\begin{cases} (z^\alpha, \pi^\alpha) \to (\bar{z}, \bar{\pi}) \text{ for the product topology } \prod_{L^{m+n}} \sigma^\infty \times \sigma^{ba} \\ \pi^\alpha \in N_{Y_j(z^\alpha)} \left(y_j^\alpha \right) \cap S \text{ for all } \alpha \in \Gamma \\ \left(\pi^\alpha \left(y_j^\alpha \right) \right)_{\alpha \in \Gamma} \text{ converges} \end{cases}$$

Then $\lim \pi^\alpha \left(y_j^\alpha \right) \geq \bar{\pi} (\bar{y}_j)$. If $\lim \pi^\alpha \left(y_j^\alpha \right) = \bar{\pi} (\bar{y}_j)$, then $\bar{\pi} \in N_{Y_j(\bar{z})} (\bar{y}_j) \cap S$.

The proof of this proposition is given in the Appendix. This result claims that the Clarke's normal cone (with external factors) has sufficient continuity properties in the space L when the individual production set has a smooth boundary.

7.3   Other Basic Assumptions

We now posit the following assumptions:

Assumption (C). For every i

(i) X_i is a $\left(\prod_{L^{m+n}} \sigma^\infty, \sigma^\infty \right)$-closed correspondence with convex values and containing 0.

(ii) For every $z \in L^{m+n}$, for every x_i in $X_i (z)$, there exists x in $X_i (z)$ such that $x_i \prec_{i,z} x$, and for every $x_i, x_i' \in X_i (z)^2$, for every $t \in (0, 1)$, if $x_i \prec_{i,z} x_i'$, then $x_i \prec_{i,z} t x_i + (1 - t) x_i'$.

(iii) The set $\Gamma_i = \left\{ (z, x_i, x_i') \in L^{m+n+2} : (x_i, x_i') \in X_i (z)^2, x_i \precsim_{i,z} x_i' \right\}$ is a $\prod_{L^{m+n}} \sigma^\infty$-closed subset of L^{m+n+2}.

(iv) The wealth function r_i is continuous on R^{1+n} and strictly increasing in the second variable. Furthermore, $\sum_{i=1}^{m} r_i \left(\pi (\omega_i), (\pi (y_j))_{j=1}^{n} \right) = \pi(\omega) + \sum_{j=1}^{n} \pi (y_j)$.

---

[5] See Guesnerie (1975).

Assumption (B). For every $\omega\prime \geq \omega$, the set
$A(\omega', z) = \{(y_j)_{j=1}^n \in \prod_{j=1}^n \partial_\infty Y_j(z) : \sum_{j=1}^n y_j + \omega' \in L_+\}$ is norm bounded.

Assumption (WSA). (Weak Survival) For all $(\pi, z, \lambda) \in PE \times R_+$, if $(y_j)_{j=1}^n \in A(\omega + \lambda\chi_M, z)$, then
$$\pi\left(\sum_{j=1}^n y_j + \omega + \lambda\chi_M\right) > 0.$$

Assumption (R). For all $(\pi, z) \in PE$, if $z \in A(\omega)$, then
$$r_i\left(\pi(\omega_i),\ (\pi(y_j))_{j=1}^n\right) > 0.$$

Assumption (C) is the natural generalization of the assumptions of Bonnisseau and Médecin (2001) to an infinite-dimensional context (see Fuentes 2011). Assumption (B) is essential for the existence of an equilibrium. It means that for every $\omega\prime \geq \omega$, the set of weakly efficient attainable production plans is relatively weakly compact, from which it follows that so is $A(\omega')$.

When the same price is offered by the producers, according to $N_{Y_j(z)}(y_j) \cap S$, Assumption WSA implies that the global wealth of the economy is strictly greater than the subsistence level. Assumption R states that the revenue functions are a way to redistribute the total wealth among the consumers and the individual revenues are above the survival level for each consumer when the global wealth is large enough to allow such redistribution. We point out that when $Y_j(z)$ is a convex subset of L for every j and every $z \in L^{m+n}$, $\omega \in intL_+$ and $0 \in Y_j(z)$, both assumptions (WSA) and (R) are satisfied.

Remark. Most papers in general equilibrium theory with infinite commodity spaces make use of a well-known assumption called *properness* since Mas-Colell (1986). This condition informally means that there is a commodity bundle v which is so desirable that the marginal rate of substitution of any other commodity for v is bounded away from zero. Properness was introduced to deal with the consequences of the emptiness of the (norm) interior of the positive cone, namely, the fact that price equilibrium functional $\overline{\pi}$ may be identically zero. We point out that the list of spaces for which the positive orthant has empty interior includes several Banach spaces with some few exceptions such as the space $\mathscr{L}_\infty(M, \mathscr{M}, \mu)$. That is why we do not need to impose any properness assumption.

7.4   Subeconomies

7.4.1   *Construction of Finite-Dimensional Economies*

Let F be a finite-dimensional subspace of L containing both χ_M and $(\omega_i)_{i=1}^m$. We denote by $\mathscr{F}$ the family of such subspaces F directed under set inclusion. For every $F \in \mathscr{F}$, we define its positive cone by $F_+ = F \cap L_+$ and its interior by $intF_+ = F \cap intL_+$ which is not empty since χ_M belongs to $intL_+$. Hence, it defines an order which allows us to endow each F with an euclidean structure such that $\| \chi_M \| = 1$ and $\left\{ \chi_M^{\perp F} \right\} \cap F_+ = \{0\}$, where $\chi_M^{\perp F}$ denotes the orthogonal space to χ_M. Hence, the dual space of F is F itself,[6] and we denote by p^F the inner product $\langle p^F, \cdot \rangle_F$.

The truncated consumption correspondence for the commodity space F is given by $X_i^F : F^{m+n} \mapsto F_+$ and defined by $X_i^F \left(z^F \right) = X_i \left(z^F \right) \cap F_+$. Analogously, the truncated production correspondence $Y_j^F : F^{m+n} \mapsto F$ is defined by $Y_j^F \left(z^F \right) = Y_j \left(z^F \right) \cap F$, and, by the definition of Y_j, one easily checks that $Y_j^F \left(z^F \right) = \left\{ y^F \in F : f_j \left(y^F, z^F \right) \le 0 \right\}$ and $\partial Y_j^F \left(z^F \right) = \left\{ y \in F : f_j \left(y, z \right) = 0 \right\} = \partial_\infty Y_j \left(z^F \right) \cap F$. Hence, $Z^F \subset Z$.

Let $S^F = \left\{ p^F \in F_+^0 : \langle p^F, \chi_M \rangle_F = 1 \right\}$, where F_+^0 denotes the positive polar cone of F_+. r_i^F is the revenue of the i-th consumer induced by r_i in the truncated economy. The relation $\succsim_{i,z^F}^F$ is the preorder induced on $X_i^F \left(z^F \right)$ by $\succsim$. We then denote the subeconomies by $\mathscr{E}^F = \left(\left(X_i^F, \succsim_{i,z^F}^F, r_i^F \right)_{i=1}^m, \left(Y_j^F \right)_{j=1}^n, \left(\omega_i \right)_{i=1}^m \right)$ for all $F \in \mathscr{F}$.

We point out that for all $F \in \mathscr{F}$, for all $z^F \in F^{m+n}$ and for all i and j, $X_i^F \left(z^F \right)$ and $Y_j^F \left(z^F \right)$ are non-empty subsets of F because of the Assumptions C(i) and P(iv) together with the fact that F is a subspace of L. We also remark that for all $F \in \mathscr{F}$ and all $\left(y_j, z \right) \in F^{m+n+1}$, $N_{Y_j^F (z)}^F \left(y_j \right) \cap S^F = \left\{ \lambda \nabla_1 f_j \left(y, z \right) \big|_{F_+^0} : \lambda \ge 0 \right\} \cap S^F$. The set of production equilibria and of weakly efficient attainable allocations in $\mathscr{E}^F$ are, respectively,

$$ \mathrm{PE}^F = \left\{ \left(p^F, z^F \right) \in S^F \times Z^F : p^F \in \bigcap_{j=1}^n N_{Y_j^F (z)}^F \left(y_j \right) \right\} $$

and

$$ A^F \left(\omega \right) = \left\{ z^F \in Z^F : \sum_{i=1}^m x_i^F \le \sum_{j=1}^n y_j^F + \omega \right\} \subset A \left(\omega \right). $$

7.4.2 Bewley's Limiting Technique and Additional Assumptions

In the paper of Bonnisseau and Médecin (2001), the consumption set is represented by a correspondence that is l.h.c. As noted in Fuentes (2011), if we assume that the correspondence X_i is l.h.c. for all i, the restriction to a finite-dimensional subspace may not be l.h.c. Hence, Bonnisseau and Médecin's theorem (smooth case) does not apply, and, thus, we cannot follow the Bewley's approach. One solution is to assume that for all i, the restriction of X_i to a finite-dimensional subspace is l.h.c.

---

[6]F and F^*, the topological dual of F, are isomorphic (See MacLane and Garret 1999, Theorem 9, p. 357).

Assumption C(v). For all i

(v) There is a finite-dimensional subspace $\bar{F} \in \mathscr{F}$, such that for any finite-dimensional subspace $F \in \mathscr{F}$ such that $\bar{F} \subset F$, the correspondence X_i^F is l.h.c. on F^{m+n}.

Another problem in assuming that the correspondence X_i is l.h.c. for all i, relies in the fact that even if there is an equilibrium in each subeconomy $\mathscr{E}^F$, we cannot prove that a limit point $\left(\left((\bar{x}_i)_{i=1}^m, (\bar{y}_j)_{j=1}^n\right), \bar{\pi}\right)$ is an equilibrium vector in the original infinite-dimensional economy. Specifically, in the Claims 3 and 4 in the proof of Theorem 1 below, it can be seen that the lower hemi-continuity of X_i is not enough to prove that, for all i, if $x_i \succsim_{i,\bar{z}} \bar{x}_i$, then $\bar{\pi}(x_i) \geq r_i\left(\bar{\pi}(\omega_i), (\bar{\pi}(\bar{y}_j))_{j=1}^n\right)$. Consequently, we cannot use the limiting argument of the Bewley type. One solution is to establish the following assumption:

Assumption C(vi). For all i

The correspondence X_i is $\left(\prod_{L^{m+n}} \sigma^\infty, f\right)$-l.h.c. on L^{m+n}, that is, if z^α $\prod_{L^{m+n}} \sigma^\infty$−converges to z in L^{m+n} and $x \in X_i(z)$, there exists a finite-dimensional subspace $\dot{F}$ such that there is a net $(x^\alpha) \subset x + \dot{F}$ with $x^\alpha \in X_i(z^\alpha)$ for all α and $x^\alpha \longrightarrow x$.

We point out that $\dot{F}$ may depend on $x \in X_i(z)$ and the net (z^α). We also note that the above Assumption implies that the correspondence X_i is $\left(\prod_{L^{m+n-1}} \sigma^\infty, \mathscr{T}\right)$-l.h.c. since the net (x^α) $\mathscr{T}$−converges to x due to the fact that it belongs to an affine finite-dimensional subspace.

When the boundary of the production set is smooth, such as in our case, if the production correspondence is l.h.c., then so is its restriction to a finite-dimensional subspace (see Remark 3 in the Appendix). Then, contrary to what is stated in Fuentes (2011), we do not need an additional assumption for the restricted production correspondences.

There are two remaining problems in the application of the Bewley technique. First, even if the original economy is supposed to satisfy the Weak Survival Assumption, this may not be true for the subeconomies. Secondly, even if the original economy is supposed to satisfy the Local Non-Satiation Assumption, we cannot say this is true in the subeconomies. Consequently, Theorem 3.1 of Bonnisseau and Médecin (2001) cannot be applied to $\mathscr{E}^F$. As we shall show later, if the commodity space F is large enough, then the economy satisfies weaker versions of Assumptions (WSA) and (LNS).

7.5   Existence of Marginal Pricing Equilibria

Now, we are ready to state the following result:

Theorem 1. *Under Assumptions (C), (P), (B), (WSA) and (R), the economy* $\mathscr{E} = \left((X_i, \succsim_{i,z}, r_i)_{i=1}^m, (Y_j)_{j=1}^n, (\omega_i)_{i=1}^m\right)$ *has a marginal pricing equilibrium.*

To compare this result with the literature, we first remark that it generalizes the one given in Shannon (1996) for the case without externalities and one producer and the one in Bonnisseau and Cornet (1990a) for the case with commodity space R^l. It also extends the main result of Bonnisseau (2002) under the particular circumstance of smooth production sets. In Fuentes (2011), the behaviour of the firms is defined through a general pricing rule. Nevertheless, the existence result uses a bounded losses assumption which is not necessary with the marginal pricing rule. Furthermore, we can suppress Assumption (PR) on the continuity of pricing rules (by Proposition 2 in this paper) and Assumption P(v) on the lower hemi-continuity of Y_j^F (See Remark 3 in the Appendix).

7.6   Proof of the Theorem

7.6.1   Equilibria in the Subeconomies

The results of this section follow the guidelines of Bonnisseau's proof of Proposition 2 (Bonnisseau 2002). The differences between our results and those of the author are due to the intrinsic differences between the finite-dimensional model without externalities (Bonnisseau and Cornet 1990a) and the one with external factors (Bonnisseau and Médecin 2001). We can observe that every subeconomy $\mathscr{E}^F$ satisfies Assumptions (P), (B), (R) and (C) (except LNS) of Theorem 3.1 of Bonnisseau and Médecin (2001). As we remarked at the end of Sect. 7.4, Assumptions (LNS) and (WSA) are not necessarily fulfilled by $\mathscr{E}^F$. The following lemma shows that each subeconomy satisfies weak versions of the survival and the local non-satiation of the preferences if F is large enough. Before stating the above result, we need to introduce the elements for its treatment. Let $\bar{\eta} > 0$ be a real number. Since $A(\omega + \bar{\eta}\chi_M, z)$ is norm bounded by Assumption (B), there exists (Schaefer and Wolf 1999, p. 25) $a > 0$ such that $a > 2\bar{\eta}$, $A(\omega + \bar{\eta}\chi_M, z) \subset B\left(0, \frac{a}{2}\right)^n$ and $A(\omega + \bar{\eta}\chi_M) \subset B\left(0, \frac{a}{2}\right)^{m+n}$. Let $\bar{r} > 2a$ such that $\{\omega + \bar{\eta}\chi_M\} + \overline{B}(0, na) \subset B(0, \bar{r})$. Let $\bar{\lambda}$ be a real number such that $\bar{\lambda} \geq 2n\bar{r} + \|\omega\|$. We point out that $\bar{\lambda}$ satisfies Lemma 4.2 of Bonnisseau and Médecin (2001) in our model.

Lemma 1. *Under Assumptions (C), (P), (B), (WSA) and (R), there exists a subspace $\hat{F} \in \mathscr{F}$ such that for all $F \in \mathscr{F}$, if $\hat{F} \subset F$, then the subeconomy $\mathscr{E}^F$ satisfies:*

(WSA^F):   *For all $\left(p^F, z^F, \lambda^F\right) \in \mathrm{PE}^F \times [0, \bar{\lambda}]$, if $\left(y_j^F\right)_{j=1}^n \in A^F\left(\omega + \lambda^F\chi_M, z^F\right)$,*

   then $\left\langle p^F, \sum_{j=1}^n y_j^F + \omega + \lambda^F\chi_M\right\rangle_F > 0$.

(LNS^F):   *For all $\left(\left(x_i^F\right)_{i=1}^m, \left(y_j^F\right)_{j=1}^n\right) \in A^F(\omega)$, and for all $\varepsilon > 0$, there exists $\left(x_i'^F\right)_{i=1}^m \in \prod_{i=1}^m \left(X_i^F\left(z^F\right) \cap B\left(x_i^F, \varepsilon\right)\right)$, such that $x_i'^F \succ_{i,z^F}^F x_i^F$ for all i.*

The proof of this lemma parallels the one given in Fuentes (2011). Just replace Assumption P by Remark 1 and Proposition 1 and Assumption PR by Proposition 2.

We recall that Bonnisseau and Médecin defined a new cone for the marginal pricing rule when there are external effects. Indeed, if we use the Clarke's normal cone (with externalities), the equilibrium may not exist due to the presence of discontinuities. However, if the individual production set is smooth, their cone coincides with the Clarke's cone.[7] The proposition below establishes that at least one equilibrium exists in the subeconomies.

Proposition 3. *Let $\bar{F}$ and $\hat{F}$ be the subspaces coming from Assumption C(v) and Lemma 1, respectively. Under Assumptions (C), (P), (B), (WSA) and (R), if we have $\bar{F} \subset F$, and $\hat{F} \subset F$, then the subeconomy $\mathscr{E}^F$ has an equilibrium $(z^F, p^F) \in Z^F \times S^F$.*

Proof. We remark that in the proof of Bonnisseau and Médecin (2001), the authors use Assumption (WSA) in Lemmas 4.2 (3) and 4.4 and in Claim 4.3. We also note that in the proof they fix belongs a parameter $\bar{t} > 0$ (p. 283). We replace it by $\bar{\eta}$ as given in paragraph before Lemma 1. For Lemma 4.2 (3) and Claim 4.3, Survival Assumption is applied only for production plans which satisfy that $\sum_{j=1}^{n} y_j + \omega + \eta \chi_M \geq 0$ with $\eta \leq \bar{\eta}$. Since $\bar{\eta} < \bar{\lambda}$ from the definition of $\bar{r}$, we have that condition (WSA^F) of Lemma 7 is enough to conclude. For Lemma 4.4, we shall prove that (WSA^F) is enough to use the deformation lemma. We now introduce the Bonnisseau and Médecin (2001)'s fundamental mathematical expressions we shall need. Let

$$\lambda_j^F : \chi_M^{\perp_F} \times F^{m+n} \longmapsto R$$
$$(s_j, z) \longmapsto \lambda_j^F(s_j, z)$$
$$\Lambda_j^F(s_j, z) = s_j - \lambda_j^F(s_j, z) \chi_M \in \partial Y_j^F(z)$$
$$X^F(z) = \sum_{i=1}^{m} X_i^F(z) + F_+ = F_+$$
$$Y_0^F(z) = -X^F(z)$$
$$\lambda_0^F : \chi_M^{\perp_F} \times F^{m+n} \longmapsto R$$
$$(s_j, z) \longmapsto \lambda_0^F(s_j, z)$$
$$\Lambda_0^F(s_j, z) = s_j - \lambda_0^F(s_j, z) \chi_M \in \partial(-F_+)$$
$$\theta^F\left((s_j)_{j=1}^{n}, z\right) = \sum_{j=1}^{n} \lambda_j^F(s_j, z) + \lambda_0^F\left(-\sum_{j=1}^{n} s_j - proj_{\chi_M^{\perp_F}} \omega, z\right) - \langle \omega, \chi_M \rangle_F$$
$$\Delta^F\left((s_j)_{j=1}^{n}, z\right) = \left\{ (p_j - p)_{j=1}^{n} \; \middle| \; \begin{array}{l} p_j \in N_{Y_j(z)}\left(\Lambda_j^F(s_j, z), z\right), j = 1, \ldots, n \\ p \in N_{-F_+}\left(\Lambda_0^F\left(-\sum_{j=1}^{n} s_j - proj_{\chi_M^{\perp_F}} \omega, z\right)\right) \cap S^F \end{array} \right\}$$
$$M_{\bar{\eta}}^F(z) = \left\{ \left((s_j)_{j=1}^{n}\right) \in \left(\chi_M^{\perp_F}\right)^n : \sum_{j=1}^{n} \Lambda_j^F(s_j, z) + \omega + \bar{\eta}\chi_M \in F_+ \right\} \text{ for every}$$
$$z \in Z_D^F$$
$$GM_{\bar{\eta}}^F = \left\{ \left((s_j)_{j=1}^{n}, z\right) \in \left(\chi_M^{\perp_F}\right)^n \times Z_D^F : \sum_{j=1}^{n} \Lambda_j^F(s_j, z) + \omega + \bar{\eta}\chi_M \in F_+ \right\}$$
$$GM_{\bar{\eta}, \alpha}^F = \left\{ \left((s_j)_{j=1}^{n}, z\right) \in \left(\chi_M^{\perp_F}\right)^n \times Z_D^F : \bar{\eta} \leq \theta^F\left((s_j)_{j=1}^{n}, z\right) \leq \alpha \right\}$$
$$\alpha = \max\left\{ \theta^F\left((s_j)_{j=1}^{n}, z\right) : \left((s_j)_{j=1}^{n}, z\right) \in \left(\bar{B}^F(0, 2a) \cap \{\chi_M^{\perp_F}\}\right)^n \times Z_D^F \right\}$$

---

[7] Bonnisseau and Médecin 2001, p. 277

where, $\overline{B}^F(0,a) = \overline{B}(0,a) \cap F$, $D^F := \overline{B}^F\left(0,\overline{\lambda}\right)^m \times \overline{B}^F(0,\overline{r})^n$ and $Z_D^F := Z^F \cap D^F$.

For λ_j^F and λ_0^F, $\sum_{j=1}^n \Lambda_j^F(s_j, z) + \omega + \eta\chi_M \geq 0$ if and only if $\theta^F\left((s_j)_{j=1}^n, z\right) \leq \eta$ (Lemma 4.3). The authors apply a deformation lemma for which it must prove that the conditions of the lemma are satisfied. One of these conditions (the one which uses Survival Assumption) requires that $0 \notin \Delta^F\left((s_j)_{j=1}^n, z\right)$ for all $\left((s_j)_{j=1}^n, z\right) \in GM_{\overline{\eta},\alpha}^F$. If it is not, then (see the proof of Lemma 4.4) there exists $\left((s_j)_{j=1}^n, z\right) \in \left(\chi_M^{\perp_F}\right)^n \times Z_D^F$ such that $\overline{\eta} \leq \theta^F\left((s_j)_{j=1}^n, z\right) \leq \alpha$ and $p \in N_{-F_+}\left(\Lambda_0^F\left(-\sum_{j=1}^n s_j - proj_{\chi_M^{\perp_F}}\omega, z\right)\right) \cap S$ such that $p \in \cap_{j=1}^n N_{Y_j^F(z)}(y) \cap S^F$. By the above result, $\sum_{j=1}^n \Lambda_j^F(s_j, z) + \omega + \alpha\chi_M \geq 0$, and it can be proved that $p\left(\sum_{j=1}^n \Lambda_j^F(s_j, z) + \omega + \alpha\chi_M\right) = 0$ contradicting Survival Assumption since $\left(\Lambda_j^F(s_j, z)\right)_{j=1}^n \in A^F(\omega + \alpha\chi_M, z)$. Therefore, Assumption (WSA^F) is enough to conclude if one proves that $\alpha \leq 2n\overline{r} + \|\omega\|$.

Since $\Lambda_j^F(s_j, z) \in \partial Y_j^F(z)$, $\Lambda_j^F(s_j, z) \notin int F_+$ (otherwise, $0 \notin \partial Y_j^F(z)$). Consequently, for $\varepsilon > 0$, there exists $\xi \in B\left(\Lambda_j^F(s_j, z), \varepsilon\right) \cap (F\backslash F_+)$ and $M' \subset M$ such that $\mu(M') \neq 0$ and $\Lambda_j^F(s_j, z)(m) = s_j(m) - \lambda_j^F(s_j, z) - \varepsilon < \xi(m) \leq 0$ for all $m \in M'$. Hence, one deduces that $\lambda_j^F(s_j, z) > -\|s_j\| - \varepsilon$. In the same way, $\Lambda_j^F(s_j, z) \notin int(-F_+)$ (otherwise, $\Lambda_j^F(s_j, z) \notin \partial Y_j^F(z)$). Hence, for $\varepsilon > 0$, $B\left(\Lambda_j^F(s_j, z), \varepsilon\right) \cap (F\backslash(-F_+)) \neq \emptyset$, from which one deduces that $\lambda_j^F(s_j, z) < \|s_j\| + \varepsilon$. Consequently, $-\|s_j\| - \varepsilon < \lambda_j^F(s_j, z) < \|s_j\| + \varepsilon$. Since the inequality is true for all $\varepsilon > 0$, one has $\left|\lambda_j^F(s_j, z)\right| \leq \|s_j\|$ for all j. On the other hand, for $\Lambda_0^F(u, z) \in \partial(-F_+)$, one easily checks that $\left|\lambda_0^F(u, z)\right| \leq \|u\|$ since $-F_+$ is convex.

Let $\left((s_j)_{j=1}^n, z\right) \in \left(\overline{B}^F(0, 2a) \cap \{\chi_M^{\perp_F}\}\right)^n \times Z_D^F$. From the above remarks and the fact that $\left|proj_{\chi_M^{\perp_F}}\omega\right| \leq \|\omega\|$, it follows that $\theta^F\left((s_j)_{j=1}^n, z\right) \leq 4na + \|\omega\| < 2n\overline{r} + \|\omega\| \leq \overline{\lambda}$, which in turn implies that $\alpha \leq 2n\overline{r} + \|\omega\|$.

For the Local Non-Satiation Assumption, we remark that it is used in Bonnisseau and Médecin (2001) only in Claim 4.6 where $z^F \in A^F(\omega)$. Consequently, condition (LNS^F) of Lemma 1 is enough to conclude, and the proof of the Proposition 3 is complete.

7.6.2 The Limit Point

Let $\left(\left((x_i^F)_{i=1}^m, (y_j^F)_{j=1}^n\right), p^F\right)_{F \in \mathscr{F}}$ be the net of equilibria of the subeconomies $(\mathscr{E}^F)_{F \in \mathscr{F}}$ given by Proposition 3. From the definition of $N_{Y_j^F(z)}^F(y_j) \cap S^F$, there exist price vectors $\left(\pi_j^F\right)_{j=1}^n \in \prod_{j=1}^n N_{Y_j(z^F)}\left(y_j^F\right) \cap S$ such that $p^F = \pi_{j|F}^F$ for all j.

Hence, we obtain the net $\left(\left(x_i^F\right)_{i=1}^m, \left(y_j^F\right)_{j=1}^n, \left(\pi_j^F\right)_{j=1}^n\right)_{F \in \mathscr{F}}$. Proposition 3 implies

that $\left(\left(x_i^F\right)_{i=1}^m, \left(y_j^F\right)_{j=1}^n\right)_{F \in \mathscr{F}} \in A(\omega)$, which is norm bounded by Assumption

(B). Hence, from the Banach-Alaoglu theorem, it remains in a $\prod_{L^{m+n}} \sigma^\infty$−compact

subset of L^{m+n}. Furthermore, the net $\left(\pi_j^F\right)_{F \in \mathscr{F}}$ belongs to S which is σ^{ba}-compact.

Consequently, there exists a subnet

$$\left(\left(x_i^{F(t)}\right), \left(y_j^{F(t)}\right), \left(\pi_j^{F(t)}\right)\right)_{t \in (T, \geq)}$$ which $\prod_{L^{m+n}} \sigma^\infty \times \sigma^{ba}$−converges to

$((\bar{x}_i), (\bar{y}_j), (\bar{\pi}_j))$. This also implies that the subnets of real numbers

$$\left(\left\langle p^{F(t)}, y_j^{F(t)} \right\rangle_{F(t)}\right) = \left(\pi_j^{F(t)}\left(y_j^{F(t)}\right)\right) \text{ and}$$

$$\left(\left\langle p^{F(t)}, x_i^{F(t)} \right\rangle_{F(t)}\right) = \left(\pi_j^{F(t)}\left(x_i^{F(t)}\right)\right) \text{ are bounded so that they can be supposed to}$$

converge.

We now prove that at least one limit point exists which in turn is a marginal pricing equilibrium of the economy $\mathscr{E}$.

Claim 1. $\bar{\pi}_1 = \bar{\pi}_2 = \ldots = \bar{\pi}_n > 0$

Proof. We first prove that $\bar{\pi}_1 = \bar{\pi}_2 = \ldots = \bar{\pi}_n \geq 0$. Let $x \in L$. There exists $F \in \mathscr{F}$ such that $x \in F$. There exists $t_0 \in T$ such that $F \subset F(t)$ for all $t > t_0$. As $p^{F(t)} = \pi_{j|F(t)}^{F(t)}$ for all j, we have that, for all $t > t_0$, $\langle p^{F(t)}, x \rangle_{F(t)} = \pi_j^{F(t)}(x)$ for all j. Without loss of generality, we denote the limit of $\langle p^{F(t)}, x \rangle_{F(t)}$ by $\bar{\pi}(x)$. Hence, $\lim \pi_j^{F(t)}(x) = \bar{\pi}(x)$ for all j. Since $ba^+(M, \mathscr{M}, \mu)$ is closed, we have the first part of the Claim. Since $\pi_j^{F(t)}(\chi_M) = 1$ for all j and $t \in T$, we have that $\bar{\pi}(\chi_M) = 1$. Therefore, the proof is complete.

Claim 2. $\left((\bar{x}_i)_{i=1}^m, (\bar{y}_j)_{j=1}^n\right) \in \prod_{i=1}^m X_i(\bar{z}) \times \prod_{j=1}^n \partial_\infty Y_j(\bar{z})$ and $\sum_{i=1}^m \bar{x}_i = \sum_{j=1}^n \bar{y}_j + \omega$

Proof. $\left(\left(x_i^{F(t)}\right)_{i=1}^m, \left(y_j^{F(t)}\right)_{j=1}^n\right) \in Z^{F(t)}$. Since $\left(z^{F(t)}\right)_{t \in (T, \geq)} \prod_{L^{m+n}} \sigma^\infty$−converges

to $\bar{z}$, we get $\bar{z} = \left((\bar{x}_i)_{i=1}^m, (\bar{y}_j)_{j=1}^n\right) \in \prod_{i=1}^m X_i(\bar{z}) \times \prod_{j=1}^n \partial_\infty Y_j(\bar{z})$ by Assumption C(i) and Proposition 1. Since $\sum_{i=1}^m x_i^{F(t)} = \sum_{j=1}^n y_j^{F(t)} + \omega$ for all $t \in T$, one obtains $\sum_{i=1}^m \bar{x}_i = \sum_{j=1}^n \bar{y}_j + \omega$.

Claim 3. For all i, if $x_i \succsim_{i,\bar{z}} \bar{x}_i$, then $\bar{\pi}(x_i) \geq r_i\left(\bar{\pi}(\omega_i), \lim \left(\pi_j^{F(t)}\left(y_j^{F(t)}\right)\right)_{j=1}^n\right)$.

Proof. See Fuentes (2011).

Claim 4. For all i, $\bar{\pi}\left(\bar{x}_i\right) = r_i\left(\bar{\pi}\left(\omega_i\right), \left(\bar{\pi}\left(\bar{y}_j\right)\right)_{j=1}^n\right)$ and for all j, $\bar{\pi}\left(\bar{y}_j\right) = \lim \pi_j^{F(t)}\left(y_j^{F(t)}\right)$.

Proof. By Proposition 2, we have $\lim \pi_j^{F(t)}\left(y_j^{F(t)}\right) \geq \bar{\pi}_j\left(\bar{y}_j\right)$ for all j. The rest of the proof is identical to the proof of Step 6 of Fuentes (2011).

From Claims 2 and 4 together with Proposition 2, one obtains $\bar{z} \in Z$, $\bar{\pi} \in \bigcap_{j=1}^n N_{Y_j(\bar{z})}\left(\bar{y}_j\right) \cap S$ and $\sum_{i=1}^m \bar{x}_i = \sum_{j=1}^n \bar{y}_j + \omega$. It only remains to show that condition a. of Definition 1. is satisfied.

Claim 5. For all i, $\bar{x}_i$ is a greater element for $\succsim_{i,\bar{z}}$ in the budget set $\left\{x_i \in X_i\left(\bar{z}\right) : \bar{\pi}\left(x_i\right) \leq r_i\left(\bar{\pi}\left(\omega_i\right), \left(\bar{\pi}\left(\bar{y}_j\right)\right)_{j=1}^n\right)\right\}$.

Proof. We have to show that for every agent i, if $x_i \succ_{i,\bar{z}} \bar{x}_i$, then $\bar{\pi}\left(x_i\right) > \bar{\pi}\left(\bar{x}_i\right)$. From Claims 3 and 4, one has $\bar{\pi}\left(x_i\right) \geq \bar{\pi}\left(\bar{x}_i\right)$. Suppose $\bar{\pi}\left(x_i\right) = \bar{\pi}\left(\bar{x}_i\right)$. From Claims 3, 4 and Assumptions (WSA) and (R), $\bar{\pi}\left(\bar{x}_i\right) = r_i\left(\bar{\pi}\left(\omega_i\right), \left(\bar{\pi}\left(\bar{y}_j\right)\right)_{j=1}^n\right) > 0$. For all $t \in (0, 1)$, we have $\bar{\pi}\left(tx_i\right) < \bar{\pi}\left(x_i\right) = \bar{\pi}\left(\bar{x}_i\right)$. For t close enough to 1, $tx_i \in X_i\left(\bar{z}\right)$ and, since preferences are continuous, $tx_i \succ_{i,\bar{z}} \bar{x}_i$. From Claim 4, we get $\bar{\pi}\left(tx_i\right) \geq \bar{\pi}\left(\bar{x}_i\right)$, a contradiction with the above inequality.

7.7 Existence of Marginal Cost Pricing Equilibria

An equilibrium as defined in Definition 1 is called marginal cost pricing equilibrium in Shannon (1996) and many other papers. The terminology has been adopted because it is suggestive even though it is not always correct as indicated earlier by Guesnerie (1990). Indeed, $\bar{\pi} \in N_{Y_j(\bar{z})}\left(\bar{y}_j\right)$ implies that $\bar{\pi}$ is proportional to the marginal cost only if the set of input combinations for producing a given level of output is convex. Marginal cost pricing equilibrium also means that every producer minimizes its costs. Bonnisseau and Cornet (1990a,b) investigated and established a formal link between the marginal pricing rule and the one of marginal cost pricing in the finite-dimensional case. We are now interested in having a marginal cost pricing equilibrium for an economy with externalities and infinitely many commodities. We must introduce both the notions of *iso-output set* and *cost functional*, for which we have to distinguish a priori between inputs and outputs. Although we follow the approach of Bonnisseau and Cornet (1990a), there appear significant drawbacks in using their technique in our economy as we shall see later.

Let I^j and O^j be partitions of the set M for the j-th producer, such that $M = I^j \cup O^j$ and $I^j \cap O^j = \emptyset$. We define the following subspaces of L.

$L^{I^j} = \{u \in L : u(m) = 0 \ \mu - a.e. \text{ if } m \notin I^j\}$

$L^{O^j} = \{u \in L : u(m) = 0 \ \mu - a.e. \text{ if } m \notin O^j\}$

For every $y_j \in L$, we denote $proj_{L^{I^j}}\left(y_j\right)$ as y_{I^j}. Note that $y_{I^j} \in L^{I^j}$ since $proj_{L^{I^j}}\left(y_j\right)$ is measurable. The same applies for $y_{O^j} = proj_{L^{O^j}}\left(y_j\right)$.

We now define the *iso-output* set: for all $(r, b, z) \in \left(L^{I^j}\right)^*_+ \times L^{O^j} \times L^{m+n}$, we let

$$Y_j (b, z) = \left\{-y_{I^j} \in L : \text{ there exists } y_j \in Y_j (z), \, y_j = y_{O^j} + y_{I^j} \text{ and } y_{O^j} = b\right\}.$$

For all $(r, b, z) \in \left(L^{I^j}\right)^*_+ \times L^{O^j} \times L^{m+n}$, we define the cost functional c_j as follows:

$$c_j (r, b, z) = \inf \left\{r (a) : a \in Y_j (b, z)\right\}$$

if $Y_j (b, z) \neq \emptyset$.

For every $(r, b, z) \in \left(L^{I^j}\right)^*_+ \times L^{O^j} \times L^{m+n}$, we denote by $\nabla_O c_j (r, b, z)$ the (Fréchet) gradient vector of c_j with respect to b. Thus, for every x in L^{O^j}, $\nabla_O c_j (r, b, z) (x) = \lim_{t \to 0} \frac{c_j(r, b + tx, z) - c_j(r, b, z)}{t}$, and hence, $\nabla_O c_j (r, b, z) \in \left(L^{O^j}\right)^*$.

As in Bonnisseau and Cornet (1990a), we separate between the first $n - 1$ producers and the $n-$th one which maximizes his profit. For the $n - 1$ first ones, we posit the following assumption:

Assumption C(vi) (P'). For $z \in L^{m+n}$

(i) There exists a partition of the set M into two non-empty subsets I^j and O^j such that $\mu \left(I^j\right) \neq \emptyset$ and $\mu \left(O_j\right) \neq \emptyset$. For every $y_j \in Y_j (z)$, $y_j (m) = y_{I^j} (m) \leq 0$ if $m \in I^j$. Furthermore, there exists $\tilde{y}_j \in Y_j (z)$ such that $y_{O^j} \leq \tilde{y}_{O^j}$ and $\tilde{y}_j (m) = \tilde{y}_{O^j} (m) \geq 0$ if $m \in O^j$.

(ii) The set $Y_j (b, z)$ is convex

(iii) The set $\Omega_j = \left\{b \in L^{O^j} : Y_j (b, z) \neq \emptyset\right\}$ is $\sigma^\infty_{L^{O^j}}$−open.

(iv) For every $r \in \left(L^{I^j}\right)^*_+$, the cost functional $c_j (r, \cdot, z)$ is $\mathcal{T}_{L^{O^j}}$−differentiable on Ω_j.

For the nth producer, we let

Assumption C(vi) (P''). The correspondence $Y_n : L^{m+n} \longmapsto L$ is convex valued.

We remark that in an economy without externalities and with R^l as commodity space, the above assumptions are the same as those in Bonnisseau and Cornet (1990a). We refer to that paper for an economic interpretation. We note that every $y_j \in Y_j (z)$ has a unique representation $y_j = y_{I^j} + y_{O^j}$ since $L^{I^j} \cap L^{O^j} = \{0\}$.

For every $\pi \in L^*_+$, we denote by π_{I^j} (π_{O^j}) the restriction of π to L^{I^j} (L^{O^j}). We now can give a precise definition of marginal cost pricing equilibrium.

Definition 2. A marginal cost pricing equilibrium of the economy $\mathscr{E}$ is an element $(\hat{z}, \hat{\pi}) = \left(\left((\hat{x}_i)^m_{i=1}, (\hat{y}_j)^n_{j=1}\right), \hat{\pi}\right)$ in $Z \times S$ such that

a. For all i, $\hat{x}_i$ is $\succsim_{i,\hat{z}}$-maximal in $\left\{ x_i \in X_i\left(\hat{z}\right) : \hat{\pi}\left(x_i\right) \le r_i\left(\hat{\pi}\left(\omega_i\right), \left(\hat{\pi}\left(\hat{y}_j\right)\right)_{j=1}^n\right)\right\}$

b'. For all $j = 1, \dots n - 1$, $\hat{\pi}_{pj}\left(-\hat{y}_{pj}\right) = c_j\left(\hat{\pi}_{pj}, \hat{y}_{oj}, \hat{z}\right)$ (cost minimization), $\hat{y}_{oj} \ge 0$ (output condition) and $\hat{\pi}_{oj} = \nabla_{oc_j}\left(\hat{\pi}_{pj}, \hat{y}_{oj}, \hat{z}\right)$ (marginal cost pricing). For $j = n$, $\hat{\pi}\left(\hat{y}_n\right) \ge \hat{\pi}\left(y\right)$ for all $y \in Y_n\left(\hat{z}\right)$ (profit maximization)

c'. $\sum_{i=1}^m \hat{x}_i \le \sum_{j=1}^n \hat{y}_j + \omega$ and $\hat{\pi}\left(\sum_{i=1}^m \hat{x}_i - \sum_{j=1}^n \hat{y}_j - \omega\right) = 0$

One easily checks that Condition c. of Definition 1 implies Condition c'. above. Condition b' says that at equilibrium every producer minimizes his cost, prices equal marginal cost and resultant production vectors are non-negative.

Lemma 2. *Let us assume that P and P' hold. Let $p \in L_+ \setminus \{0\}$, let $y_j = y_{pj} + y_{oj} \in \partial_\infty Y_j\left(z\right)$ such that $p\left(-y_{pj}\right) = c_j\left(p_{pj}, y_{oj}, z\right)$ and $p_{oj} \le \nabla_{oj} c_j\left(p_{pj}, y_{oj}, z\right)$. Then $p \in N_{Y_j(z)}\left(y_j\right)$.*

Proof. The proof is a direct transcription of the proof of Lemma 2 (a) in Bonnisseau and Cornet (1990a) since, in this point, there are not relevant differences when considering externalities and infinitely many commodities.

The next proposition is the key argument of the proof of Theorem 2.

Proposition 4. *Let (z, π) be a MPE of $\mathcal{E}$ such that $y_{oj} = y_{oj}^+$ for all j. Then (z, π) is a MCPE of $\mathcal{E}$ if Assumptions P, P' and P'' hold.*

The proof of this proposition is given in the Appendix. This shows the relationship between the two notions of marginal pricing equilibrium and marginal cost pricing equilibrium under the particular circumstance that, at marginal pricing equilibrium, all outputs are non-negative. We remark that in the paper of Bonnisseau and Cornet (1990a), they show the relationship between the two notions of equilibrium also in the case $y_{oj} \ne y_{oj}^+$. We refer to the Appendix for more details on this subject.

A sufficient condition for $y_{oj} = y_{oj}^+$ is that the price system is (punctually) strictly positive.

Lemma 3. *Let $z \in L^{m+n}$, let $Y_j : L^{m+n} \mapsto L$ be a correspondence satisfying Assumption P and P'(i). Let $y_j \in \partial_\infty Y_j\left(z\right)$ such that $\nabla_1 f_j\left(y_j, z\right) \in \mathcal{L}_1^{++}$. Then, $y_{oj} = y_{oj}^+$.*

Proof. Suppose that $y_{oj} \ne y_{oj}^+$. Hence, $y_j' = y_{pj} + y_{oj}^+ > y_j = y_{pj} + y_{oj}$ and $y_j' \in \partial_\infty Y_j\left(z\right)$ by Assumptions P'(i) and free disposal. Consequently, $\nabla_1 f_j\left(y_j, z\right)\left(y_j - y_j'\right) < 0$ since $\nabla_1 f_j\left(y_j, z\right)$ is in the quasi-interior of $\mathcal{L}^+$. Since $\nabla_1 f_j$ is $\left(\sigma^\infty \times \prod_{L^{m+n}} \sigma^\infty\right)$-continuous by Assumption P, there exists an σ^∞-open neighbourhood $U\left(y_j\right)$ of y_j, such that $\nabla_1 f_j\left(y_j'', z\right)\left(y_j - y_j'\right) < 0$ for all $y'' \in U\left(y_j\right)$. Let $y_j^\kappa = \kappa y_j' + \left(1 - \kappa\right) y_j$ such that $\kappa > 0$. For all $\kappa \in \left(0, 1\right)$, $y_j' > y_j^\kappa > y_j$, so that $y_j^\kappa \in \partial_\infty Y_j\left(z\right)$; and for κ close enough to 0, $y_j^\kappa \in U\left(y_j\right)$. Consequently, $\nabla_1 f_j\left(y_j^\kappa, z\right)\left(y_j - y_j'\right) < 0$ which implies that $\left(y_j - y_j'\right) \in \text{int}\left[\nabla_1 f_j\left(y_j^\kappa, z\right)\right]^o =$

$\text{int}T_{Y_j(z)}\left(y_j^\kappa\right)$ (Clarke 1983, Theorem 2.4.7). Since the set of vectors hypertangent to $Y_j(z)$ at y_j is non-empty, $\left(y_j - y_j'\right)$ is hypertangent to $Y_j(z)$ at y_j (Clarke 1983, Theorem 2.4.8). Consequently, $y_j^\kappa + \varepsilon\left(y_j - y_j'\right) + \varepsilon a \chi_M \in Y_j(z)$ for all $\varepsilon > 0$ small enough and a suitably chosen $a > 0$. Let us take $\varepsilon < \kappa$, then $y_j^\kappa + \varepsilon\left(y_j - y_j'\right) \in \text{int}Y_j(z)$. Since $y_j^\kappa + \varepsilon\left(y_j - y_j'\right) = (\kappa - \varepsilon)y_j' + (1 - \kappa + \varepsilon)y_j$, one has that $y_j^\kappa + \varepsilon\left(y_j - y_j'\right) > y_j$ a contradiction.

Actually, the above proof shows a stronger result than the statement of the lemma: y_j is efficient. We remark that the proof parallels that of Proposition 2 in Bonnisseau and Créttez (2007) for the finite-dimensional case. The only difference is that we use Theorem 2.4.8 of Clarke (1983) instead of Theorem 2.5.8 of Clarke's book.

We posit an additional assumption before stating the main result of this section.

Assumption SPP (Strictly Positive Prices). For all j, if $(z, \pi) \in A(\omega) \times \cap_{j=1}^n N_{Y_j(z)}\left(y_j\right) \cap S$, then $\nabla_1 f_j\left(y_j, z\right) \in \mathscr{L}_1^{++}$.

From Assumption P(iv), $\nabla_1 f_j\left(y_j, z\right) \in \mathscr{L}_1^+$ for all $y_j \in \partial_\infty Y_j(z)$. Hence, Assumption SPP only requires that the common price vector π, which is given by the marginal pricing rule of each producer, be strictly positive when the allocation is feasible and weakly efficient. Assumption SPP is weaker than Assumption P(4) in Shannon (1996) where it is required that $\nabla_1 f_j\left(y_j, z\right) \in \mathscr{L}_1^{++}$ for all $y_j \in Y_j(z)$.

Theorem 2. *Under Assumptions (C), (P), (P'), (P"), (WSA), (R) and (SPP), the economy $\mathscr{E} = \left((X_i, \succsim_{i,z}, r_i)_{i=1}^m, (Y_j)_{j=1}^n, (\omega_i)_{i=1}^m\right)$ has a marginal cost pricing equilibrium.*

Proof. The proof follows immediately from Theorem 1, Assumption SPP and Proposition 4.

Acknowledgements Earlier versions of this paper were presented at the Seminario de Funciones Generalizadas, Universidad de Buenos Aires (2013); the Primer Workshop en Economía Matemática, Universidad de San Andrés (2014); the XLIX Reunión Anual de la Asociación Argentina de Economía Política, Universidad Nacional de Posadas (2014); and the 15th SAET Conference on Current Trends in Economics, University of Cambridge (2015). I would like to thank their audiences. I also wish to thank Juan José Martínez and two anonymous reviewers whose comments and suggestions have improved the quality of this paper. Mistakes and other shortcomings are, of course, entirely my own.

Appendix

Proof of Proposition 1

Before proving Proposition 1, we show that, given Assumption P, for every $z \in L^{m+n}$ and every j, $\overline{\text{int} Y_j(z)}^8 = Y_j(z)$. Let y belong to $Y_j(z)$. If y belongs to $\text{int} Y_j(z)$, then y belongs to $\overline{\text{int} Y_j(z)}$. If y belongs to $\partial_\infty Y_j(z)$, then for all $\varepsilon > 0$, $y - \frac{\varepsilon}{2}\chi_M$ belongs to $\text{int} Y_j(z)$ by free disposal. Consequently, $B(y, \varepsilon) \cap \text{int} Y_j(z) \neq \varnothing$ for all $\varepsilon > 0$, and thus, y belongs to $\overline{\text{int} Y_j(z)}$.

We now prove that the correspondence $Y_j : L^{m+n} \mapsto L$ is $\left(\prod_{L^{m+n}} \sigma^\infty, \mathscr{T}\right)$-l.h.c. From the above result and Lemma 14.21 in Aliprantis and Border (1994), it is enough to prove that $\text{int} Y_j : L^{m+n} \mapsto L$ is $\left(\prod_{L^{m+n}} \sigma^\infty, \mathscr{T}\right)$-l.h.c. Let $y \in \text{int} Y_j(z)$ and let z^α be a net which $\prod_{L^{m+n}} \sigma^\infty$−converges to z. Since f_j is $\sigma^\infty \times \prod_{L^{m+n}} \sigma^\infty$-continuous, there exists $\alpha_0 \in \Gamma$ such that $\alpha > \alpha_0$ implies $f_j(y, z^\alpha) < 0$. Hence, there exists a net $y^\alpha (= y) \in \text{int} Y_j(z^\alpha)$ for all α and $y^\alpha \to y$.

The weak* closeness of Y_j is immediate from Assumption P(ii).

Remark 3. Given Assumption P, if the correspondence Y_j is convex valued, then $\overline{\text{int} Y_j(z)} = Y_j(z)$ without free disposal requirement (Schaefer and Wolf 1999, p. 38, 1.3). On the other hand, we can repeat the argument made above to show that the correspondence $Y_j^F : F^{m+n} \mapsto F$ is l.h.c.

Proof of Proposition 2

We omit the index j in order to simplify the notation. We first state the following Lemma:

Lemma 4. *For a given $\bar{z} \in L^{m+n}$, let $T_{Y(\bar{z})}(\bar{y})$ be the Clarke tangent cone of $Y(\bar{z})$ at $\bar{y}$. Let $v \in T_{Y(\bar{z})}(\bar{y})$ and $\delta > 0$. There exist weak* open neighbourhoods of $\bar{z}$ and $\bar{y}$, $W^{\bar{z}}$ and $W^{\bar{y}}$, respectively, such that for all $\varepsilon > 0$, for all $z \in W^{\bar{z}}$ and for all $y \in W^{\bar{y}} \cap B(\bar{y}, \varepsilon)$, $v + \bar{y} - y - \delta\chi_M \in T_{Y(z)}(y)$.*

Proof. Given $\bar{z} \in Z$, we have to prove that $\nabla_1 f(\bar{y}, \bar{z})(v + \bar{y} - y - \delta\chi_M) \leq 0$. Let $0 < \alpha < \frac{\delta \nabla_1 f(\bar{y}, \bar{z})(\chi_M)}{2(\|v\| + \varepsilon + \delta)}$. From Assumption P(v), there exists a $\prod_{L^{m+n+1}} \sigma^\infty$−open neighbourhood of $(\bar{z}, \bar{y})$, $U^{\bar{z}} \times U^{\bar{y}}$, such that for all $(z, y) \in U^{\bar{z}} \times U^{\bar{y}}$, $|\nabla_1 f(y, z) - \nabla_1 f(\bar{y}, \bar{z})| < \alpha$. Let us consider the following σ^∞−open neighbourhood of $\bar{y}$,

$$V^{\bar{y}} = \left\{ y \in L : |\nabla_1 f(\bar{y}, \bar{z})(\bar{y} - y)| < \frac{\delta \nabla_1 f(\bar{y}, \bar{z})(\chi_M)}{2} \right\}$$

---

[8]For $z \in L$, $\text{int} Y_j(z) = \{y \in L : f_j(y, z) < 0\}$.

Let $W^{\bar{y}} = U^{\bar{y}} \cap V^{\bar{y}}$, $W^{\bar{z}} = U^{\bar{z}}$ and $\varepsilon > 0$. For all $(y, z) \in W^{\bar{y}} \cap B(\bar{y}, \varepsilon) \times W^{\bar{z}}$,

$$
\begin{aligned}
\nabla_1 f(y, z)(v + \bar{y} - y - \delta \chi_M) &= (\nabla_1 f(y, z) - \nabla_1 f(\bar{y}, \bar{z}))(v + \bar{y} - y - \delta \chi_M) \\
&\quad + \nabla_1 f(\bar{y}, \bar{z})(v + \bar{y} - y - \delta \chi_M) \\
&< \alpha(\|v\| + \varepsilon + \delta) \\
&\quad + \nabla_1 f(\bar{y}, \bar{z})(v) + \nabla_1 f(\bar{y}, \bar{z})(\bar{y} - y) - \delta \nabla_1 f(\bar{y}, \bar{z})(\chi_M) \\
&< \frac{\delta \nabla_1 f(\bar{y}, \bar{z})(\chi_M)}{2} + \nabla_1 f(\bar{y}, \bar{z})(v) \\
&\quad + \frac{\delta \nabla_1 f(\bar{y}, \bar{z})(\chi_M)}{2} - \delta \nabla_1 f(\bar{y}, \bar{z})(\chi_M) \\
&= \nabla_1 f(\bar{y}, \bar{z})(v) \leq 0
\end{aligned}
$$

We now proceed to the proof of Proposition 2. Let $(z^\alpha, \pi^\alpha)_{(\Gamma, \leq)}$ be a net of $Z \times S$, $\prod_{L^{m+n}} \sigma^\infty \times \sigma^{ba}$–converging to $(\bar{z}, \bar{\pi})$. Let $v \in T_{Y(\bar{z})}(\bar{y})$ and $\delta > 0$. There exist $\varepsilon > 0$ and $\alpha_0 \in \Gamma$ such that for all $\alpha > \alpha_0$, $y^\alpha \in B(0, \varepsilon)$. We note that $\|y^\alpha - \bar{y}\| < \varepsilon + \|\bar{y}\| = \varepsilon'$. Hence, for all $\alpha > \alpha_0$, $y^\alpha \in B(\bar{y}, \varepsilon')$. From the above lemma, there exist weak*-open neighbourhoods of $\bar{z}$ and $\bar{y}$, $W^{\bar{z}}$ and $W^{\bar{y}}$, respectively, such that for $\varepsilon' > 0$ and all $\alpha > \alpha_0$, $(y^\alpha, z^\alpha) \in W^{\bar{y}} \cap B(\bar{y}, \varepsilon') \times W^{\bar{z}}$ and $v + \bar{y} - y^\alpha - \delta \chi_M \in T_{Y(z^\alpha)}(y^\alpha)$.

Since $\pi^\alpha \in N_{Y(z^\alpha)}(y^\alpha)$, $\pi^\alpha(v + \bar{y} - y^\alpha - \delta \chi_M) \leq 0$ for all $\alpha > \alpha_0$. Passing to the limit, we obtain $\bar{\pi}(v) + \bar{\pi}(\bar{y}) - \lim_\alpha \pi^\alpha(y^\alpha) - \delta \leq 0$. Since $0 \in T_{Y(\bar{z})}(\bar{y})$, $\bar{\pi}(\bar{y}) \leq \lim_\alpha \pi^\alpha(y^\alpha) + \delta$, and since this inequality holds true for all $\delta > 0$, we have $\bar{\pi}(\bar{y}) \leq \lim_\alpha \pi^\alpha(y^\alpha)$.

Let $v \in T_{Y(\bar{z})}(\bar{y})$. If $\lim_\alpha \pi^\alpha(y^\alpha) = \bar{\pi}(\bar{y})$, then $\bar{\pi}(v) \leq 0$. Consequently, $\bar{\pi} \in N_{Y(\bar{z})}(\bar{y}) \cap S$ since $\pi^\alpha \in S$ for all α.

Proof of Proposition 4

We first state and prove the following lemma, which is used in the proof of Proposition 4. To simplify, we suppress index j.

Lemma 5. Let $p_I \in (L^I)^*_+$, then there exists $\hat{p}_I \in L^*_+$ such that $\hat{p}_I(x) = p_I(x^I)$ if $x \notin L^O$ and $\hat{p}_I(x) = 0$ if $x \in L^O$.

Proof. Let $p_I \in (L^I)^*_+$. By a classical extension theorem, there exists a functional $\tilde{p}_I \in L^*_+$, and hence, a measure $\tilde{v}_I \in ba^+(M, \mathcal{M}, \mu)$ such that $\tilde{p}_I(x) = \int_{m \in M} x(m)\, d\tilde{v}_I(m)$ and $p_I(x) = \tilde{p}_I(x)$ for all $x \in L^I$, since L^* and $ba(M, \mathcal{M}, \mu)$ are isometrically isomorphic (Dunford and Schwarz 1958). We now define the measure $\hat{v}_I$ as:

$$
\hat{v}_I(A) = \begin{cases} \tilde{v}_I(A^I) & \text{if } A \subsetneqq O \\ 0 & \text{otherwise} \end{cases}.
$$

One easily checks that $\hat{v}_I \in ba^+(M, \mathcal{M}, \mu)$ which is identified with a functional $\hat{p}_I \in L_+^*$. Take $x \notin L^O$. There exists $M' \subset I$ such that $\mu(M') \neq 0$ and $x^I(m) \neq 0$ for all $m \in M'$. Consequently, $\hat{p}_I(x) = \hat{p}_I(x^I) + \hat{p}_I(x^O) = \int_{m \in M} x^I(m) \, d\hat{v}_I(m) + \int_{m \in M} x^O(m) \, d\hat{v}_I(m) = \int_{m \in I} x^I(m) \, d\hat{v}_I(m) = \int_{m \in I} x^I(m) \, d\tilde{v}_I(m) = \tilde{p}_I(x^I) = p_I(x^I)$. If $x \in L^O$, $\hat{p}_I(x) = \int_{m \in M} x^O(m) \, d\hat{v}_I(m) = 0$.

Remark 4. The above lemma can be rewritten in terms of the subspace $(L^O)^*$ as follows: for every $p_O \in (L^O)_+^*$, there exists a functional $\hat{p}_O \in L_+^*$ such that $\hat{p}_O(x) = p_O(x^O)$ if $x \notin L^I$ and $\hat{p}_O(x) = 0$ if $x \in L^I$.

First, we claim that for all $t > 0$, $-y_{Ij}$ does not belong to the relative interior of $Y_j(y_{Oj} + t\chi_{Oj}, z)$. Otherwise, $y_j \in \text{int} Y_j(z)$. We also note that for all $t > 0$, the relative interior of $Y_j(y_{Oj} + t\chi_{Oj}, z)$ is non-empty. Finally, since for all $t > 0$, $Y_j(y_{Oj} + t\chi_{Oj}, z)$ is convex, $\cup_{t>0} \text{int} Y_j(y_{Oj} + t\chi_{Oj}, z)$ is open, non-empty and convex (Schaefer and Wolf 1999, p. 38, 1.2).

Since $-y_{Ij} \notin \cup_{t>0} \text{int} Y_j(y_{Oj} + t\chi_{Oj}, z)$, there exists a continuous linear functional $p_{Ij} \in (L^{Ij})_+^*$ such that $p_{Ij}(-y_{Ij}) \leq p_{Ij}(a) \, \forall a \in \cup_{t>0} \text{int} Y_j(y_{Oj} + t\chi_{Oj}, z)$,[9] whence $p_{Ij}(-y_{Ij}) \leq p_{Ij}(a')$ for all $a' \in \cup_{t>0} Y_j(y_{Oj} + t\chi_{Oj}, z)$. Consequently, $p_{Ij}(-y_{Ij}) = c_j(p_{Ij}, y_{Oj} + t\chi_{Oj}, z)$ since $-y_{Ij} \in Y_j(y_{Oj}, z)$ for all $t > 0$. By the above lemma, we can extend the functional p_{Ij} to an element of L_+^*—denoted by p_{Ij} as well—such that $p_{Ij}(\xi) = 0$ for all $\xi \in L^{Oj}$. Let $p_{Oj} = \nabla_{Oj} c_j(p_{Ij}, y_{Oj}, z) \in (L_+^{Oj})^*$. We also extend p_{Oj} to L_+^*—denoted by p_{Oj} as well—such that $p_{Oj}(\xi) = 0$ for all $\xi \in L^{Ij}$. Consequently, by Lemma 2, $p_j = p_{Ij} + p_{Oj} \in N_{Y_j}(z)(y_j)$. Since, (z, π) is a marginal pricing equilibrium, $\pi = \lambda p_j$ for some $\lambda > 0$. Hence, $\pi_{Ij} = \lambda p_{Ij}$ and $\pi_{Oj} = \lambda p_{Oj} = \lambda \nabla_{Oj} c_j(p_{Ij}, y_{Oj}, z) = \nabla_{Oj} c_j(\lambda p_{Ij}, y_{Oj}, z) = \nabla_{Oj} c_j(\pi_{Ij}, y_{Oj}, z)$. Consequently, $\pi_{Ij}(-y_{Ij}) = c_j(\pi_{Ij}, y_{Oj}, z)$ and $\pi = \pi_{Ij} + \nabla_{Oj} c_j(\pi_{Ij}, y_{Oj}, z) \in N_{Y_j(z)}(y_j)$. Hence, conditions a., b'. and c'. of Definition 2 are satisfied.

Remark. We point out that Bonnisseau and Cornet show that if (z, π) is a marginal pricing equilibrium, then there exists a vector $(w_j)_{j=1}^n \in L^n$ (our notation) defined as $w_j = y_{Ij} + y_{Oj}^+$, such that $\left((x_i)_{i=1}^m, (w_j)_{j=1}^n, \pi\right)$ is a marginal cost pricing equilibrium. A significant difference between our approach and theirs is that in their case, $\left((x_i)_{i=1}^m, (w_j)_{j=1}^n\right) \in \Pi_{i=1}^m X_i \times \Pi_{j=1}^n Y_j$, while in ours, if $z \in Z$, $\left((x_i)_{i=1}^m, (w_j)_{j=1}^n\right)$ may not be in $\Pi_{i=1}^m X_i \left((x_i)_{i=1}^m, (w_j)_{j=1}^n\right) \times \Pi_{j=1}^n Y_j \left((x_i)_{i=1}^m, (w_j)_{j=1}^n\right)$ since the sets are not comparable. This justifies Assumption SPP.

---

[9] Let us suppose that $p_{Ij}(-y_{Ij}) \geq p_{Ij}(a)$ for all $a \in \cup_{t>0} \text{int} Y_j(y_{Oj} + t\chi_{Oj}, z)$. For any $t > 0$ and a sufficiently large $\alpha > 0$, we have $-y_{Ij} + \alpha \chi_{Ij} \in \text{int} Y_j(y_{Oj} + t\chi_{Oj}, z)$ by free disposal condition. Hence, $p_{Ij}(-y_{Ij}) \geq p_{Ij}(-y_{Ij} + \alpha \chi_{Ij})$, a contradiction.

Another important difference with the above paper is that, even if $\left((x_i)_{i=1}^m, (w_j)_{j=1}^n\right)$ belongs to $\Pi_{i=1}^m X_i \left((x_i)_{i=1}^m, (z_j)_{j=1}^n\right) \times \Pi_{j=1}^n Y_j \left(\left((x_i)_{i=1}^m, (z_j)_{j=1}^n\right)\right)$, we cannot prove that $\nabla_{O^j} c_j (\pi_{Ij}, y_{O^j}, z) = \nabla_{O^j} c_j (\pi_{Ij}, y_{O^j}^+, \quad z)$ as they did, since the argument they constructed does not work in Fréchet derivatives in infinite-dimensional spaces. Consequently, in the present context, $\nabla_{O^j} c_j (\pi_{Ij}, y_{O^j}, z) = \nabla_{O^j} c_j (\pi_{Ij}, y_{O^j}^+, z)$ whenever $y_{O^j} = y_{O^j}^+$ which also justifies the Assumption SPP.

References

Aliprantis, C., Border, K.: Infinite Dimensional Analysis. Springer, Berlin (1994)

Aumann, R.: Existence of competitive equilibria in markets with a continuum of traders. Econometrica **34**, 1–17 (1966)

Balder, E.: More on equilibria in competitive markets with externalities and a continuum of agents. J. Math. Econ. **44**, 575–602 (2008)

Beato, P.: The existence of marginal cost pricing equilibria with increasing returns. Q. J. Econ. **97**, 669–688 (1982)

Beato, P., Mas-Colell, A.: On marginal cost pricing with given tax-subsidy rules. J. Econ. Theory **37**, 356–365 (1985)

Bewley, T.: Existence of equilibria in economies with infinitely many commodities. J. Econ. Theory **4**, 209–226 (1972)

Bonnisseau, J.M.: Existence of equilibria in economies with externalities and nonconvexities. Set Valued Anal. **5**, 209–226 (1997)

Bonnisseau, J.M.: The marginal pricing rule in economies with infinitely many commodities. Positivity **6**, 275–296 (2002)

Bonnisseau, J.M., Cornet, B.: Existence of marginal cost pricing equilibria in an economy with several non convex firms. Econometrica **58**, 661–682 (1990)

Bonnisseau, J.M., Cornet, B.: Existence of marginal cost pricing equilibria: the nonsmooth case. Int. Econ. Rev. **31**, 685–708 (1990)

Bonnisseau, J.M., Créttez, B.: On the characterization of efficient production vectors. Econ Theory **31**, 213–223 (2007)

Bonnisseau, J.M., Médecin, M.: Existence of marginal pricing equilibria in economies with externalities and non-convexities. J. Math. Econ. **36**, 271–294 (2001)

Brown, D.: Equilibrium analysis with non-convex technologies. In: Hildenbrand, W., Sonnenschein, H. (eds.) Handbook of Mathematical Economics, vol. 4, pp. 1963–1995. North-Holland, Amsterdam (1991)

Clarke, F.: Optimization and Nonsmooth Analysis. Wiley, New York (1983)

Cornet, B.: Existence of equilibria in economies with increasing returns. In: Cornet, B., Tulkens, H. (eds.) Contributions to Operations Research and Economics: The XXth Anniversary of CORE, pp. 79–97. The MIT Press, Cambridge (1990)

Cornet, B., Topuzu, M.: Existence of equilibria for economies with externalities and a measure space of consumers. Econ. Theory **26**, 397–421 (2005)

Dunford, N., Schwarz, J.: Linear Operators, Part I. Wiley-Interscience, New York (1958)

Fuentes, M.: Existence of equilibria in economies with externalities and non-convexities in an infinite dimensional commodity space. J. Math. Econ. **47**, 768–776 (2011)

Guesnerie, R.: Pareto-optimality in nonconvex economies. Econometrica **43**, 1–29 (1975)

Guesnerie, R.: First best allocation of resources with non convexities in production. In: Cornet, B., Tulkens, H. (eds.) Contributions to Operations Research and Economics: The XXth Anniversary of CORE, pp. 99–143. The MIT Press, Cambridge (1990)

Hotelling, H.: The general welfare in relation to problems of taxation and of railway and utility rates. Econometrica **6**, 242–269 (1938)

Laffont, J.: Decentralization with externalities. Eur. Econ. Rev. **7**, 359–375 (1976)

Laffont, J.: Effets externes et théorie économique. Monographies du Séminaire d'Econométrie. Editions du CNRS, París (1977)

Laffont, J.J., Laroque, G.: Effets externes et théorie de l'équilibre general. Cahiers du Séminaire d'Econométrie, vol. 14. Editions du CNRS, París (1972)

MacLane, S., Garret, B.: Algebra, 3rd edn. AMS Chelsea Publishing, New York (1999)

Mantel, R.: Equilibrio con rendimientos crecientes a escala. An. Asoc. Argentina de Economía Política **1**, 271–283 (1979)

Mas-Colell, A.: The price equilibrium existence problem in topological vector lattices. Econometrica **54**, 1039–1054 (1986)

Mas-Colell, A., Zame, W.: Equilibrium theory in infinite dimensional spaces. In: Hildenbrand, W., Sonnenschein, H. (eds.) Handbook of Mathematical Economics, vol. 4, pp. 1836–1898. North-Holland, Amsterdam (1991)

Mas-Colell, A., Whinston, M., Green, J.: Microeconomic Theory. Oxford University Press, New York (1995)

Noguchi, M., Zame, W.: Competitive markets with externalities. Theor. Econ. **1**, 143–166 (2006)

Rustichini, A., Yannelis, N.: What is perfect competition. In: Khan, M.A., Yannelis, N. (eds.) Equilibrium Theory in Infinite Dimensional Spaces. Springer, New York (1991)

Schaefer, H.H., Wolf, M.P.: Topological Vector Spaces, 2nd edn. Springer, New York (1999)

Shannon C.: Increasing returns in infinite-horizon economies. Rev. Econ. Stud. **64**, 73–96 (1996)

Villar, A.: Curso de Microeconomía Avanzada. Un enfoque de equilibrio general. Antoni Bosch, Barcelona (1997)

Chapter 8
On Optimal Growth Under Uncertainty: Some Examples

Adriana Gama-Velázquez

Abstract The one-sector model of optimal growth with uncertainty in the production has been thoroughly characterized in a general way. In particular, it has been shown that the stochastic production function can be replaced by a transition probability that maps inputs into random outputs. Such replacement is relevant since it relaxes the sufficient conditions to obtain desirable conditions on the solutions to the model. Nonetheless, the model with the stochastic production represented by a transition probability lacks of examples that lead to closed-form solutions. This paper provides a revision on how to rewrite the model, its advantages, and three novel examples with explicit solutions to the optimal consumption policy and value function. These three examples assume a logarithmic utility function for every period and change the (explicit) distribution of the random production process. In the three cases, a linear optimal policy of the consumption is obtained. Examples like these allow us to do more economic analysis on applications that attain such forms, such as comparative statics.

Keywords Optimal growth • Uncertainty • Stochastic technology • Transition probability • Second-order stochastic dominance • Bellman equation • Optimal consumption policy • Value function

8.1 Introduction

In the 1960s, important deterministic models of optimal growth were developed, for instance, by Cass (1965) and Koopmans (1965). Big part of the importance of such models relies on the existence and stability results of the optimal consumption policies. Nonetheless, the robustness of these deterministic models is threatened by any perturbation such as unexpected macroeconomic shocks. Deterministic models always predict monotonic optimal sequences of capital stocks, but in reality, optimal sequences of capital stocks are non-monotonic due to such macroeconomic shocks.

A. Gama-Velázquez (✉)
El Colegio de México, Centro de Estudios Económicos, Camino al Ajusco 20,
Pedregal de Santa Teresa, Tlalpan 10740, Mexico City, Mexico
e-mail: agama@colmex.mx

© Springer International Publishing Switzerland 2016   147
A.A. Pinto et al. (eds.), *Trends in Mathematical Economics*,
DOI 10.1007/978-3-319-32543-9_8

Since some applications cannot be explained by deterministic models, authors like Mirrlees (1965),[1] Mirman (1971, 1972, 1973), Brock and Mirman (1972, 1973), Amir (1997), and Li (1988) introduce uncertainty via the production function. In this way, the spectrum of problems that can be explained through optimal growth models, now with uncertainty, is broaden. Similarly, with the incorporation of models with uncertainty, it is possible to test the robustness of the results obtained in deterministic models. In other words, the inclusion of uncertainty in the production process helps to better understand fluctuations in the economy.

Specifically, Brock and Mirman (1972) and Amir (1997) consider the problem where the decision-maker maximizes her expected utility function in an infinite horizon subject to a given technology that dictates how much output will be available the next period. The reason why the utility is uncertain this time is that the production process is stochastic, which differentiates these models from deterministic models of optimal growth. This means that the available output for the next period depends not only on the available input left by the planner in the current period, like in deterministic models, but it is also affected by a random shock represented by a random variable, which can be interpreted as an economic shock.

The authors prove that under particular but not stringent assumptions, which are summarized in the next section, the optimal consumption is a unique policy that leads to a desirable value function; specifically, the latter is monotonic, concave, and differentiable. Thus, the solution of the dynamic problem is characterized in a general and convenient way, without having to know precisely the functional forms of the primitives. This characterization is desirable in any growth model, hence its importance; nonetheless, the literature in economic growth with uncertainty lacks of particular examples with closed-form solutions that illustrate the general results in the existing models. This paper intends to fill this gap by providing examples that illustrate such results. For this purpose, a relevant literature review on the model is first presented, in order to contextualize the problems and clarify the notation.

In deterministic models, several examples with particular utility and production functions that lead to closed-form solutions are provided; thus, this paper wants to reach a counterpart for the uncertainty case. In particular, the primary goal of this study is to find conditions on the utility function and on the transition probability of the random output such that explicit solutions can be found.

The following section provides the reader with a standardized version of the model under study, as well as with the main results existing in the current literature concerning this paper. Section 8.3 presents the novel results of this paper: three examples with a particular utility form (logarithmic) and three different transition probabilities describing the random outcome that lead to explicit solutions of the optimal consumption policy. Section 8.4 presents a final discussion of the results, and the Appendix shows the details for their obtention.

---

[1] As cited by Brock and Mirman (1972).

8.2   The Model

The model studied in this paper was widely characterized by Brock and Mirman (1972), and it is described in this section. Brock and Mirman (1972) take the well-known models in economic growth of Cass (1965) and Koopmans (1965) and generalize them by adding a stochastic nature to the production process. These models are generalized in a similar way, by Mirrlees (1965) and Mirman (1971, 1972, 1973). In particular, the growth model with uncertainty under study considers a decision-maker who maximizes her expected utility for infinitely many periods subject to a stochastic production function. In other words, the decision-maker faces a dynamic problem over an infinite horizon with uncertainty in the production. In this model, the output does not depend only in the current input of the agent like in the Cass–Koopmans model; now, the output is random and depends on the available input and a random shock represented by a random variable.

Specifically, the aim of the model studied in this paper is to characterize the optimal consumption policy and value function of the agent that solves the following problem

$$\sup E \sum_{t=0}^{\infty} \delta^t u(c_t) \text{ subject to } x_{t+1} = f(x_t - c_t, r) \text{ and } 0 \le c_t \le x_t, \ t = 0, 1, \ldots,$$

$$(8.1)$$

for a given initial capital stock $x_0 \ge 0$. The variables x_t and c_t stand for capital stock and consumption in period t, respectively, u is the utility function of the agent in any given period, r is a random shock, and $\delta \in [0, 1)$ is the agent's discount factor. Since the planner faces a stochastic production for the next period, she maximizes her expected utility over her lifetime, which is represented by E in the setting of the problem.

Brock and Mirman (1972) provide conditions such that the optimal consumption policy that solves problem (8.1) is well behaved. In other words, such that the solution to the dynamic problem satisfies desirable conditions analogous to those of the optimal growth model under certainty. Specifically, under their conditions, the optimal consumption policy is unique and the marginal propensities of consumption lie between zero and one. In addition, the value function is concave, monotonic, and differentiable.

Such conditions are summarized as follows. The utility function in any period, u, satisfies the standard assumptions, that is, $u' > 0$, $u'' < 0$, and $u'(0) = \infty$ (to guarantee an interior solution). The key assumptions in Brock and Mirman (1972) are those imposed on the stochastic technology. It has to be strictly increasing and concave for every level of input and for every possible realization of the random variable or shock in the production. Formally, the random output is given by a production function f that satisfies the properties in Definition 1, which is a generalization of Brock and Mirman (1972) conditions, and it is due to Amir (1997).

The simplifying notation, which does not require a subindex t, is as follows: $y \geq 0$ represents the input for a given period,[2] and $x' \geq 0$ is the random output for the next period.

Definition 1. x' and y are related by

$$x' = f(y, r), \tag{8.2}$$

where r is a random variable with support $[\alpha, \beta] \subset [0, \infty)$, and f satisfies:

i) $f(\cdot, r)$ is strictly increasing for every $r \in [\alpha, \beta]$;
ii) $f(\cdot, r)$ is strictly concave for every $r \in [\alpha, \beta]$;
iii) there exists $\bar{y} > 0$ such that $f(y, r) < y$ if $y > \bar{y}$ for every $r \in [\alpha, \beta]$.

Intuitively, Definition 1 requires the production function to have decreasing returns to scale in a global sense, i.e., for all levels of input and for all random shocks. This is a direct consequence of asking strict concavity of f for all the possible inputs and shocks [part ii)]. Such condition on the returns to scale may be too restrictive since some production processes present decreasing returns to scale in average but local increasing returns to scale, i.e., for some level of inputs and shocks. To broaden the scope of Brock and Mirman (1972) approach, Amir (1997) relaxes Definition 1 and obtains the same results. First, Amir (1997) represents the technology in a slightly different way. Instead of using a function f, the author does it through a transition probability that takes the current input and maps it into the distribution of the next period's output. Hence, the agent's problem now looks like problem (8.3).

$$\sup E \sum_{t=0}^{\infty} \delta^t u(c_t) \text{ subject to } x_{t+1} \sim q(\cdot | x_t - c_t) \text{ and } 0 \leq c_t \leq x_t, \ t = 0, 1, \ldots,$$
$$\tag{8.3}$$

where all the variables are defined as before and q denotes the transition probability that maps the investment level $(x_t - c_t)$ into the random output for the next period.

Next, using second-order stochastic dominance, the assumptions of concavity and monotonicity in Definition 1 are imposed on the integral of the output distribution instead of on the distribution itself. Such conditions on the stochastic production are formally specified in Definition 2.

Definition 2. x' and y are related by

$$x' \sim q(\cdot | y) \tag{8.4}$$

---

[2]Hence, $y = x - c$, where x is the capital stock and c stands for consumption, both for the given period.

where q is a transition probability from $[0, \infty)$ and the associated distribution function, F, satisfies:

i) $\int_0^a F(t|y)dt$ is decreasing in y for every $a \geq 0$;
ii) $\int_0^a F(t|y)dt$ is convex in y for every $a \geq 0$;
iii) there is a function $h : [0, \infty) \to [0, \infty)$ such that $h(0) = 0$, $h(\bar{y}) = \bar{y}$, and $h(y) < y$ if $y > \bar{y}$, for some $\bar{y} > 0$, and $E(x'|y) = \int_0^\infty t\,dF(t|y) \leq h(y)$ for every $y \geq 0$.

Notice that the representation of the future (and stochastic) output in both Definitions 1 and 2 is equivalent in the sense that they provide information on the distribution of the future output given the current input. Moreover, Amir (1997) formalizes this argument by showing that a representation of the stochastic production in Eq. (8.2) can be also represented in terms of Eq. (8.4) and vice versa.

Amir (1997) also shows that Definition 1 is stronger than Definition 2, since each part of the former implies its respective part in the latter. Hence, Definitions 1 and 2 can be used indistinctly when studying a particular problem, but Definition 2 broadens the set of applications that we can characterize without knowing the primitives. In the first definition, monotonicity and concavity are imposed on the production function for every level of output and for every realization of the random variable or shock. In the second one, monotonicity and concavity are imposed on the integrals of the distribution of the random outcome, which means that the random output has to be monotonic and concave only in average but can be non-monotonic or non-concave for some levels of inputs or realizations of the random shock. In other words, increasing returns of the technology are locally acceptable as long as in average, the returns are decreasing. This assumption is more realistic than asking for decreasing returns in a global sense.

Thus, Amir (1997) replaces Definition 1 by Definition 2, which is weaker, and shows that the growth model under study still has a unique solution with marginal propensities of consumption between zero and one and value function concave, monotonic, and differentiable.

Long and Plosser (1983) present an example in the many-sector optimal growth model with uncertainty in the production with closed-form solutions. In particular, such uncertainty is expressed through the production function, like in problem (8.1), but no specific transition probability is provided.[3] Hence, a counterpart for problem (8.3) wants to be found, i.e., specific primitives of the problem (utility function and transition probability) that lead to closed-form solutions.

The following section presents three examples with a logarithmic utility function and three different transition probabilities that allow us to find explicit solutions to

---

[3]In the example of Long and Plosser (1983), the agent chooses n goods plus leisure time in order to maximize her expected utility subject to her time constraint and production possibilities. Her utility function is additively separable and logarithmic in every choice variable. The technology is Cobb–Douglas, except that at every period, it is multiplied by a positive random shock that follows a time-homogeneous Markov process.

the dynamic model. Such applications with explicit solutions allow us to do more analysis on them, for instance, to do comparative statics with respect to the relevant parameters.

8.3 Examples

The one-sector model of optimal growth with certainty literature exhibits plenty of specific applications where it is possible to find closed-form solutions. The same does not happen in the literature of the model with uncertainty in the production, in particular, when such uncertainty is represented by a transition probability. This section pretends to fill such gap by providing three specific illustrations of the model with explicit solutions.

In the deterministic case, Amir et al. (1991) state that the logarithmic utility function, $u(c) = \ln c$, leads to an explicit optimal consumption policy that is linear in the state variable, when the production process follows the Cobb–Douglas form $f(x) = x^{\alpha}$, where $\alpha > 1$, c is the consumption, and x is the input or state variable. Similarly, one gets optimal policies that are linear in the stock when the utility function is Cobb–Douglas or exponential, $u(c) = c^{\beta}$ or $u(c) = -e^{-c}$, respectively, and the production function is linear, with $\beta > 0$.

We now turn to the stochastic case. The following examples assume that the utility function of the agent is logarithmic, which satisfies the standard desirable conditions on a utility function (strictly increasing and strict concavity), and combine it with three different stochastic technologies that lead to linear optimal policies of consumption.

The notation is consistent with that of Sect. 8.2. Hence, c, x, and r represent consumption, available output, and the random shock in the current period, respectively; x' represents the available output for the next period. Recall that the latter depends on how much of the good remained from the previous period, $x-c$, and a random shock, r; thus, x' is the random stock or random outcome for the next period. In the frame of this growth model, the stock $x - c$ is interpreted as the savings of the consumer, which in this case are affected by an economic shock. In other fields, like the ones that study the extraction of resources, $x-c$ is interpreted as the available stock for the next period; the random shock helps in modeling stochastic reproduction of natural resources or animals, such as the reproduction of fish.

Example 1. Consider an agent that maximizes her expected utility over an infinite number of periods. The utility function of the agent, u, is logarithmic for every period, $u(c) = \ln c$, and the random output for the next period is given randomly by

$$x' = (x - c)/r,$$

where r is a random variable uniformly distributed between zero and one. Thus, the associated cumulative distribution function (cdf) for the future production is[4]

$$F(t|x-c) = \begin{cases} 1 - \frac{x-c}{t} & \text{if } 0 \le x - c \le t, \\ 0, & \text{otherwise.} \end{cases}$$

Notice that $0 \le x - c$ holds by the setting of the problem.
The Bellman equation of this problem is

$$V(x) = \max_{0 \le c \le x} \ln c + \delta \int_{x-c}^{\infty} V(t) dF(t|x-c),$$

which becomes

$$V(x) = \max_{0 \le c \le x} \ln c + \delta(x-c) \int_{x-c}^{\infty} \frac{V(t)}{t^2} dt \qquad (8.5)$$

after substituting the transition probability represented by F.

Since the utility function of the agent is logarithmic, it is natural to think that the value function takes the same functional form. Making this guess and following the standard method to solve dynamic problems, we find the solution to this problem; the details are in the Appendix. The optimal consumption policy, $c(x)$, is linear in the capital stock and is given by

$$c(x) = (1 - \delta)x,$$

and the value function $V(x)$ is, as we guessed, logarithmic

$$V(x) = A \ln x + B,$$

where

$$A = \frac{1}{1 - \delta} \text{ and } B = \frac{\ln(1 - \delta)}{1 - \delta} + \frac{\delta(1 + \ln \delta)}{(1 - \delta)^2}.$$

The explicit solution to this problem allows us to do some comparative statics. For instance, we can analyze how consumption changes when the agent is more patient, i.e., when the discount factor is bigger. Notice that such change is negative, $\partial c(x)/\partial \delta = -x < 0$; hence, a more patient planner will consume less in the first period. Since she is more patient, she will save more input for future periods. □

---

[4]To compute the cdf, notice that $F(t|x-c) = \text{Prob}(x' \le t|x-c) = \text{Prob}\left(\frac{x-c}{r} \le t\right) = \text{Prob}\left(r \ge \frac{x-c}{t}\right)$. The first equality follows by definition of cdf; the second one, by substituting $x' = (x-c)/r$; and the last one, by rearranging terms. Finally, its explicit form is obtained using the fact that the random variable r is uniformly distributed between zero and one.

   Given the results of the first example, we can suspect that a random shock with linear probability density function (pdf) together with a logarithmic utility function leads to closed-form solutions. Example 2 solves the one-sector optimal growth model with such specifications.

Example 2. Let us continue working with $u(c) = \ln c$ and the random output given by $x' = (x - c)/r$, like in Example 1. This time, suppose that r is a random variable with pdf given by $f(r) = 2r \cdot 1(0 \leq r \leq 1)$. Following the same steps as in Example 1, the corresponding transition probability represented by F is now

$$F(t|x - c) = \begin{cases} 1 - \left(\frac{x-c}{t}\right)^2 & \text{if } 0 \leq x - c \leq t, \\ 0, & \text{otherwise.} \end{cases}$$

   The specific Bellman equation of this problem, after substituting the transition probability, becomes

$$V(x) = \max_{0 \leq c \leq x} \ln c + 2\delta(x - c)^2 \int_{x-c}^{\infty} \frac{V(t)}{t^3} dt. \tag{8.6}$$

Since the utility function is logarithmic, we again guess that $V(x) = A \ln x + B$; then, the optimal consumption policy coincides with that in Example 1. See the Appendix for the details. The value function differs from that of the previous example because the random outcome is different and so is its expected value. Specifically, the solution is given by the optimal consumption

$$c(x) = (1 - \delta)x$$

and value function

$$V(x) = A \ln x + B,$$

with

$$A = \frac{1}{1 - \delta} \text{ and } B = \frac{\ln(1 - \delta)}{1 - \delta} + \frac{\delta(1/2 + \ln \delta)}{(1 - \delta)^2}.$$

   Notice that the value of the parameter A coincides with that in Example 1, but not the parameter B, which explains the difference between the value functions of the examples. In particular, the value function of Example 2 is lower with respect to that in Example 1 by $\frac{\delta}{2(1-\delta)^2}$.

   The comparative static result in Example 1 still holds given that the optimal consumptions coincide.   □

   In general, an explicit solution of the dynamic model can be achieved whenever the utility function of the planner is logarithmic and the pdf of the shock that affects the future production is linear, i.e., whenever the pdf in Example 2 is given by $f(r) = nr \cdot 1(0 \leq r \leq 1)$, where n is a natural number. The last example formalizes this result.

Example 3. Consider an expected utility maximizer agent whose utility function in every period is logarithmic, $u(c) = \ln c$; there is an infinite number of periods. Such agent faces the random output $x' = (x - c)/r$, where r is a random variable with pdf $f(r) = nr \cdot 1(0 \leq r \leq 1)$ and n is a natural number. Under this generalization, the transition probability function becomes

$$F(t|x - c) = \begin{cases} 1 - \frac{n}{2}\left(\frac{x-c}{t}\right)^2, & \text{if } 0 \leq x - c \leq t, \\ 0, & \text{otherwise.} \end{cases}$$

and the Bellman equation,

$$V(x) = \max_{0 \leq c \leq x} \ln c + n\delta(x - c)^2 \int_{x-c}^{\infty} \frac{V(t)}{t^3} dt. \tag{8.7}$$

Guessing the value function $V(x) = A \ln x + B$, we get that the optimal consumption policy is (see the Appendix for the details)

$$c(x) = \frac{(2 - n\delta)}{2}x,$$

and the value function is as guessed above with

$$A = \frac{2}{2 - n\delta} \text{ and } B = \frac{2}{2 - n\delta}\left[\frac{n\delta}{2(2 - n\delta)} - \frac{2}{2 - n\delta}\ln 2 + \ln(2 - n\delta) + \frac{n\delta}{2 - n\delta}\ln(n\delta)\right].$$

Finally, notice that the comparative statics in Examples 1 and 2 can be generalized. The change of the optimal consumption with respect to the discount factor is negative, $\partial c(x)/\partial \delta = -nx/2 < 0$ (recall that n is a natural number), i.e., a more patient consumer (larger δ) will consume less today in order to save more for tomorrow, because she values more her future consumption. □

8.4 Conclusions

In this paper, we study the one-sector model of optimal growth with an infinite number of periods and stochastic production. This is a model where the planner maximizes her expected utility, since her output for the next period is unknown. Specifically, in the current period, the planner saves some input to produce output for the next period, but such production is affected by an exogenous and random shock, which can be interpreted as a macroeconomic or a natural shock. A common application of these models with stochastic production is the extraction of resources, where the reproduction of the natural resources is stochastic, for instance, the reproduction of fish.

Works like Brock and Mirman (1972) and Amir (1997) provide sufficient conditions on the primitives of the model under study, to obtain desirable solutions, in particular, to get a unique optimal consumption policy with marginal propensities of consumption between zero and one, as well as a concave, monotonic, and differentiable value function. It is natural to ask what particular characteristics on the transition probability defining the random production and on the utility function lead to explicit optimal policies, so that more research can be done on the behavior of these explicit results. Nonetheless, there is a lack of examples with specific transition probabilities that map inputs into random outputs that this paper tries to fill. The opposite happens in the deterministic literature, where many examples with closed-formed solutions have been written; thus, a counterpart to the deterministic model is presented in this work.

This paper offers three novel examples. In all of them, the planner maximizes her expected utility over an infinite horizon, and her utility is logarithmic in every period. Her available outcome for the next period is equal to the savings of the current period affected by a random shock; hence, the production process is stochastic. In particular, this uncertainty is mathematically expressed by the savings divided by a random variable, which takes three different distributions, one for every example.

In the first example, the random variable affecting the production is uniformly distributed; in the second one, its pdf is given by $f(r) = 2r \cdot 1(0 \leq r \leq 1)$. The third and last example is a generalization of the second one; the pdf of the random variable is $f(r) = nr \cdot 1(0 \leq r \leq 1)$, where n is a natural number.

The first two examples lead to the same optimal policy, which is linear in the current capital stock x. This optimal policy is decreasing in the discount factor, which is an expected result given that a decision-maker who is more patient will be willing to consume less in the first period in order to consume more in the future. The value functions in such examples are logarithmic, and they differ only by the constant. In the second case, the value function is lower by $\frac{\delta}{2(1-\delta)^2}$.

The second example suggests that a random variable with linear pdf leads to linear solutions whenever the utility function is logarithmic in every period; thus, the problem is solved for a more general linear pdf (detailed before). Again, the optimal policy is linear in the capital stock and decreasing with respect to the discount factor; the value function is still logarithmic.

Appendix

Solution to Example 1

Since the utility function of the agent is logarithmic in every period, a good guess for the value function is that it is logarithmic as well. That is, $V(x) = A \ln x + B$; hence, we substitute this guess into Eq. (8.5) and solve for the optimal policy, and if we are able to find values for the constants A and B consistent with our guess, we have a solution. This solving method is standard in dynamic programming.

Specifically, following the steps described above, we get that the Bellman equation of this problem is

$$V(x) = \max_{0 \le c \le x} \ln c + \delta(x - c) \int_{x-c}^{\infty} \frac{A \ln t + B}{t^2} dt. \tag{8.8}$$

Let us solve the integral first, starting by distributing it

$$\int_{x-c}^{\infty} \frac{A \ln t + B}{t^2} dt = A \int_{x-c}^{\infty} \frac{\ln t}{t^2} dt + B \int_{x-c}^{\infty} \frac{1}{t^2} dt; \tag{8.9}$$

integrating by parts, we get

$$\int_{x-c}^{\infty} \frac{\ln t}{t^2} dt = \frac{\ln(x - c)}{x - c} + \int_{x-c}^{\infty} \frac{1}{t^2} dt. \tag{8.10}$$

Notice that, as will be explained below, the following equalities hold

$$\int_{x-c}^{\infty} \frac{A \ln t + B}{t^2} dt = A \frac{\ln(x-c)}{x-c} + (A + B) \int_{x-c}^{\infty} \frac{1}{t^2} dt$$
$$= A \frac{\ln(x-c)}{x-c} + (A + B) \frac{1}{x-c}$$
$$= \frac{(\ln(x-c)+1)A+B}{x-c};$$

the first equality follows from substituting identity (8.10) into (8.9); the second one, by integrating the second term in the right hand side of the first equality; and the last one, by reducing the expression algebraically.

Hence, substituting the last equality in Eq. (8.8), the Bellman equation becomes

$$V(x) = \max_{0 \le c \le x} \ln c + \delta[(\ln(x - c) + 1)A + B], \tag{8.11}$$

with first-order condition

$$\frac{1}{c(x)} - \frac{\delta A}{x - c(x)} = 0,$$

which leads the optimal policy for consumption and value function, respectively

$$c(x) = \frac{x}{1 + \delta A} \text{ and } V(x) = \ln\left[\frac{x}{1 + \delta A}\right] + \delta\left[\left(\ln\left(\frac{\delta A x}{1 + \delta A}\right) + 1\right)A + B\right].$$

Notice that the latter is obtained by plugging the optimal consumption policy in (8.11) and can be simplified into

$$V(x) = (1 + \delta A) \ln x - (1 + \delta A) \ln(1 + \delta A) + \delta A(1 + \ln \delta A) + \delta B. \tag{8.12}$$

To obtain the explicit values of the parameters A and B and complete our solution, we equalize the value function obtained in Eq. (8.12) with our guess. Specifically, we equalize the coefficients in our equations and solve for A and B. This means that

$A = (1 + \delta A)$ and $B = -(1 + \delta A) \ln(1 + \delta A) + \delta A(1 + \ln \delta A) + \delta B$. Solving this system of two equations and two variables, we get

$$A = \frac{1}{1 - \delta} \text{ and } B = \frac{\ln(1 - \delta)}{1 - \delta} + \frac{\delta(1 + \ln \delta)}{(1 - \delta)^2}.$$

Hence, we have a linear optimal policy in the capital stock given by

$$c(x) = (1 - \delta)x$$

and value function

$$V(x) = A \ln x + B,$$

with A and B defined above.   □

Solution to Example 2

To solve this problem, we proceed in a similar way as in Example 1, starting with our logarithmic guess of the value function, $V(x) = A \ln x + B$. Based on this guess and the distribution of the random production, we get the following Bellman equation [by substituting the guess in Eq. (8.6)]

$$V(x) = \max_{0 \le c \le x} \ln c + 2\delta(x - c)^2 \int_{x-c}^{\infty} \frac{A \ln t + B}{t^3} dt. \tag{8.13}$$

The integral in Eq. (8.13) can be expressed as follows

$$\int_{x-c}^{\infty} \frac{A \ln t + B}{t^3} dt = A \int_{x-c}^{\infty} \frac{\ln t}{t^3} dt + B \int_{x-c}^{\infty} \frac{1}{t^3} dt; \tag{8.14}$$

integrating by parts, we have

$$\int_{x-c}^{\infty} \frac{\ln t}{t^3} dt = \frac{\ln(x - c)}{2(x - c)^2} + \frac{1}{2} \int_{x-c}^{\infty} \frac{1}{t^3} dt. \tag{8.15}$$

Then,

$$\int_{x-c}^{\infty} \frac{A \ln t + B}{t^3} dt = A \frac{\ln(x-c)}{2(x-c)^2} + \left(\frac{A}{2} + B\right) \int_{x-c}^{\infty} \frac{1}{t^3} dt$$
$$= A \frac{\ln(x-c)}{2(x-c)^2} + \left(\frac{A+2B}{4}\right) \frac{1}{(x-c)^2}$$
$$= \frac{(2 \ln(x-c) + 1)A + 2B}{4(x-c)^2}.$$

The first equality follows from substituting Eq. (8.15) in (8.14); the second one, by solving the integral; and the third one, by reducing algebraically.

Substituting the last expression in Eq. (8.13), we get the following Bellman equation

$$V(x) = \max_{0 \le c \le x} \ln c + \frac{1}{2}\delta[(2\ln(x - c) + 1)A + 2B],$$

with first-order condition

$$\frac{1}{c(x)} - \frac{\delta A}{x - c(x)} = 0.$$

Then, the optimal consumption in the current period and the value function are, respectively,

$$c(x) = \frac{x}{1 + \delta A} \text{ and } V(x) = \ln\left(\frac{x}{1 + \delta A}\right) + \frac{1}{2}\delta\left[\left(2\ln\left(\frac{\delta xA}{1 + \delta A}\right) + 1\right)A + 2B\right].$$

We are looking for parameters A and B such that $V(x) = A\ln x + B$; given the previous result, we are looking for A and B such that

$$A\ln x + B = \ln\left(\frac{x}{1 + \delta A}\right) + \frac{1}{2}\delta\left[\left(2\ln\left(\frac{\delta xA}{1 + \delta A}\right) + 1\right)A + 2B\right],$$

which can be simplified into

$$A\ln x + B = (1 + \delta A)\ln x - (1 + \delta A)\ln(1 + \delta A) + \delta A\left(\ln \delta A + \frac{1}{2}\right) + \delta B.$$

The previous equality implies that $A = \frac{1}{1-\delta}$ and $B = \frac{\ln(1-\delta)}{1-\delta} + \frac{\delta(1/2+\ln\delta)}{(1-\delta)^2}$.

Notice that the optimal consumption policy is the same for both Examples 1 and 2. Nonetheless, the value function is smaller in the current Example by $\frac{\delta}{2(1-\delta)^2}$.   □

Solution to Example 3

To solve this problem, we follow the methodology in Examples 1 and 2; hence, we now skip some technical details. After guessing that the value function is logarithmic, $V(x) = A\ln x + B$, substituting it into the Bellman equation (8.7) and solving for the integral, we get

$$V(x) = \max_{0 \le c \le x} \ln c + \frac{n}{4}\delta\left[(2\ln(x - c) + 1)A + 2B\right], \tag{8.16}$$

with the corresponding first-order condition

$$\frac{1}{c(x)} - \frac{n}{2} \frac{\delta A}{x - c(x)} = 0.$$

From the previous equation, we get that the optimal consumption is

$$c(x) = \frac{2x}{2 + n\delta A};$$

and by (8.16), the value function becomes

$$V(x) = \ln\left(\frac{2x}{2 + n\delta A}\right) + \frac{n}{4}\delta\left[\left(2\ln\left(\frac{n\delta Ax}{2 + n\delta A}\right) + 1\right)A + 2B\right].$$

Recall that $V(x) = A \ln x + B$ (from our initial guess); hence, equalizing the last two identities and doing the algebra, we get

$$A = \frac{2}{2 - n\delta} \text{ and } B = \frac{2}{2 - n\delta}\left[\frac{n\delta}{2(2 - n\delta)} - \frac{2}{2 - n\delta}\ln 2 + \ln(2 - n\delta) + \frac{n\delta}{2 - n\delta}\ln(n\delta)\right].$$

Then, the solution can be reduced to

$$c(x) = \frac{(2 - n\delta)}{2}x \text{ and } V(x) = A \ln x + B,$$

with A and B defined above. □

Acknowledgements The author gratefully acknowledges Rabah Amir for many useful comments, and CONACyT and SEP for their financial support while working on this project.

References

Amir, R.: A new look at optimal growth uncertainty. J. Econ. Dyn. Control **22**, 67–86 (1997)

Amir, R., Mirman, L., Perkins, W.: One-sector nonclassical optimal growth: optimality conditions and comparative dynamics. Int. Econ. Rev. **32**, 625–644 (1991)

Brock, W., Mirman, L.: Optimal economic growth and uncertainty: the discounted case. J. Econ. Theory **4**, 479–513 (1972)

Brock, W., Mirman, L.: Optimal economic growth and uncertainty: the no discounting case. Int. Econ. Rev. **14**, 560–572 (1973)

Cass, D.: Optimum growth in an aggregate model of capital accumulation. Rev. Econ. Stud. **32**, 233–240 (1965)

Koopmans, T.: On the concept of optimal economic growth. In: Pontificiae Academiae Scientiarum Scripta Varia 28, pp. 225–300. Academiae Vaticana (1965)

Li, L.: A stochastic theory of the firm. Math. Oper. Res. **13**, 447–466 (1988)

Long Jr, J., Plosser, C.: Real business cycles. J. Polit. Econ. **91**, 39–69 (1983)

Mirman, L.: Uncertainty and optimal consumption decisions. Econometrica **39**, 179–185 (1971)

Mirman, L.: On the existence of steady-state measures for one-sector growth models with uncertainty technology. Int. Econ. Rev. **13**, 271–286 (1972)

Mirman, L.: The steady-state behavior of a class of one-sector growth models with uncertain technology. J. Econ. Theory **6**, 219–242 (1973)

Mirrlees, J.: Optimum accumulation under uncertainty. Dissertation, Cambridge University (1965)

Chapter 9
Fundamental Principles of Modeling in Macroeconomics

Samuel Gil Martín

Abstract Modern macroeconomic research has been increasingly concerned with its capability of establishing sound positive and normative conclusions in uncertain environments. The Lucas' critique imposes the requirement to do so from solid microeconomic principles which alter in a significant way the overall notion of fundamental value behind any convincing theory. The necessity to bring these general issues to dynamic settings to test alternative hypotheses introduces new tools of analysis which are required to understand the current literature, both theoretical and empirical. In this article we explore the notion of fundamental taken as given the level of total resources and show how it extends when different degrees of uncertainty are introduced.

Keywords Stationarity • Risk • Uncertainty • Recursive utility

9.1 Introduction

The Lucas' critique has brought about a consensus about the necessity to found macroeconomics on sound microeconomic principles. The paths of the main aggregates such as inflation, employment, and asset pricing cannot be accounted for unless the driving interests of the economic actions are properly set. For such purpose, the additive expected utility representation of preferences has been the conventional cornerstone of theoretical research. Implicitly, this elegant, parsimonious, and versatile representation construction has emerged from a requirement of consistency and rationality of expectations. However, the theory faces somewhat serious difficulties and drawbacks when confronted with data and experimental evidence. The Allais (1953) paradox, obtained by revealed preference, could be resolved by a relatively simple transformation of the original model. The Ellsberg (1961) paradox puts forward a meaningful difference between different orders of uncertainty from a behavioral standpoint. While Ramsey and Savage propose that

S. Gil Martín (✉)
Facultad de Economia, Universidad Autónoma de San Luis Potosí, Avenida Pintores s/n, Col.
Burócratas del Estado, San Luis Potosi 78213, Mexico
e-mail: samuel.gilmartin@gmail.com

© Springer International Publishing Switzerland 2016
A.A. Pinto et al. (eds.), *Trends in Mathematical Economics*,
DOI 10.1007/978-3-319-32543-9_9

163

personal decisions subject to uncertain outcomes reveal subjective likelihoods of events, the Ellsberg paradox shows that this is not indeed a representative case. The positive and normative implications are likely to be more drastic than those related to the Allais' paradox, as long as the existence of an incontrovertible model is called into question.

Welfare theorems establish conditions of efficiency and separation of the logic of market efficiency and distribution. Specifically, the first welfare theorem imposes the requirement of market completeness to achieve an efficient outcome which permits economic agents full insurance against idiosyncratic risk. By transforming the original model by a multiplicative stochastic discount factor, this outcome may be expressed as a risk-neutral pricing measure. The efficient market hypothesis states that asset prices are fully informative: they cannot possibly embed any forecastable component because their fluctuations are purely random.

Unfortunately, there are some facts that can hardly be reconciled with some empirical evidence. Financial markets have experienced over the last 30 years a tremendous increase in the bulk of data available that has made it possible to test the theories developed under the umbrella of the dynamic stochastic general equilibrium paradigm. The equity premium puzzle (Mehra and Prescott, 1985), as well as many others that highlight possible and substantial drawbacks in the standard macroeconomic theory, has built upon standard models that make use of the rational expectations hypothesis.[1] As long as the role of financial markets is to provide liquid assets to allocate and monitor capital, efficiency requires production of information that ultimately shall be reflected in prices. Information technology has remarkably reduced transaction costs, including information costs and data storage. In addition, during the period 1960–2012, the share of the financial industry to GDP in the USA has more than doubled, from 2.5 to 5.5 %, having grown six times faster than GDP in the last 30 years (Boi et al. 2012). The TFP has also suffered important alterations during the last 30 years, breaking a long period of stability. Both facts are most probably behind the polarization of wealth observed worldwide over a whole generation.

Tables 9.1 and 9.2 capture a tension observed between some indices of prosperity and poverty that synthetize the extent of these paradoxes. The development of macroeconomic theory in recent years is seemingly bound to be dealing with issues related with complexities of this sort. However, and because of the same reasons, the main questions which gave rise to the birth of economics remain open or have been reformulated: the difficulties to explain the capacity of a community to generate wealth and abundance are essentially associated to its ability to formulate a theory of value generated by agents' decisions susceptible to be confronted with data.

---

[1] See Brunnermeier (2001) and Easley and O'Hara (2003) among many surveys available in the literature.

Table 9.1 Figures of population change and income are measured in annual rates, and life expectancy is measured in years

	0–1000	1000–1800	1800–2000	2000–2011
Population	0.02	0.17	0.86	1.36
Income per capita	−0.00	0.00	1.07	2.23
Life expectancy	28–24	24–26	26–66	67.2 (2009)

Source: Maddison (2001) and own

Table 9.2 Consumption measured in terms of exosomatic energy (approximation) per person and day

Fire	Agriculture	1900	2000
2	4	14	30–40

Source: Common and Stagl (2005) and Gómez Romero (2010)

Table 9.1 shows that growth is an astonishingly contemporary phenomenon, but at the same time, it is well known that total factor productivity (TFP) is not driven by accumulation of physical factors.

On the contrary, Table 9.2 gives an opposite view of this piece of evidence. The necessary consumption of subsistence (2000–2500 kcal per capita and day approx) is known as endosomatic consumption and is shared by any living being on this planet according to their biological needs. Consumption, as measured in units of exosomatic primary energy, consists of those goods that are obtained by means of technologies, from the use of fire onward, that allow a transformation of energy, raw materials, and information from the environment into final goods.

This chapter revisits and provides a conceptualization of economic actions that give rise to fundamental implications in the theory of value. In doing so, we explore some methods of analysis aimed to embed preferences in a general macroeconomic theory described by complexity and an extraordinary number of different kinds of interactions. In this way, this essay aims to provide a further debate on whether and how likely the circular flow of income is able or not to generate welfare and production under increasingly uncertain and interrelated contexts. The difficulties found extensively in the literature observationally equivalence can be restated conversely, namely, problems to distinguish fundamentals from frictional value driven by strategic and informational spillovers.

The rest of this work is organized as follows. Sections 9.2 and 9.3 deal with standard representations of preferences and some of their extensions within the probabilistically sophisticated paradigm. Section 9.4 introduces the notion of Knightian uncertainty and gives a hint of some of its macroeconomic implications. Section 9.5 extends these concepts to a dynamic framework. In Sect. 9.6 it is provided an empirical application that explores the boundaries of the macroeconomic research, allowing a generalization of the concept of stationarity. We suggest that different alternative specification of the Brownian motion, while obeying the basic stationarity conditions, can potentially affect the fundamental components of value. Section 9.7 concludes.

9.2   Decision Under Risk

9.2.1   The Economic Space

Macroeconomic models are formulated in terms of an underlying state of the
world. An economic model requires an accurate specification of (1) agents and
their preferences defined over a set of admissible economic actions, (2) resources
which are available for the society as a whole, and (3) a characterization of the
market setting. While the economic fundamentals are defined by the parameters of
preferences and resources, some models emphasize the effects of market frictions
on resource allocation. Let us start this section by making precise the meaning of
the state of an economy.

The economic world, in all its complexity, is represented at the most abstract
and general level by a standard probability space $(\Omega, \mathscr{F}, P)$. An element ω of ω
gives a full description of the economy. Overall, agents living in uncertain models
need not know the true value of ω; in other words they are not expected to think
of a complete account of the world before making a decision. Static theories are
constructed by identifying the universe Ω with a fixed metric state space S. Time
can be introduced by defining an analogue space $(T, \mathscr{B})$, where the time set $T \subset \mathbb{R}$
is either an interval of (continuous models) or a discrete subset set, and $\mathscr{B}$ is its Borel
sigma algebra. A dynamic model is constructed by means of the standard product
space $(S \times T, \mathscr{F} \otimes \mathscr{B})$. Very often the model is closed through an exponential
discounting, though it is not a necessary requirement. We will deal with dynamic
models in Sect. 9.5, and it is the case that many of the major insights can be obtained
by considering finite space states.

An element $s \in S$ comprises those specific indicators which are relevant
to characterize the performance of the circular flow of income. A model is a
probability measure on the measurable space $(S, \mathscr{F})$. $\mathscr{F}$ is a standard sigma
algebra of events, that is, a family of subsets of S closed by numerable unions,
intersections, and pairwise complements. Its elements are called measurable sets or
events. Any economy is populated by agents whose preferences are order relations
over a stipulated set of feasible actions. An information structure is given by
those information sets of agents who takes active part of the economic life. For
instance, in perfectly competitive environments, one expects that the price system
alone conveys enough information as to coordinate an efficient market outcome. In
general, an action is a measurable map $c : S \to \mathscr{C}$, where $\mathscr{C}$ stands for a convex
set of consequences. We will thoroughly assume that S and $\mathscr{C}$ are Polish spaces, in
order to guarantee convenient regularity conditions of choice under high degrees of
uncertainty.

The size of the action space will depend essentially on the operative constraints
of the economy and those that affect the decision-maker (DM) itself. In a purely
deterministic consumption model, S is a singleton, and households' actions, $c \in \mathscr{C}$,
reduce to the choice of an element of the consumption space. Under the usual

conditions of existence and smoothness, an ordinal preference relation can be represented by a smooth utility function u on $\mathscr{C}$. This index will survive in more sophisticated environments as a fundamental building block in macroeconomics.

In stochastic settings, every act is a measurable function $c : S \to \mathscr{C}$ that generates a partition (equivalence class) on the phase state S, denoted by a family of subsets $c^{-1}(x) \subset S, x \in \mathscr{C}$. Measurability implies that this family is a subsigma algebra of $\mathscr{F}$ or equivalently a closed family of events closed under arbitrary unions, intersections, and complementaries containing the empty set and Ω. Measurable mappings can be approximated as a limit of a sequence of simple functions, that is, acts giving rise to finite partitions of events. Formally, a simple act has the form

$$c = \sum_{s=1}^{n} 1_{c^{-1}(x_s)}(s) \cdot x_s, \tag{9.1}$$

where $x_s \in \mathscr{C}$ is a constant act which equals the outcome in state $c^{-1}(x_s)$, 1_A is the indicator function defined for any event, and n is an integer. A cardinal representation of preferences $\succsim$ is a continuous functional $V : \mathscr{A} \to \mathbb{R}$, such that $c \succsim c' \Leftrightarrow V(c) \geq V(c')$ (see Appendix A).

The Bayesian theory of decision-making has been widespread applied both to rational expectation equilibrium models and to game theoretical approaches. The Bayesian paradigm presumes the existence of a single $P : \mathscr{F} \to [0, 1]$ associated to our original measurable space. In this case, $(\Omega, \mathscr{F}, P)$ constitutes a probability space. Each act c may be unambiguously identified with a distribution $\pi := P \circ c^{-1}$. Probabilistically sophisticated environments of this kind make it possible the association of a DM's feasible set with a (convex) subset of the space of distributions over the course of actions. We denote this set as $\Pi(\mathscr{C})$, which in mathematical terms is the dual space of $\mathscr{A}$. Its elements are commonly referred to as lotteries. Since $E1_{c^{-1}(x_s)} = \pi_s = P(x_s)$, in finite settings a lottery can be represented as a vector of payoffs together with its corresponding vector of probabilities. Accordingly, (9.1) can be written $c = (x; \pi)$.

9.2.2 Expected Utility Theory

When the consequences of individual actions are uncertain, the phase state S cannot be longer defined by a single element. The first attempt to axiomatize decision-making under risk imposes individuals a single model (probability measure) of reference: Von Neumann and Morgenstern (1944) provide a representation of preferences by means of a linear functional, unique up to affine transformations of the utility index u, of this general form:

$$V(c) = Eu(c) = \int u(c) \, dP. \tag{9.2}$$

In particular, when S is finite, and considering (9.1),

$$V(c) = Eu(c) = Eu\left(\sum_{s=1}^{n} 1_{c^{-1}(x_s)} \cdot x_s\right) = \sum_{s=1}^{n} E1_{\{c=x_s\}} \cdot u(x_s) = \sum_{s=1}^{n} \pi_s \cdot u(x_s).$$

From (9.2) it follows that $U(\delta_x) = u(x)$, so riskless decision-making becomes a particular case of decision under risk.

In many macroeconomic models, it is convenient to confine the set of consequences $\mathscr{C}$ to a subset of the real line. Most data are encoded in aggregate terms, and empirical applications need to justify a methodology based entirely on macroeconomic variables. The VNM preferences represented in (9.2) is a first step in approaching theory to data. It will be often convenient to rewrite (9.2) in units of this aggregator: assuming that u is strictly increasing, and making use of the invariance of the utility functional to increasing transformations, we can rewrite (9.2) in terms of the certainty equivalent associated to a consumption lottery

$$m^P(c) := u^{-1}E^P u(c) = u^{-1}(V(c)) \tag{9.3}$$

Savage (1954) have addressed much of the criticisms launched at the EU theory by pointing toward the cognitive roots of the probability distribution P. Savage postulated an order relation on the space of events ($\mathscr{F}$) according to its likelihood of occurrence. This approach gives rise to a weaker version of the expected utility theory, known as subjective expected utility theory (SEU).

9.2.3 A Static LQ Model

The linear quadratic (LQ) model offers the main insights introduced so far. Consider a quadratic one-period utility $u : \mathbb{R}^d \times \mathbb{R}^{d'} \supset \mathscr{A} \times S \to \mathbb{R}$, which depends on a variable of control, say $c \in \mathscr{A} = \mathbb{R}^d$ that can alter the behavior of the state $X \in S$:

$$u(c, X) = -\frac{1}{2}c'c - X'\Phi X. \tag{9.4}$$

The cost matrix Φ is a definite positive and without loss of generality, a symmetric matrix. We may interpret this problem as an attempt to achieve the maximum target with the possibly minimum energy input X. The following decomposition is standard in macroeconomic theory and asset pricing when it comes to bringing these insights into a dynamic economy. Assume that $X = X^c$ has finite second moments: let a^c and $\sigma\sigma'$ be, respectively, its expectation and covariance matrix; then the state variable X can be (a.s.) uniquely expressed as

$$X = X^c = a + Hc + \sigma B. \tag{9.5}$$

$a \in \mathbb{R}^n, H \in \mathbb{R}^{n \otimes d}, and \sigma \in \mathbb{R}^{n \otimes \ell} \neq 0$ are parameters, and B is a standard Gaussian random noise valued in $\mathbb{R}^\ell$; each of its components are interpreted as a risk factors that synthetize all the information contained in the probability space $(\Omega, \mathscr{F}, P)$, in terms of a reference model P. The optimal rule for the maximization problem

$$V(c) = \max_{c \in \mathbb{R}^d} Eu(c, X^c)$$

subject to (9.5) can be obtained by direct substitution; straightforward computations yield the optimal plan for our benchmark model:

$$c = \left(I_d + H'\Phi H\right)^{-1} H'\Phi a. \tag{9.6}$$

Let $u = u(c, X^c)$ and note that it is quadratic in B and c

$$- 2u = c' \left(I_d + H'\Phi H\right) c + 2 \left(a + \sigma B\right)' \Phi Hc + \left(a + \sigma B\right)' \Phi \left(a + \sigma B\right)$$
$$= a'\Phi a + b(c, B) + 2a'\Phi\sigma + B'\sigma'\Phi\sigma B, \tag{9.7}$$

where

$$b = b(c, B) = c' \left(I_d + H'\Phi H\right) c + \left(a + \sigma B\right)' \Phi Hc + a'\sigma\Phi B. \tag{9.8}$$

Consider the utility index φ with a constant coefficient of absolute risk aversion α. The certainty equivalent associated with φ, as defined in (9.3), can be written as follows:

$$m(c) := -\alpha^{-1} \ln E e^{-\alpha u(c, X^c)}. \tag{9.9}$$

Let us define

$$A := I_s + \alpha\sigma'\Phi\sigma = V\Lambda V', \tag{9.10}$$

where V is the orthogonal matrix ($V^{-1} = V'$ because A is symmetric) formed row-wise by its eigenvectors. Λ is the diagonal matrix composed by the eigenvalues, repeated according to its degree of multiplicity, and they are clearly strictly greater than one. Then one can write the CE explicitly, by using Eqs. (9.7)–(9.10):

$$\alpha m(c) = -\ln(2\pi)^{-\frac{1}{2}} \int_{-\infty}^{\infty} e^{-\frac{\alpha}{2}(a'\Phi a + b(c,B) + \alpha B'\sigma'\Phi\sigma B) - \frac{\|B\|^2}{2}} dB$$
$$= \frac{\alpha}{2}a'\Phi a - \ln(2\pi)^{-\frac{1}{2}} \int_{-\infty}^{\infty} e^{-\frac{1}{2}(\alpha b(c,B) + B'AB)} dB. \tag{9.11}$$

The exponential term of the integrand is a polynomial with a quadratic component which depends on a symmetric definitive positive matrix. By analogy to the one-dimensional case [see Backus et al. (2004, Appendix 9)], one may apply a change of variable of the form $w = V\Lambda^{1/2}V'B + \alpha\gamma = A^{1/2}B + \alpha\gamma$, making it possible[2] to avoid the crossed terms and write the CE in closed form:

$$\alpha b\,(c, B) + B'AB = \alpha\beta\,(c) + \|w\|^2.$$

This relation combined with (9.8) yields

$$\beta = \alpha\,\|\gamma\|^2 + c'\left(I_d + H'\Phi H\right)c + d'\Phi Hc$$
$$\gamma = A^{-\frac{1}{2}}\sigma'\,(a + Hc)$$

This permits us to write

$$\beta\,(c) = \beta_0 + \beta_1'c + c'\Psi c, \tag{9.12}$$

so that the parameters of the model have been restated according to

$$\Psi = I_d + H'\left(\Phi + \alpha^2\sigma A^{-1}\sigma'\right)H \tag{9.13a}$$
$$\beta_0 = a'\sigma A^{-1}\sigma'a, \quad \beta_1 = H'\left(I_n + \alpha^2\sigma A^{-1}\sigma'\right)a \tag{9.13b}$$
$$w = A^{1/2}B + A^{-1/2}\sigma'\,(a + Hc).$$

Next we can solve the second integral of (9.11):

$$\int_\infty^\infty e^{-\frac{1}{2}(\alpha a'\sigma B + B'AB)}\,dB = \det\,(A)^{-\frac{1}{2}}\,e^{-\frac{1}{2}\beta(c)}\int_{-\infty}^\infty e^{-\frac{\|w\|^2}{2}}\,dw.$$

Note by (9.10) that $\det\,(A) = \det\,(\Lambda)$ and let $\beta_1 \geq 0$, with equality if $\alpha\Sigma = 0_n$, stand for the natural logarithm of such determinant. Thus, by virtue of (9.11)–(9.13), we find that the CE is equivalent (affine) to $\beta\,(c)$

$$m\,(c) = -\frac{1}{2}\left(\det\,(\Lambda) + a'\Phi a + \beta\,(c)\right). \tag{9.14}$$

By $\max m\,(c)$ or $\beta\,(c)$, it is immediate to recover the optimal decision rule:

$$c = \Psi^{-1}\lambda = \left[I_d + H'\left(\Phi + \alpha^2\sigma A^{-1}\sigma'\right)H\right]^{-1}H'\left(I_n + \alpha^2\sigma A^{-1}\sigma'\right)a. \tag{9.15}$$

This solution coincides with the SEU model when $\alpha\sigma = 0$, namely, when there is either no risk or the DM is neutral in the face of it.

---

[2] The factor α on the right hand of the last equality has been introduced to simplify slightly the computations.

9.2.4 Information Structure

Fundamentals may lack to provide a good characterization of the behavior of an economic system when information is privately held due to strategic externalities. There is a vast number of examples covered in the literature where the presence of transaction costs and asymmetric information transforms trading within the marketplace into a game. There are many empirical findings that detect a predictable component in both the short-run and the long-run movements of asset prices. In particular, the latter effects may help to explain the relevant role of variables such as information, liquidity (the costs associated to immediate execution of investment), and uncertainty in the macroeconomic realm (Easley and O'Hara 2003).

Savage (1954) and Anscombe and Aumann (1963) provided an axiomatization in which individuals face *horse* (second order) lotteries that can be compounded, in agreement with the independent axiom proper of EU decision-making (see the Appendix). This behavior is at odds with a bulk of experimental evidence of ambiguity aversion, even under the extreme case where inside information reveals virtually nothing, e.g., a coin (Ellsberg 1961). We have seen that a decision with first-order uncertainty (risk) can be conceptualized by identifying the admissible set of actions with the space of probability measures over consequences, $\mathscr{A}$ or $\Pi\,(\mathscr{C})$. We can view this operation as one endowed of a cognitive nature [see, for instance, Brunnermeier (2001, Chap. 1)]. Indeed the information set $\mathscr{F}$ can be viewed as a partition over the phase space Ω, denoted by $[\cdot]$, which satisfies the following conditions for each $\omega \in \Omega$:

Axiom (Axiom of Truth). $\omega \in [\omega]$

Axiom (Axiom of Introspection). $\omega' \in [\omega] \Leftrightarrow [\omega'] = [\omega]$

The first condition is clear; introspection can be positive ($\Rightarrow$) or negative ($\Leftarrow$). By positive introspection individuals rule out the existence of a state of the world[3]; thus, $z \in [\omega'] \cap [\omega]^c$ whenever $\omega' \in [\omega]$. For instance, assume the axiom of truth holds and let ω^o mean that we are able to forecast the lack of any macroeconomic shock—demand or supply sided, within a given lapse of time, say a quarter. On the contrary, let $[\omega^d] \cup [\omega^s]$, denote that we are able to forecast a shock of either category within the same period, where the superscripts d and s stand for demand and supply, respectively. Assume $[\omega^d] = [\omega^s]$, that is, the impossibility of discernment between the two sources of noise. By positive introspection we are able to discard from $[\omega^o]$ any state of the world included in $[\omega^d]$ so long as $[\omega^s]$ and $[\omega^o]$ are disjoint subsets. Let now $[\omega^r]$ mean that we are currently going through a recession and $[\omega^*] = \{\omega^*\}$ that such event has a permanent component. This fatal event can be ruled out from $[\omega]$ by negative introspection.

A converse argument allows one to relate an arbitrary partition to a knowledge operator, defined as $\mathscr{K}\,(G) := \{\omega : [\omega] \in G\}$. One does not need to impose any

---

[3]The superscript c denoted the complementary of a set.

axiom on the underlying information structure [·] of $\mathscr{K}$ in order to obtain the following properties whose interpretation is obvious:

K1	$\mathscr{K}(\Omega) = \Omega$
K2 Monotonicity	$\mathscr{K}(G) \subset \mathscr{K}(H)$ if $G \subset H$
K1	$\mathscr{K}(G) \cap \mathscr{K}(H) = \mathscr{K}(G \cap H)$

Axiom of truth can be restated as $\mathscr{K}(A) \subset A$. Positive (negative) introspection may be stated, in terms of the knowledge operator, as knowing what one does (not) know, respectively, $\mathscr{K} \subset \mathscr{K}^2$ and $\mathscr{K}^c \subset \mathscr{K} \circ \mathscr{K}^c$.

With the axioms of truth and introspection, a partition and its corresponding knowledge operator uniquely define a sigma algebra $\mathscr{F}$.

9.3 Alternative Explanations to Expected Utility

SDGE models impose, by dealing with expected utility, implausible parameter constraints on individual patterns of behavior, once confronted against experimental evidence. In the next sections, we will explore alternative specifications of preferences that give rise to substantial changes in the concept of fundamentals and thereby in the theory of value as a whole. Furthermore, theoretical progress is subject to an observational equivalence issue. Recall that in probabilistically sophisticated models, by virtue of (9.1), an economic act may be identified with a pair of vectors $(x; \pi)$ that describe outcomes and probabilities.

9.3.1 Probabilistically Sophisticated Models

The Allais' paradox (1953) can be illustrated in the following example: consider the lotteries

$$c_1 = 1_{\{c=1\}} \qquad c_2 = .33 \cdot 1_{\{c=1+\varepsilon\}} + .66 \cdot 1_{\{c=0\}} + .01 \cdot 1_{\{c=0\}}$$
$$c_1' = .34 \cdot_{\{c=1\}} + .66 \cdot 1_{\{c=0\}} \qquad c_2' = .33 \cdot 1_{\{c=1+\varepsilon\}} + .67 \cdot 1_{\{c=0\}}.$$

Since the utility index is unaltered by scaling and translation, we lose no generality putting $u(0) = 0$ and $u(1) = 1$. Allais reported that, for $\varepsilon = 1/24$, individuals would majoritarily decide $c_1 \succ c_2$ and $c_2' \succ c_1'$. However, this fact contradicts SEU, because $U(c_1) > U(c_2)$ means $1 > .33 \cdot u(1 + \varepsilon) + .66$; equivalently, $U(c_1') = .34 > .33 \cdot u(1 + \varepsilon) = U(c_2')$. The interpretation of this result is based on the fact that, on facing uncertain choices, a DM tends to put more weight on bad outcomes than they do on good outcomes.

A possible way to accommodate this piece of evidence consists of substituting the axiom of independence by the so-called axiom of betweenness; for $c, c' \in \Pi(\mathscr{C}), \exists \lambda \in (0, 1)$:

Axiom (BET Betweenness). $c \succ c' \Rightarrow c \succ (c, c')_\lambda \succ c'.$

This axiom grants that the family of convex linear combinations of any two given lotteries is ordered according with its two extremes. Weighting utility (WU) and rank-dependent preferences (RDP) are classes of preferences that accommodate the behavior patterns observed by Allais. WU makes use of the following axiom instead of independence:

Axiom (w-SUB Weak Substitution). $a \sim b \Rightarrow \exists c \in \Pi(\mathscr{C}), \exists \alpha = \alpha(\lambda), \lambda \in (0, 1) : (a, c)_\alpha \sim (b, c)_\lambda.$

Clearly IND is a particular case of w-SUS when $\alpha = \lambda$. A key attribute of this axiom is the imposed independence between α and c. The SEU theory implies that if the axioms ORD, w-SUB, and continuity (a weakest version of continuity applies for the discrete case) hold, a twisted representation of preferences is possible, namely,

$$V(c) = \int u(c)\,\theta(c)\,dP =: \int u(c)\,dP^\theta, \qquad (9.16)$$

where

$$P^\theta(A) := \frac{\int 1_A \theta(c)\,dP}{\int \theta(c)\,dP}$$

RDP operates through the following class of transformations of the reference model:

$$\int u(c)\,dg(P), \qquad (9.17)$$

where $g : [0, 1] \to [0, 1]$ is a continuous and strictly increasing transformation of full range.

A major problem with these specifications is that they both violate the first-order criterion of stochastic dominance. For example, consider two numbers $y \geq x$ and $x, y \in u(A) = [0, 1]$ and set $\theta(1 - \lambda) + \theta(\lambda) < \theta(1) = 1$. Clearly, δ_y first order dominates stochastically (FOSD) to the binary lottery $(x, y)_\lambda$. When x and y are close enough, $y < \theta(1 - \lambda)y + \theta(\lambda)x$. Any violation of the domination principle poses similar problems to those involved in nontransitive preferences: if an individual with continuous preference relation happens to display a cycle $a \succsim b \succsim c \succsim a$, it would be profitable to offer her c in exchange for a, after which she can be offered the alternative b in exchange for $a - \varepsilon$, with $\varepsilon > 0$ (Machina 1982). Seemingly, intransitivity would seem to be at odds with continuity of preferences before (Quiggin 1982) reconciled transitivity with FOSD. Different modification and weighting functions have been proposed in the literature: see, for instance, Chew (1983, 1989), Dekkel (1986), and Fishburn (1989). Kahneman and Tversky (1992) proposed a cumulative prospect theory to avoid the violation of first-order stochastic dominance.

Fishburn proposed a unifying theory that comprises continuous preference relations characterized by the axiom BET. To this aim, this author defined a bilinear utility functional $\phi : \mathscr{A} \times \mathscr{A} \to \mathbb{R}$, defined on a convex set of feasible alternatives $\mathscr{A} \subset \Pi(\mathscr{C})$. This utility functional satisfies the *skew-symmetric* condition $\phi(p, q) = -\phi(q, p)$. Preference representation stems from the relation $p \succ q \Leftrightarrow \phi(p, q) > 0$, in such a way that it is possible to relax the transitivity assumption.

9.3.2 The Aggregator

Chew-Dekel preferences provide a generalization of expected utility that accommodates the Allais' paradox in line with the considerations made in the last subsection. Their model is established by means of a risk aggregator $f : \mathbb{R}^2 \supset \mathscr{C} \times \mathbb{R} \to \mathbb{R}$, which satisfies the following fixed-point requirement:

$$m = Ef(c, m) \tag{9.18}$$

Epstein and Zin (1989) show that the special case $f(c, m) = u(c) + m$ is equivalent to the axiom of betweenness. It is commonly assumed that f satisfies the following properties:

a. sure consequences correspond to their own certainty equivalent: $f(c, c) = c$;
b. first-order stochastic dominance: f is increasing in its first argument
c. risk aversion: f is concave in its first argument.

The next property is technical and gives rise to parsimonious settings:

d. linear homogeneity.

Properly speaking, only the two first conditions define a certainty equivalent. Risk aversion is usually stated to guarantee the second-order conditions of optimality that guarantee the existence of equilibria. However, specific contexts have been reported where economic agents would be prone to seeking risk or ambiguity (Heath and Tversky 1991). The fourth hypothesis is technical and is stated for the sake of analytical convenience: with the assumption of homogeneity, the SEU model reduces to

$$f(c, m) = m\left[\frac{1}{\alpha}\left(\left(\frac{c}{m}\right)^\alpha - 1\right) + 1\right] \tag{9.19}$$

with an explicit certainty equivalent that can be straightforwardly recovered from (9.18)

$$m(c) = (Ec^\alpha)^{1/\alpha} \tag{9.20}$$

Chew and Dekel proposed the following extension of the EU, known as weighted utility (WU)

$$f(c, m) = m \left[\frac{1}{\alpha} \left(\left(\frac{c}{m} \right)^{\alpha+\beta} - 1 \right) + 1 \right],$$

which gives

$$m(c) = \left(Ec^{\alpha+\beta} \right)^{1/\alpha}. \tag{9.21}$$

This relation is easily interpreted if we apply the change of measure

$$d\bar{P} := \frac{c^{\beta}}{Ec^{\beta}} dP$$

and rearrange so as to obtain (9.20): under the transformed model by a multiplicative factor known as the Radon-Nikodym derivative of the twisted model $\bar{P}$ with respect to P:

$$m^{\bar{P}}(c) = \left(E^{\bar{P}} c^{\alpha} \right)^{1/\alpha}.$$

Under this version, individuals primarily focus on the impact of bad outcomes and allocate them a greater weight than EU does. Similar extensions have been provided in the literature, such as rank-dependent preferences as mentioned in the previous subsection. Alternative kinds of exotic behavior can be explained in parsimonious settings, such as disappointment aversion and, in dynamic settings, habits [see Backus et al. (2004)]. Both have the potential to act as macroeconomic propagation mechanisms with a considerable macroeconomic impact.

Disappointment aversion utility (DAU) extends SEU so that its corresponding aggregator g takes the form

$$g(c, m) = f(c, m) + \frac{\delta}{\alpha} \left(c^{\alpha} m^{1-\alpha} - m \right) 1_{\{c<m\}},$$

where f is the SEU aggregator given in (9.19). With $\delta = 0$, $g = f$. If $\delta > 0$, the DM places more weight to those events worse than the CE. One can prove to satisfy the relation (9.21) with (Backus et al, 2004)

$$\bar{P} = \frac{1 + \delta 1_{\{c<m\}}}{1 + \delta E^{P} 1_{\{c<m\}}} P.$$

In other words, both models are observationally equivalent.

Macroeconomic theory most commonly deals with finite variance distributions. Random variables with finite second moments $\int X^2 dP < \infty$ are naturally endowed with a norm which endows the space of mean square integrable random variables with the properties of a Hilbert space, usually denoted as $L^2(\Omega, \mathscr{F}, P)$. These variables admit a decomposition, used extensively in the sequel, of the form $X = a + \sigma B$, where B is a random variable with zero mean and unitary variance. The free parameters a and σ stand, respectively, for the mean and standard deviation. With these notions at hand, we are able to provide an insightful small-risk approximation for $m(c)$ for the Gaussian case.

Example 1 (Certainty Equivalent with Gaussian Returns with a CRRA Utility Index (Backus et al. 2004)). Consider a risky payoff c log-normally distributed, so $\log c = \mu + \sigma \varepsilon$ with $\varepsilon \sim \mathcal{N}(0, 1)$. As a particular case of the LQ model of Sect. 9.3.3, we deduce this well-known expression for the mean value of c:

$$Ec = e^{\mu + \frac{\sigma^2}{2}}.$$

If the utility index has exponent $\alpha = 1 - \gamma$, where γ is the parameter of risk aversion, the SEU certainty equivalent can be written as

$$m^{\mathrm{SEU}}(c) = e^{\mu + \frac{(1-\gamma)\sigma^2}{2}}.$$

The risk premium of c, defined as the ratio between Ec and $m(c)$ expressed in logs, will be proportional, under the stated assumptions, to the Arrow-Pratt coefficient of risk aversion and the standard deviation:

$$\varrho^{\mathrm{SEU}} := \log \frac{Ec}{m^{\mathrm{SEU}}(c)} = \frac{1}{2}\gamma\sigma^2.$$

In the WU model, a similar operational method of calculus can be applied to compute the certainty equivalent which yields

$$m^{\mathrm{WU}}(c) = e^{\mu + \frac{(1-\gamma')\sigma^2}{2}}, \varrho^{\mathrm{SEU}} = \frac{1}{2}\gamma'\sigma^2, \text{ with } \gamma' = \gamma + 2\beta.$$

Neither of these models has a linear term in the quadratic form that defines the log of the certainty equivalent, contrary to the DAU case. The techniques involved to compute the CE are similar to those involved in the Black and Scholes (1973) formula. Let Φ be the standard Gaussian distribution and

$$\left[m^{\mathrm{DAU}}(c)\right]^\alpha = e^{\alpha\mu + \frac{(1-\gamma)\alpha^2\sigma^2}{2}} + \delta\left[e^{\alpha\mu + \frac{(1-\gamma)\alpha^2\sigma^2}{2}} \Phi\left(\frac{\ln m - \mu - \alpha\sigma^2}{\sigma}\right) - \Phi\left(\frac{\ln m - \mu}{\sigma}\right)\right]$$

$$\varrho^{\mathrm{SEU}} = \frac{1}{2}\gamma'\sigma^2, \text{ with } \gamma' = \gamma + 2\beta.$$

In all the three preference models described above, it is relatively easy to test the sensitivity to small risks. Consider the CE of the outcome $c = 1 + \varepsilon$, where the small-risk ε is a random variable with null expectation and positive standard deviation σ. By obtaining second-order approximations and taking expectations, it is straightforward to estimate the certainty equivalents relative to these three models for $Q = $ SEU,WU,DAU:

$$m^Q (c) \simeq 1 - \varrho^Q.$$

A remarkable feature of DAU preferences is that it places proportionally greater risk aversion to small risks than it places to large risks, although this trend can be easily reversed under a further generalization. These equations make it possible to make simple estimates of the equity premium.

9.3.3   Differentiability

Machina (1982) suggests an interesting local interpretation of SEU preferences with no need to resort to independence. Let $\mathscr{D} = \mathscr{A} - \mathscr{A}$ the space of directions of the form $v = c' - c$. To define the Frechet derivative, $\mathscr{D}$ must be endowed with a norm. The weak topology, the minimal topology such that the maps $f \to \int f dP$ are continuous, is a most natural framework of analysis. An alternative for a norm is given by $\|v\| := \int d \, |v|$. A utility functional $V(c)$ is *Frechet differentiable* if there exists a linear functional $D(c; \cdot)$ on $\mathscr{D}$, which satisfies the following equality:

$$\delta V (c) := V (c + v) - V (c) = D (c; v) + o (\|v\|). \tag{9.22}$$

The Riesz's representation theorem assures the existence of a mapping $J : \mathscr{A} \times \mathscr{D} \to \mathbb{R}$, such that $D(c, v) = - \int J (c, v) dP$. Let $u (x, \cdot) := J [(-\infty, x], \cdot]$. By way of substitution in (9.22), and proceeding with integration by parts, we obtain the following relation[4]:

$$\delta V (c) = \int u (\cdot, c) \, dv + o (\|v\|). \tag{9.23}$$

This equation says that the marginal utility along the trajectory along v equals the expected value of the difference in utility $c + v$ and c.

Example 2 (Example of Local Quadratic Utility Functional (Machina 1982, p. 295)).

$$V (c) = \int R (\cdot, c) \, dP + 1/2 \left(\int \sigma dP \right)^2$$

induces local utility index $u (x, c) = E^P R (x, c) + 1/2 \left[E^P \sigma (x, c) \right]^2$, and a local utility.

---

[4]See Machina (1982, p. 294 and Appendix 1).

Let $\pi(t) := P(c(t))$. To compare local and global behaviors, Machina utilizes the notion of paths $v(t;s) := c(t) - c(s)$, whose norm is differentiable in the parameter $t \in [0, 1]$. Let $v(t) := V(c(t))$. Taking into account (9.23) and the fact that $o'(0) = 0$,

$$v'(t) = \frac{d}{dt} \int u(\cdot, c(s)) \, dv(\cdot; t).$$

Total variation will be determined by the fundamental theorem of calculus:

$$\Delta U := v(1) - v(0) = \int_0^1 \frac{d}{dt} \int u(\cdot, p(s)) \, dp(\cdot, t) \, ds.$$

This expression states that global behavior equals the sum of each marginal displacement of preferences along the path.[5] While RDP theory is not generally Frechet differentiable, the most part of the analysis of the Sect. 9.3.3 can be exploited through the Gateaux derivative. We say that V is Gateaux differentiable in p along the direction $\delta \in \mathscr{D}$

$$DV(c, \delta) = \lim_{h \to 0^+} \frac{V(c + \delta) - V(c)}{h} = \int u(\cdot, c) \, d\delta.$$

If $DV(c, \delta)$ is well defined for each admissible δ, we say that $u(x, c)$ is the Gateaux derivative in c.

9.4   Knightian Uncertainty

SEU, Chew-Dekel preferences, DAU, and the related research on behavioral economics covered in Sect. 9.3 are built upon probabilistically sophisticated models, namely, models with a unique admissible scenario P. A major critique of this approach is that agents are given more information than an econometrician facing misspecification issues (Sargent 2001, Hansen, 2007). In the spirit of early literature initiated by Keynes (1921) and Knight (1921), the Ellsberg's paradox (1961) went one step further than Allais' in that agents reveal concerns for model robustness; in other words, people are concerned about the consequences of ignoring the true model. Uncertainty is made of social interactions which, as opposed to risk, convey alternative scenarios.

Extensions of SEU require a representation of the utility functional by means of a nonlinear operator. In general terms, we will consider two broad classes of models which have been conveniently axiomatized. These two classes of models include

---

[5]See Karni and Schmeidler (1991, Sects. 5 and 7).

most of the extensions known to expected utility. The Ellsberg's paradox can be illustrated through a DM faced with a choice problem where one ball is to be drawn from an urn containing three balls, one of which is unambiguously red, while the others may be black or white. The following bets are experimentally confronted to a population DM with the aim to reveal their preferences: here $S = \{R, B, N\}$ denote, respectively, the states for each color. Actions cannot be possibly described by lotteries, since there is one for each possible model $P \in \mathcal{P}$.

$$c_1 = 1_R \qquad\qquad c_2 = 1_B$$
$$c_1' = 1_{R \cap N} = c_1 + 1_N \quad c_2' = 1_{B \cap N} = c_2 + 1_N.$$

According to the sure-thing principle, the preference relation should be invariant if there is no change in the course of action. However, the majority revealed $c_1 \succ c_2$ and $c_2' \succ c_1'$. One of the most immediate and remarkable effects of this paradox consists of the expansion of the size of the space state so as to include the epistemic reality. Those beauty contests that arise through a myriad of transactions sensitive to market microstructures and strategic interactions can be included in the phase state and therefore in the fundamental components of asset value.

Reasoning by analogy to risky choice, acts of order n can be defined as elements of $S_n = \Pi^n(\mathcal{C}) \setminus \Pi^{n-1}(\mathcal{C})$. These acts represent beliefs (second order) as well as beliefs about what the other agents' beliefs. Under the hypothesis that X is a Polish space, it can be shown that $\Pi^n(X)$ is itself Polish. This construction involves no major technical difficulties once the state S_n is embedded with the vague topology, i.e., the coarsest topology making the functionals $\mu \to \int g d\mu$ continuous for any continuous g. For each S_n, we consider the Borel σ-algebra $\mathcal{G}_n$ generated by the vague topology and define n-th order acts as the $\mathcal{G}_n$-measurable mappings[6] $c : S_n \to \mathcal{C} \subset \mathbb{R}$. We will denote the set of n-th order acts as $\mathcal{A}_n = \mathcal{M}(S_n, \mathcal{C})$ of measurable maps from S_n to the space of consequences and the set of admissible actions by $\mathcal{A} := \bigcup_{n \geq 0} \mathcal{A}_n$.

9.4.1 Smooth Preferences

Klibanoff et al. (2005), 2007 have axiomatized smooth preferences with second-order uncertainty. Thimme and Völkert (2012) have generalized this setting to arbitrary orders of uncertainty: let $Pc^{-1} =: p \in S_n, n > 0$ be stand for a n-th order belief. With the notation introduced above, we can define the n-th order certainty equivalent $m_n : S_n \to \mathbb{R}$ as follows:

$$m_n(p) = u_n^{-1}\left(\int_{S_{n-1}} u_n dp\right), \tag{9.24}$$

---

[6]The inclusion of the set of consequences in the real line may be interpreted as measured in terms of numéraire units.

The certainty equivalent corresponding to

$$m(c) = \lim_{n \to \infty} m_n^{\mu_n} \left(\cdots m_2^{\mu_2} \left(m_1^{\mu_1} (c) \right) \right) \tag{9.25}$$

whenever such limit exists. Conditions of existence of expression (9.25) are the standard conditions of measurability if we impose a finite degree of uncertainty or exponential decay.

Second-order uncertainty is commonly referred to as ambiguity. It may be interpreted as a fundamental component of the epistemic reality. In this case, KMM show that decision-maker is ambiguity averse (lover) when the function $u_2 \circ u_1^{-1}$ concave (convex); if such a function is linear, then lotteries can be compounded and preferences are ambiguity neutral (SEU). Note that under homothetic preferences, $u_2 \circ u_1^{-1}(x) = x^{\alpha_2/\alpha_1}$. This implies that ambiguity aversion implies that the Arrow-Pratt coefficient of risk aversion corresponding to utility index $u_2(1 - \alpha_2)$ is greater than that of u_1.

Example 3. Assume consequences lie in the unitary interval $\mathscr{C} = [0,1]$, and consider binary actions $c(S) = \{\varepsilon, 1\}$. First-order beliefs can be represented in this simple case by the probability of success $p = P(c = 1)$. Under homothetic preferences,

$$u_n(c) = \frac{c^{1-\sigma_n} - 1}{1 - \sigma_n}.$$

It is convenient to normalize $u_1^0(\varepsilon) = 0$ and $u_n^0(1) = 1$, by imposing the translation $u_n^0 = u_n + 1$. In this case, it is clear that

$$\varepsilon = (1 - \alpha)^{\frac{1-\alpha}{\alpha}} \to e^{-1} \text{ as } 1 - \alpha := \sigma_1 \to 1.$$

This fact implies $m_1^p(c,p) = p^{\frac{1}{\alpha}}$.

Assume the agent is able to compound all uncertainty of order three or higher. Second-order beliefs are characterized by a density $\varphi(p)$. If p is uniformly distributed and $\beta := 1 - \alpha_2$, then

$$V(c) = m(c)^\beta = \int_0^1 p^{1/\alpha} dp = \frac{\alpha}{1 + \alpha}.$$

9.4.2 Variational Preferences

Maccheroni et al. (2006); (MMR henceforth) have provided an axiomatic basis of a class of preferences that comprise the multiple prior preferences (MPP) of Gilboa and Schmeidler (1992) and robust control theory (RCT) as particular cases. This class of preferences admits this representation, unique after affine normalization of the utility indices u and h,

$$V(c) = \min_{P \in \Pi} \int u(c)\, dP + \lambda h(P), \tag{9.26}$$

where $h : \Pi(S) \to \overline{\mathbb{R}}$ is a convex functional which in addition is lower semicontinuous and grounded, i.e., $\inf h = 0$. It is given by the following integral representation:

$$h(P) = \sup_{c \in \mathscr{A}} \left\{ u(m(c)) - \int u(c)\, dP \right\}. \tag{9.27}$$

MPP corresponds to the special case where $P \subset \Pi$ is a convex subset of priors, and

$$h(P) = \delta_{\mathscr{P}}(P) = \begin{cases} 0 \text{ if } P \in \mathscr{P} \\ \infty \text{ otherwise} \end{cases}$$

The size of $\mathscr{P}$ measures the concerns for robust misspecification. This special characterization conveys a maxmin interpretation of a malevolent nature that plays a zero-sum game against the DM in a dynamically consistent way (see Appendices B and C):

$$V(c) = \min_{P \in \mathscr{P}} \int u(c)\, dP \tag{9.28}$$

A special and important case corresponds to Schmeidler (1989) CEU preferences, when

$$\mathscr{P} = \text{Core}(v) = \{P \in \Pi : P \geq v\}$$

for a given capacity[7] v. The utility functional is in this case expressed by means of a Choquet integral defined as

$$V(c) = \min_{P \in \text{Core}(v)} \int u(c)\, dP = \int u(c)\, dv. \tag{9.29}$$

Example 4 (RDP Preferences Quiggin (1982)). Let $z := u(c) \succsim \in CEU$, κ a permutation and $\Gamma_\kappa = \{z : z_{\kappa(1)} \geq \cdots \geq z_{\kappa(n)}\}$ is convex cone. Each $\kappa(i)$ induces an equivalence class through the set of acts $u^{-1}(\Gamma_s)$. Let

$$\pi_s = v\left(\bigcup_{i=1}^{s} \kappa(i)\right) - v\left(\bigcup_{i=1}^{s-1} \kappa(i)\right).$$

---

[7] A capacity is a positive, increasing, and nonadditive function $v : \mathscr{F} \to [0, 1]$. The condition of nonadditivity makes the difference between a probability measure and a capacity.

Note that $\sum \pi_s = 1$. In this case, (9.21) can be expressed as

$$V(c) = \mathscr{V}z = \int z dv = \sum_{s=1}^{n} u\left(c_{\kappa(s)}\right) \pi_s \quad \forall z \in \Gamma_\kappa.$$

Accordingly, a belief system can be characterized by the convex hull of the system of measures (see Appendix C): $\partial \mathscr{V}(0) = \text{co}\{P_s : s \in \Sigma\}$. Concavity of $\mathscr{V}$ is an equivalent condition of supermodularity[8] of v. In such a case, $\text{Core}(v) = \text{Core}(\mathscr{V})$, finding an interesting link with the Shapley characterization in the theory of cooperative games.

9.4.3 Robust Control

Robust control theory corresponds to the particular case of variational preferences (9.26) where the functional h equals the index of relative entropy

$$R(P\| Q) := E^Q \ln Q/P. \tag{9.30}$$

R measures the distance between two probability measures; here Q is interpreted as the DM's reference model. When $\lambda = \infty$, the model reduces to SEU. This model appeared in many applications well before the MMR axiomatization, within the general domain of engineering and operational research. The DM acts as if he were involved in a zero-sum game, played against a malevolent nature whose size is measured according to the degree of ambiguity aversion.

Consider the LQ model of Sect. 9.2 and allow distortions of a given model of reference P given by a law of motion (9.5). A set of feasible scenarios which captures the DM's concerns of misspecification is assumed to be described by a parameter $\theta \in \Theta$, valued in $\mathbb{R}^n$.

$$X = X^c = a + Hc + \sigma(B + \theta). \tag{9.31}$$

The optimal plan is obtained by altering the mean proportionally to θ as compared to (9.6):

$$c^\theta = \left(I_d + H'\Phi H\right)^{-1} H'\Phi(a + \sigma\theta). \tag{9.32}$$

Let $u = u(c, X^c)$ and write

$$-2u = a'\Phi a + b\left(c, B^\theta\right) + 2a'\Phi\sigma B^\theta + B^{\theta'}\sigma'\Phi\sigma B^\theta, \tag{9.33}$$

---

[8]A function $v : 2^S \to \mathbb{R}$ is said to be supermodular if for any subsets $A, B \subset S$, we have $v(A) + v(B) \leq v(A \cup B) + v(A \cap B)$.

where the noise relative to the altered model is transformed into $B^\theta := B + \theta$ and $b\left(c, B^\theta\right)$ was defined in Eq. (9.8) that corresponds to the following objective function:

$$\min_{\theta \in \mathbb{R}^n} Eu\left(c\right) + 1/2\lambda\theta'\theta.$$

The agent is faced with an evil nature (much of it belongs to a societary realm indeed) which is to choose the worst possible scenario θ. By definition, $E^Q B^\theta = 0$, and the objective function corresponds with the entropy ambiguity index (9.30). Let $y := a + \sigma B$ rearrange (9.10):

$$-2u = c'\left(I_d + H'\Phi H\right)c + (y + \sigma\theta)'\Phi(y + \sigma\theta)$$

by taking derivatives with respect θ and equaling to zero, one obtains

$$\theta = -\left(\sigma'\Phi\sigma\right)^{-1}\sigma'\Phi y.$$

Next consider the utility index φ with a constant coefficient of absolute risk aversion α. The certainty equivalent associated with φ, as defined in (9.3), can be expressed in the following way:

$$m_\theta\left(c\right) := -\alpha^{-1}\ln E^\theta e^{-\alpha u(c)} \tag{9.34}$$

This certainty equivalent can be expressed as $\beta\left(c\right)$ as defined in Eqs. (9.12)–(9.13). Substituting into (9.34) one finds that when $\alpha = \lambda^{-1}$, both approaches coincide. In other words, the parameter λ expresses a fundamental attribute of preferences as it controls for the magnitude of the DM's opponent. The connection between concerns for robustness and risk sensitivity may be observed by approximating the certainty equivalent of an exponential utility under the LQ model.

9.5 Time

The insights developed in the previous sections can be extended within a dynamic framework under the general label of recursive preferences. The universal set Ω has been identified with a phase state S which in general is a Polish space. However, in numerous applications, this set is assumed to be finite, apparently without much loss of generality. The general method of analysis consists of applying the logic of construction developed in static settings to a dynamic model endowed with a generalized space state, S^T, made up of sample paths. The Skorohod topology[9] is a

---

[9] See, for instance, Jacod and Shiryaev (2003).

cornerstone to derive the asymptotic properties of general stochastic processes. In continuous time, it is the space of all right continuous with left limit paths. These functions are commonly known as the acronym càdlàg for their initials in French.

A simple and insightful example is given by the tree

$$\mathscr{T}_1 = \{\pm 1\}^{\{0\}} =: B = 1_{\{B=1\}} + 1_{\{B=-1\}}$$

as a representation of the probability space $(S, \mathscr{F}, P)$, where states take the two possible values $B \in S = \{\pm 1\}$ and $\mathscr{F} = 2^S$. An obvious reference model P is given by $p := P(\{\pm 1\}) = 1/2$ so that $E^P B = 0$ and the state B has unitary variance. B can thus be interpreted in terms of pure noise suffered in a given lapse of time, say 1 year. The exponent in the first defining equality of $\mathscr{T}_1$ illustrates that we can construct a dynamic model $\mathscr{T}$ endowed with a general time set T (containing 0 as its initial point) by using $\mathscr{T}_1$ as a generator. Rescale our original binary model by a factor $h = 1/n$, where n is, without loss of generality, an integer, in such a way that the null expectation and unitary standard deviation are preserved under n autoconvolutions when increments are independent. Accordingly, define the family of approximations

$$B_{h,1} = h^{1/2}B * \cdots * h^{1/2}B = B_h * \cdots * B_h.$$

The central limit theorem states that asymptotically, as $h > 0$ tends to zero, B_h converges in probability to a standard normal distribution. This logic can be extended to more arbitrary time sets $\{0, 1, \cdots T\}$, giving rise to a Brownian motion (BM) $B_t = \lim_{h \downarrow 0} B_{h,t}$.

The election of the binomial tree $\mathscr{T}_1$ as the generator of the Brownian motion is immaterial because this process acts as a general class of basin attractors of more general distributions with finite variance that is within the Euclidean space $L^2(\Omega, \mathscr{F}, P)$. The BM satisfies the conditions of stationarity and permits a parsimonious integral representation of dynamic process that can be thought of as generated by static models defined by states of the form $X = a + \sigma B$, as in Eq. (9.5). The constants a and σ equal, respectively, the expected value and the variance of X. Alternatively, we may be interested in multiplicative decompositions of the form $Y = e^X$. By virtue of the central limit theorem, the Gaussian example becomes fundamental cornerstone within the class of stationary processes with finite second moments $L^2(\Omega, \mathscr{F}, P)$. The boundaries of macroeconomic research are constructed under this paradigm.

9.5.1   The Koopmans' Aggregator

Koopmans (1960) formalized the concept of dynamically consistent preferences by means of a representation which can be viewed as the dynamical counterpart of the static aggregator defined in Sect. 9.3.1, which is derived trivially as the outcome of a stationary path with a constant continuation value. In general, for a discrete-time model with $0 \leq t \leq T \leq \infty$, we have

$$V_t := V\left(c^t\right) := f\left(u\left(c_t\right), V\left(c^{t+1}\right)\right), \tag{9.35}$$

where $f\left(\cdot\right)$ stands for the dynamic aggregator, and preferences are defined over consumption paths $c^t = \{c_n\}_{n=t}^T$. Expression (9.35) generalizes the linear version $V_t = u\left(c_t\right) + \beta V_{t+1}$, widely applied in macroeconomic applications, associated with the time-additive utility

$$V_t = \sum_{n=t}^T \beta^t u\left(c_n\right), 0 \leq T \leq \infty. \tag{9.36}$$

The aggregator defined in (9.35) is imposed under a condition of stationary preferences that allow a preference representation based on a contemporaneous utility index. This decomposition is clearly nonunique, since preferences remain unchanged under any strictly increasing transformation of the continuation value: let g be any such transformation, so that we can rewrite the continuation value as $\tilde{V} = g\left(V\right)$ and (9.36) as

$$\tilde{V}_t = \tilde{f}\left(u_t, \tilde{V}_{t+1}\right) = g\left[f\left(u_t, g^{-1}\left(\tilde{V}_{t+1}\right)\right)\right]. \tag{9.37}$$

When c is scalar and preferences are monotonic, we can measure utility in terms of the consumption index through the transformation $\tilde{V} = u^{-1}\left(V\right)$.

Patterns of intertemporal preference are captured in the aggregator: it is straightforward to see that the marginal rate of substitution between periods t and $t+1$ (the stochastic discount factor, abbreviated SDF) is given by

$$\text{MRS}_{t+1}^t = \frac{f_{2,t} f_{1,t+1}}{f_{1,t}}. \tag{9.38}$$

The derivative with respect of the second argument of the continuation value at time $t, f_{2,t}$, is the discount factor, which, as deduced form (9.37), is invariant under increasing transformations. For instance, in the standard model (9.36), the SDF (9.38) equals, in the linear case,

$$M_h^t = \frac{\beta^h u'\left(c_{t+h}\right)}{u'\left(c_t\right)}. \tag{9.39}$$

Two extensions of this kind of intertemporal preferences can be found in the literature. Uzawa (1968) offered a continuous version of the linear case $f\left(u, v\right) = u - \rho\left(u\right) v$ with risk, so that the corresponding utility functional adopts the von Neumann-Morgenstern representation

$$V\left(c^t\right) = E_t\left(\int_t^\infty e^{-\int_t^s \rho(c_\tau)d\tau} u\left(c_s\right) ds \middle| \mathscr{F}_t\right), \tag{9.40}$$

where $c^t : \mathbb{R}_+ \rightarrow C$ belongs to the set of admissible consumption streams as of time t and $\mathscr{F}_t$ stands for the information set dated at t. The approximation is made clear if we put $\beta^h = e^{-\rho h}$. The discrete model with a time window $h \geq 0$ becomes

$$V_{t+h} - V_t = h u_t + \left(e^{-\rho h} - 1\right) V_t,$$

where u_t stands for current utility. Taking limits as h approaches zero gives us the dynamic equation $\dot{V} = f(u, V) := u - \rho V$.

The second extension takes into account the Koopmans' axiomatization which leads to the aggregator representation: let $\precsim_t$ denote preferences over c^t. These are the three conditions needed to derive (9.35):

Axiom (History Independence). $\precsim_t$ *do not depend on* $c^s, s < t$.

Axiom (Future Independence). $\precsim_t$ *do not depend on* $c^s, s > t$

Axiom (Stationarity). *For any* $t, \precsim_t := \precsim^*$.

Epstein and Zin (1989) relax the axiom of future independence, giving rise to an aggregator lacking a current utility index of the form $V' = f(c, U)$, which is necessarily equivalent to (9.35) formulation when c_t is scalar.

9.5.2 Random Settings

Expression (9.40) involves a conditional expectation. It is natural to assume that news arrive as time goes by, so we need to be more accurate on this issue. Dynamic random settings are usually formulated by endowing $(\Omega, \mathscr{F})$ with a filtration $\mathbb{F} = (\mathscr{F}_t : t \in T)$ to model the process of information updating as new public events as well as idiosyncratic signals are realized. In continuous time, $\mathbb{F}$ is typically made up of a right continuous sequence[10] of increasing sigma algebras $\mathscr{F}_t$. Once a model P is introduced in the analysis, an additional technical hypothesis states that the set of null sets of $\mathscr{F}$ is contained in $\mathscr{F}_0$ and thus in every $\mathscr{F}_t$. A decision tree is represented by a tree $\mathscr{T} = \Omega = S^T$, which is generally characterized by a set of *nodes* related themselves by ordered pairs (*edges*) uniquely defined. Without loss of generality, we fix an initial condition at $X_0 = x \in S$. We consider that the tree ends at a stopping time[11] $\tau : \Omega \rightarrow T$ which can possibly be infinite.

If S and T are finite sets, a probability model boils down to a multinomial tree. In this case, given preferences, resources, and the market conditions, a model is a probability measure P defined over the whole set of terminal nodes or trajectories $\sigma \in S^T$. Extensions to continuous versions of this model of either S or T can be

---

[10] In discrete settings, the right continuity assumption holds trivially.

[11] For it to be a stopping time, it needs to be progressively measurable, that is, $\mathscr{F}_t$-measurable at each $t \in T$.

applied with additional regularity conditions. The dynamic structure of the model imposes a natural filtration of information made up of a sequence of increasingly finer σ-algebras $\mathbb{F} = (\mathscr{F}_t)$. This object suffices to complete the description of random systems defined in the previous subsection. Formally, such description is established by means of a filtered space $(\Omega, \mathscr{F}, \mathbb{F}, P)$. $P_t : \mathscr{F}_t \to [0, 1]$ is obtained by using the Bayesian updating rule: associate to each path a σ its history of length t, $\sigma^t \in \mathscr{F}_t$. If $P(\sigma^t) > 0$, we can define a stochastic process of conditional probabilities

$$P_t(\sigma) = P(\sigma|\mathscr{F}_t) = \frac{P(\sigma)}{P(\sigma^t)}. \tag{9.41}$$

The conditional expectation of a stochastic process X_t is a stochastic process defined as the projection of X_t on the subspace of random variables generated by $(\Omega, \mathscr{F}_t, P)$. In symbols, for any $t \geq 0$ and $h > 0$, we define

$$E_t X_{t+h} = E^{P_t} X_{t+h}.$$

The projection property is known as the law of iterated expectations and implies that alternatively, if we impose the terminal condition $P = P_\infty$, we can recover a model from a prespecified set of conditional probabilities: for any finite interval $[0, t]$, establish a finite partition of lengths h_s, possibly unitary and possibly random and progressively measurable stopping times: let the end points of this partition be denoted by $t_s > 0$. Put $t = t_k$,

$$P_{t-h_k}(\sigma) = \frac{P(\sigma)}{P(\sigma^{t-h_k})},$$

if P_h^t stands for the h-period ahead forecast that allows us to write the forward equation as follows:

$$P_t := \frac{P(\sigma)}{P(\sigma^{t_s})} = \frac{P(\sigma)}{P(\sigma^{t-h_k})} \cdots \frac{P(\sigma^{t_s+1})}{P(\sigma^{t_s})} =: P_{h_k}^{t-h_k} \cdots P_{h_s+1}^{t_s} \cdots P_{h_1}^0 = \prod_{s=1}^{k} P_{h_s}^{t_{s-1}}. \tag{9.42}$$

This equality corresponds with a Markov property (semigroup structure) when the family $v(\sigma) := P_h^t$ does not depend on t. Otherwise the process would follow trivially the Markov property by including the time in the state space. By doing so, we can rewrite the last expression without the reference point, included in the space state if necessary, so

$$P_t = \prod_{s=1}^{k} P_{h_s} \tag{9.43}$$

This forward equation has as a counterpart established in terms of the state space and makes use of the standard decomposition between a predictable and a pure stochastic part. The Kolmogorov extension theorem (e.g., Oksendal 2003) grants the existence of a continuous stochastic process that embeds the family of probability distributions described in (9.43), as long as it is invariant under permutations. The predictable component a_t equals the conditional expectation at t of the expected change of the state, namely,

$$\Delta X_t = a_t(X_t) + \sigma_t(X_t)\,\varepsilon_t, \quad X_0 = x \tag{9.44}$$

where X_t belongs to S and ε_t is a noise term. The function a_t is called the drift parameter and represents the local mean, whereas σ_t is a parameter of scale. We assume that $\mathscr{F}_t = \sigma(\varepsilon^t)$ where ε^t is the history of shocks as of date t. The following axioms define stationarity of the model allowing parsimonious representations that often work as operative restrictions necessary to match theory and data in a meaningful way.

Axiom (S1 Independence). $E(\varepsilon_t|\mathscr{F}_{t-1}) = 0.$

Axiom (S2 Stationarity). ε_t's are identically distributed.

Equation (9.44) admits backward integral solution of the form

$$X_t = x + \sum_{j=0}^{t} a_j(X_j) + \sigma_j(X_j)\,\varepsilon_j. \tag{9.45}$$

Expressions (9.43)–(9.45) are equivalent representations. The following example is a particular case that has become standard far beyond economic applications.

Example 5 (ARIMA (p, d, q)). This family of linear processes can be represented as

$$\Phi(L)\,\Delta^d Y_t = \mu + \Psi(L)\,\sigma_t\varepsilon_t, \tag{9.46a}$$

$$\Phi(L) = \sum_{i=0}^{p} \phi_i L^i, \Psi(L) = \sum_{j=1}^{q} \Psi_j L^j \tag{9.46b}$$

with $t = 0, 1, 2, \ldots$ where (p, d, q) are positive integers and ε_t is a standard noise term iid distributed which is factorized by a scale factor σ_t. Under conditions of stationarity and invertibility, (resp.) the polynomials Φ and Ψ have roots outside the (closed) unitary circle. To avoid trivial computations, we can assume that both characteristic polynomials have no common roots. Invertibility means that we can express (9.46) as

$$X_t = \Delta^d Y_t = \frac{\mu}{\Phi(1)} + \sum_{i=0}^{\infty} \theta_i \sigma_{t-i}\varepsilon_{t-i}, \tag{9.47}$$

which is a special case of (9.46) if $P_0 = 1_{\{x\}}$. The solution can be restated: let $\lambda_t := \rho^t$,

$$X_t = \lambda_t x + (1 - \lambda_t) \frac{\mu}{\Phi(1)} + \sum_{i=1}^{t} \theta_i \sigma_{t-i} \varepsilon_{t-i}. \tag{9.48}$$

This equation states the law of motion of the process starting at x, and its ex ante convergence toward the stationary state given by $\lim_{t \to \infty} EX_t = \mu/\Phi(1)$. The term $M_t = \sum_{i=0}^{t} \theta_i \varepsilon_{t-i}$ satisfies trivially the martingale property

$$E(M_t | \mathscr{F}_{t-1}) = M_{t-1}.$$

The martingale term in (9.48) is a purely stochastic additive term that encodes the permanent changes to the system.

Alternatively, we can express (9.48) by successive increments[12]: put $a = \Phi(1)$ and write

$$X_t = x + \sum_{i=0}^{t-1} \Delta X_i = x + \sum_{i=0}^{t-1} (\mu - aX_i + \sigma_i \varepsilon_i) = x + t\mu - a \sum_{i=0}^{t-1} X_i + \sum_{i=0}^{t-1} \sigma_i \varepsilon_i. \tag{9.49}$$

This solution can also be written in continuous time, as noted earlier. The deterministic version captures, as one might expect, the drift-driven component of the general model:

$$\dot{X} = -bX + \mu, \tag{9.50a}$$

$$X_0 = x. \tag{9.50b}$$

This system has a solution which coincides with the terms of (9.49) excluding the martingale. Abusing slightly of notation, define $\lambda(t) = e^{-bt}$, so that the integral solution to the dynamic system (9.50) can be expressed as

$$X(t) = \lambda(t) x + (1 - \lambda(t)) \frac{\mu}{b} = x + (1 - \lambda(t)) \left(\frac{\mu}{b} - x \right) = x + \int_0^t (-bX(s) + \mu) \, ds.$$

The first equality is the general solution of the dynamic system and the second equality follows from rearrangement. The third equality states the fundamental theorem of calculus. The last integral can be solved so that

$$X(t) = x + t\mu - b \int_0^t X(s) \, ds. \tag{9.51}$$

---

[12]Note the relation $\Delta X_t = X_t - X_{t-1} = \Delta^d Y_t - \Delta^d Y_{t-1} = \Delta^{d+1} Y_t$.

The equivalence between (9.49) and (9.51) is made apparent through the relation $1 - a = e^{-b}$. Introduce a variable time window $0 < h \leq 1$, and define the corresponding increments by $\delta_h X_t := X_{t+h} - X_t$, so that the scaled system (9.51) becomes

$$\delta_h X_t = \left(e^{-bh} - 1 \right) X_t + \mu h. \tag{9.52}$$

Now dividing this expression by h and taking limits as $h \to 0$, we recover (9.50). In the next section, we will see that those processes that admit decompositions between a forecastable element and a martingale correspond roughly speaking to the broadest family of integrable processes known as semimartingales. As it turns out, the noise term $\varepsilon_{t,h} = B_{t+h} - B_t = \delta_h B_t$ that should be added to the system (9.52) has variance h so that the sum of h^{-1} copies of independent random variables has unitary variance[13] as required.

AR models of higher order can be resolved keeping in mind these considerations. For instance, let X_t follow an AR(2) process so that we can write the characteristic polynomial $\Phi(L)$ in terms of its roots $\lambda_1, \lambda_2 \in \mathbb{C}$, which obeys the stationarity condition $|\lambda_i| > 0$ [see Samorodintsky and Taqqu (1994)]

$$\Phi(L) = \sum_{i=0}^{2} \phi_i L^i = (\lambda_1 \lambda_2)^{-1} (L - \lambda_1)(L - \lambda_2)$$

and

$$\Phi(L)^{-1} = \frac{\lambda_1 \lambda_2}{\lambda_1 - \lambda_2} \left(\frac{\lambda_1^{-1}}{1 - \lambda_1^{-1} L} - \frac{\lambda_2^{-1}}{1 - \lambda_2^{-1} L} \right) =: \sum_{n=0}^{\infty} \zeta_n L^n.$$

This expression allows us to recover the coefficients of the Taylor expansion of $\Phi(L)^{-1}$:

$$\zeta_n = \frac{\lambda_1 \lambda_2}{\lambda_1 - \lambda_2} \left(\lambda_1^{-n-1} - \lambda_2^{-n-1} \right).$$

When Φ has complex conjugate roots of the form $\rho e^{\pm i\mu}$, it can be seen that

$$\zeta_n = \rho^{-n} \frac{\sin(n+1)\mu}{\sin \mu}.$$

---

[13]For instance, suppose that time units (t) are measured in years, so 1-day window corresponds to $h = 1/365$.

9.5.3 Recursive Utility

In order to illustrate how ambiguity can be embedded in a dynamic framework, consider that this economy is populated by a representative agent (Guidolin and Rinaldi 2013, Sect. 2.4) whose discount rate is zero and with a risk-neutral utility index, so that

$$V(c) = \min_{P \in \mathscr{P}} E^P c.$$

If this basic structure of the environment is a common knowledge, the natural setup at $t = 2$, $P = P_2$ assigns a uniform probability of $1/4$ to each of the four terminal nodes. A model is thus fully described by the conditional probability in period 1, according with the Bayes' rule (9.39). The whole structure of the model can be comprised in a parameter of autocorrelation ϕ, with $|\phi| \leq 1$, if we assume

$$P_1(\sigma) = P_1(X_1, X_2) = \frac{1}{2}(1 + \phi) 1_{\{X_1 = X_2\}} + \frac{1}{2}(1 - \phi) 1_{\{X_1 \neq X_2\}}.$$

Consider an asset $1_{\{X_2=1\}}$ which pays one unit of consumption in the state $X_2 = 1$. Note that this asset can be built by a trivial combination of two Arrow-Debreu contingent assets. It can be seen (Backus et al. 2004) that recursive and ex ante valuation differs, in contrast with the standard consumption CAPM with dynamically consistent preferences (no trade theorems) under dynamically consistent preferences provided that there exists a full range of contingent claim commodities. Heuristically, intertemporal preferences are ordering relations which do not depend on t (we provide a more formal definition in the next subsection). Assume that σ^t differs from $\sigma^{t'}$ only from a given random time τ. Dynamically consistent preferences satisfy $\sigma \succsim_\tau \sigma' \Leftrightarrow \sigma \succsim_t \sigma$ for any $t \leq \tau$. Dynamic consistency is closely related to consequentialism

Axiom (Consequentialism). $\sigma \succsim_t \sigma' \Rightarrow \sigma \succsim_s \sigma'$ $t \geq s$ $\forall \sigma, \sigma' \in \mathscr{F}_t$.

Axioms of independence, constrained to $\Pi(S)$ and consequentialism, are equivalent conditions of dynamic consistency. Therefore, under ambiguity, dynamic consistency appears to be a quite restrictive assumption. In addition, independence and continuity imply consequentialism and dynamic consistency [Karni and Schmeidler (1991)], a result that links both axioms within a general Bayesian framework.

A remarkable feature of random setups is that the continuation value associated to an arbitrary act behaves as a stochastic process. Since the widest class of integrable process is semimartingales, it is natural to construct models within this family of decomposable processes. Further stationarity constraint restrictions will allow to consider tractable representations capable of providing with empirical conclusions.

Based on these considerations, Epstein and Zin (1989) deal with recursive representations of preferences expressed in terms of the certainty equivalent of the one-period ahead continuation value. Let $V_t = V_t^c$ and write the certainty equivalent of Sect. 9.2.3 as $m(V_t)$. Let $P_t = P_{|\mathscr{F}_t}$ and $m_t := m^{P_t}$

$$V_t = f(c_t, m_t(V_{t+1})). \tag{9.53}$$

This formulation generalizes SEU preferences while keeping the hypothesis of dynamic consistency. The aggregator f combines current utility stemming from consumption and the CE corresponding to the continuation value. A particular case is provided by the CES aggregator with a constant elasticity of substitution $\rho = (1 - \alpha)^{-1}$:

$$V(c, m) = (c^\alpha + \beta m^\alpha)^{1/\alpha}, \tag{9.54}$$

where $0 \leq \beta < 1$ $(0^0 \equiv 1)$ is the discount factor. Expected utility corresponds to the certainty equivalent $m_t = u^{-1}E_t u$ and a linear aggregator (9.36): $V_t = u_t + \beta E_t V_{t+1}$. This equality can be solved forward giving rise to the well-known time-additive specification (see also representation 9.40):

$$V_t = E_t \sum_{n=t}^{\infty} \beta^n u(c_n), \tag{9.55}$$

sometimes it is convenient to express value in terms of consumption units, as mentioned in Sect. 9.3. In the isoelastic case (9.54), the certainty equivalent obeys the equality

$$m_t(c) = (E_t c^\alpha)^{1/\alpha}$$

$$V_t := u^{-1}(U_t) = (u_t + \beta E_t U_{t+1})^{1/\alpha} = (u_t + \beta (m_t(V_{t+1}))^\alpha)^{\frac{1}{\alpha}} =$$

$$\left(c_t^\alpha + E_t \sum_{n=t+1}^{\infty} \beta^n c_n^\alpha\right)^{1/\alpha} = \left(E_t \sum_{n=t}^{\infty} \beta^n c_n^\alpha\right)^{1/\alpha}.$$

An alternative fundamental model was provided by Kreps and Porteus (1978) and extended by Epstein and Zin in dynamic contexts (1989). The certainty equivalent $m_t = u^{-1}E_t u$ remains unchanged, and the aggregator, measured in consumption units, takes the form

$$V(c, m) = (c^\alpha + \beta m^\delta)^{1/\alpha},$$

which has been introduced in Sect. 9.3.2 [see also Eq. (9.21)].

9.6 Extensions

Before concluding, we explore three further generalizations: an introduction to continuous-time frameworks, the notion of discounting, and finally, stationarity with an empirical application to financial markets.

9.6.1 The State in Continuous Time

Analogous decompositions to those considered on previous sections can be considered in continuous time as well. The BM describes the pure stochastic component of the process. Discrete models can be encoded in continuous-time settings, and conversely, continuous-time models can be constructed from simpler settings via the concept of generator that we have introduced in Sect. 9.5. Stochastic processes of interest are $\mathscr{F}_t$-measurable random vectors evolving over time. The condition of $\mathscr{F}_t$-measurability defines the optional or progressively measurable stochastic processes. The functions a_t, $and\sigma_t$ are measurable and satisfy two technical conditions that grant existence of a continuous representation ($h \downarrow 0$) of Eq. (9.44): contraction and lack of explosion times, which under the Markov property can be stated as a single one, in terms of the existence of a positive constant D such that

$$|a(x) - a(y)| + |\sigma(x) - \sigma(y)| \leq D|x - y|. \tag{9.56}$$

By identifying B with B_1, static models can be extended to a dynamic framework whose phase state evolves according to

$$dX_t = a_t dt + \sigma_t dB_t. \tag{9.57}$$

$$X_0 = x \in S, \text{ given.}$$

This equation can be stated for vector processes of multiple dimensions. In the general case, $X_t = (X_{1,t}, \ldots, X_{d,t})$ and $B_t = (B_{1,t}, \ldots, B_{s,t})$ are characterized by two predictable processes: $a_t = a_t(X)$, valued in $\mathbb{R}^d$, is known as drift, and it may be interpreted as the local direction of change or the transient component in the flow of the system. The volatility is captured by $\sigma_t = \sigma_t(X_t)$.

The backward solution of this equation admits the Ito representation

$$X_t = x + \int_0^t a_s ds + \int_0^t \sigma_s dB_s. \tag{9.58}$$

This formula highlights a generic observer's ability to split the process into a forecastable and a pure stochastic component which are the second and last elements of the right hand side of (9.58). Another general interpretation is that of a system subject to a random flow, such as a river.

9.6.2 Continuous-Time and Endogenous Discounting

The generalization of the ideas presented so far to a continuous model requires an adequate accommodation of the control space. The set of admissible acts $\mathscr{A}$ is a closed subset of a Banach separable (Lattice) space. Let preferences be defined over the set of admissible consumption paths $t \mapsto c_t$ whose continuation value $V_t = V_t^c$ is given, at any date $t \in [0, T]$,

$$V_t = E_t \int_t^T e^{-\int_t^s \rho_\tau d\tau} u\left(c_s, \rho_s\right) ds, \tag{9.59}$$

for a feasible discount factor ρ. The most remarkable fact of these preferences is the separability of the utility index with respect to the cumulative factor. This property defines V_t as recursive utility, which is a special case of variational utility, not necessarily characterized by non-separable indices. Geoffard (1996) shows that the linear preference aggregator defined is the Legendre transform of the function u, that is, $Df(c, \cdot) = Du^{-1}(c, \cdot)$ and

$$f(c, m) = \sup_\rho \left\{ u(c, \rho) + \rho m \right\}.$$

The first-order condition of this equation $u_\rho = m$ defines implicitly a solution $\rho = \rho(c, m)$ and admits a recursive formulation of the continuation value in terms of the drift:

$$\lim_{h \downarrow 0} \frac{V_{t+h} - V_t}{h} = -f(c_t, V_t). \tag{9.60}$$

The extension to uncertain decisions is immediate. Consider the standard discrete-time model ($h = 1$); for each time length $h = 1/n > 0$, we can define the family of CE $m_{h,t}$. Under the rescaled model, the phase state is assumed to evolve according to an extended version of (9.45):

$$X_t = x + \sum_{j=0}^{nt} a_j(X_j) + \sigma_j(X_j)\varepsilon_j. \tag{9.61}$$

Assuming (local) differentiability one can compute, from (9.53) and assuming $m_{h,t}(V_t) = V_t$,

$$\frac{d}{dh} m_{h,t}(V_{t+h})\bigg|_{h=0} = \lim_{\varepsilon \downarrow 0} \frac{m_{h,t}(V_{t+h}) - V_t}{h} = -f(c_t, V_t). \tag{9.62}$$

The aggregator expresses the forecastable component of the continuation value. The unpredictable component imposes further conditions in the noise structure in order to apply standard decompositions in a Hilbert (Euclidean) space of the form

$L^2 (\Omega, \mathscr{F}, P)$. The common practice among scholars is to restrict the action space to the family of square integrable processes under the norm $\|\cdot\|$ defined in the act space $\mathscr{A}$, that is,

$$E \int_0^T \|c_t\|^2 < \infty.$$

The following assumption is very much connected with the differential analysis made in Sect. 2.5.

As in Epstein and Duffie (1992), the Gateaux derivative of the CE at V in the direction z with compact support is assumed to exist:

$$Dm(V, z) = \lim_{h \downarrow 0} \frac{m(V + hz) - m(V)}{h}.$$

The CE is said to be smooth at certainty if for any $x \in \mathbb{R}$, there exists a twice continuously differentiable real-valued function $K(x, \cdot)$ such that

$$Dm^P (1_x, z) = \int K(x, y) \, dP^z (y) = E^P K(x, z).$$

where $P^z (A) = P(z \epsilon A)$. $K(x, \cdot)$ is known as the local gradient of the CE m. By definition, $Dm^P (1_x, 1_y) = y$, which implies $K(x, y) = y$ and[14] $K_y (x, \cdot) = 1$. For the expected utility representation, assume u to be twice differentiable, $m = u^{-1} Eu$ and by direct calculations (Epstein and Duffie, 1992), find

$$K(x, y) = \frac{u(x)}{u'(y)}.$$

Assume the noise is generated by a BM, so that $\mathscr{F}_t = \sigma(B^t)$. Suppose the existence of an economic agent who faces an optimization problem along the interval $[0, T]$ with zero terminal condition. It is natural to suppose that the utility process follows a diffusion of the form

$$dV_t = -\mu_t dt + \sigma_t dB_t \tag{9.63}$$

with a terminal condition[15] $V_T = 0$. From (9.62), the intertemporal aggregator measures the local expected variation of the CE at V_t in its own direction of change, namely,

---

[14]The subindex in K stands for partial derivatives.

[15]Imposing alternative terminal conditions such as $V_T = U_T (X_T)$ for some given function U_T does not change the argument. Neither does it so to assume that T is a stopping time $T : \Omega \to [0, \infty]$. The extension to an infinite control problem needs further regularity conditions discussed in Appendix C of Duffie, Epstein (1992), written by Skiadis.

$$-f_t^c = -f\left(c_t, V_t^c\right) = Dm\left(V_t^c, \frac{d}{dh}\left(V_{t+h}^c | \mathscr{F}_t\right)\Big|_{h=0}\right).$$

Using continuity of m, we can interchange the defining limits of the two derivatives that appear in the right hand side of the last equation. Let

$$\Sigma_t := \sigma_t \sigma_t' = \frac{d}{dt}[V]_t$$

by applying and equaling coefficients in the Ito Lemma, one finds

$$\mu_t^c = \mu_t\left(V_t^c\right) = f_t^c + \frac{1}{2}A\left(V_t^c\right)\Sigma_t^c \tag{9.64}$$

$A(x) := K_y(x, x)$ stands for the variance multiplier of the certainty equivalent. Backward integration of (9.63) gives, for any $t \in [0, T]$,

$$E\left(V_T | \mathscr{F}_t\right) = V_t + E\left(\int_t^T \mu_s ds | \mathscr{F}_t\right).$$

The terminal condition allows a forward representation of the continuation value

$$V_t = -E\left(\int_t^T \mu_s ds | \mathscr{F}_t\right).$$

9.6.3   Stationarity

We have discussed so far the forecastable component of the continuation value, and other economic processes, and introduced some keys to analyze its properties. This exercise is relevant insofar as it captures one of the fundamental elements of value. Nothing instead has been said about the characterization of the pure stochastic component of processes. Before concluding, let us observe that the noise structure implied by the BM, which is by nature continuous, can be completed by adding an exogenous discontinuous structure described by a Poisson distribution.[16] Both classes of distributions are stable by convolution (sum of random variables), and both have finite variance.

There is another class of stable distibututions, which display fat tails and consequently do not have a convergent variance when the number of events spanning the whole sample space diverges to infinite. In order to avoid technical complications, we can deal with integrable process, so they have a finite mean and the concept of martingale is preserved. This concept is very useful because

---

[16] See, for instance, Hansen (2012).

of its connection with that of permanent shock, which is essential to discuss asset pricing. This means that we can extend the space of processes to $L^\alpha (\Omega, \mathscr{F}, P)$ for $1 \leq \alpha \leq 2$.

Self-similar processes are invariant in distribution under a homothetic scaling of space and time (Samorodintsky and Taqqu 1994, Chap. 7). The interest of this kind of processes relies on their practical ability to deal with complex relations. In particular, it is possible to extend the central limit theorems and thereby the concept of stationarity in models displaying long-range correlations as observed in financial markets as well as in a vast array of complex systems.

A process X_t is said to be self-similar with index $H > 0$ if for any $a > 0$, the finite-dimensional distributions of the process X_{at} and that of a^H coincide. Examples of self-similar processes are the Brownian motion ($H = 1/2$) and the α-Lévy-stable motion ($H = 1/\alpha > 1/2$). Both process can be represented by a random measure $\overline{X}$ defined over the Borel space $(T, \mathscr{B}_T)$ that satisfies

$$X_t = \int_0^t \overline{X} (ds) = \overline{X} [0, t] \qquad (9.65)$$

at any date t. More general stable and self-similar processes can be constructed by means of representations of the form

$$X_t = \int_{-\infty}^{\infty} g_{\alpha, H, t} (s) \overline{X} (ds).$$

As an example, Fig. 9.1 depicts some members of the family of functions $g_t (s) = \ln |t - x| + \ln |x|$, for different values of t, which gives rise to the log-fractional stable motion.

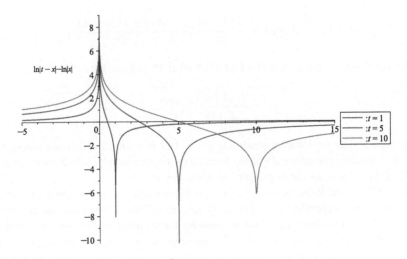

Fig. 9.1 Stationarity and long memory. Source: own

Parameters α and H stand, respectively, for the coefficients of stability and similarity. A Fractional Brownian Motion (FBM), is a Gaussian process $B_{H,t}$ with zero mean. Fractional white noise is defined to be $B_{H,t+k} - B_{H,t}$ (if $H = 1/2$ the noise reduces to iid white noise). Fractional white noise exhibits long-range memory; that is, as the lags k tend to infinity, the asymptotic behavior of the autocorrelation structure of the process is given by

$$E\left(\Delta B_{H,t+h} \Delta B_{H,t} | \mathscr{F}_t\right) = 2H\left(2H - 1\right) k^{2(H-1)}$$

whose variance and covariance are given by

$$EB_{H,t}^2 = t^{2H}$$

$$EB_{H,t} B_{H,s} = 1/2 \left(t^{2H} + t^{2H} - |t - s|^{2H}\right)$$

Processes with lower H have a greater volatility than those processes with a higher H.

The fractional differenced ARIMA processes with parameters (p, d, q) or equivalently, FARIMA processes take the general form

$$\phi\left(L\right)\left(1 - L\right)^d X_t = \Psi\left(L\right) \varepsilon_t, |d| < 1/2$$

Gil Martin and Rege (2010) fit

$$Y_t = \left(1 - L\right)^d X_t = \phi\left(L\right)^{-1} \Psi\left(L\right) \varepsilon_t.$$

A power series expansion of the fractional operator gives

$$\left(1 - L\right)^d = 1 - dL + \frac{d\left(d - 1\right)}{2!} L^2 + \frac{d\left(d - 1\right)\left(d - 1\right)}{3!} + \cdots$$

The autocorrelation function of a FARIMA $(0, d, 0)$ process tends to

$$\frac{\Gamma\left(1 - d\right)}{\Gamma\left(d\right)} |t|^{2d-1}$$

for large n. The stationarity condition $|d| < 1/2$ provides a relationship between the differencing parameter d and the long-memory parameter λ which allows one to infer the lack of long-range memory in when $d = 0$.

Gil Martin and Rege (2010) have used two approaches to investigate the presence of long-range dependence in the daily returns of the Portuguese index PSI20 (1992–2010). Our analysis found an estimate for the Husrst exponent greater than one half, suggesting the presence of long-range dependence in the stock prices. This suggests that Portuguese financial markets do not allocate investment funds efficiently and that they generate endogenously critical phenomena (persistence).

The quadratic trend produces higher values of the Hurst exponent, although they may not be an appropriate fit, as concluded after an alternative semi-parametric approach corroborates our preliminary results.

The bulk of applied research about long-term dependence has not yet shed sufficient light on their underlying causes. In addition, to the eye, short-memory processes appear to be hardly distinguishable from a white noise. A fractal or a multi-fractal series suggests the action of interacting systems generating positive feedbacks. This problem is recurrent in macroeconomics, especially since the 1980s, where systemic interactions between aggregate supply and demand called for new tools of analysis. During 'normal' periods in which those systems operate rather independently, the 'low-scale' information that becomes operative is that giving rise to short memory. Suddenly, when signs of distress occur, high-range waves of information will eventually dominate the markets, and time series displays long-term dependence. Financial markets then collide with the real side of the economy, as new propagation mechanisms trigger, producing unknown events prior to the process of revulsion.

Figure 9.2 shows the movement of the Hurst exponent 3500 days with a 1-day moving window under both linear and quadratic trends to estimate the exponent. Both settings exhibit similar behavior. The Hurst exponent based on the quadratic trend lies above 0.5 implying an unequivocal presence of long-memory asset returns which does not seem compatible with any specification of noise based solely on Brownian motions.

9.7 Conclusion

The requirement of sound microeconomic principles imposes a revision of concepts that affects substantially the notion of economic action itself, once ambiguity is accounted for. The development and evolution of macroeconomic theory ultimately continue, and it will certainly continue to be entirely determined by events that take place within the flow of circular income. Overall, this flow glues together those processes which intervene actively in the procurement of human and social needs. Economic interactions are generated endogenously and, under given conditions, they can drive the course of actions. Under plausible circumstances, the epistemic realm is suggested to enlarge the space state over which fundamentals are defined.

One of the most relevant developments of modern history has been the transnational economic development fueled by the division of labor and the subsequent increase in the average income. This process has not been engined neither by primary sources of energy or matter nor by the intensive use of human labor, or at least data shows us that this need not be the case. On the other hand, the evidence shows the presence of high-order ambiguity which highlights a price system unable to internalize production costs. In addition, the intertemporal distribution of income is the link between production and consumption. Evidence shows that consumption distribution, as measured in units of primary energy (see Sect. 9.1), is highly unequally distributed, at a geographical as well as an individual level, and there is also evidence of a higher tan unitary income elasticity of appropriation of primary energy. This problem, known as the Jevons' paradox, seems at odds with the observation that the intensity of physical factors does not explain growth unless we are willing to introduce additional variables in standard SDGE models. This work provides insights as long as the reduced state space S must be expanded so as to include an epistemic realm which is captured by the original universe defined at the beginning of Sect. 9.2. Let the economic and financial system be denoted, respectively, by the symbols $e, f \in \Omega$. The circular flow of income can be expressed as a net of relations (transactions): $CFI \equiv e \leftrightarrows f$. Let $X = (A, Z)$, where A stands for technology, a given set of transformation techniques of resources (raw materials, primary energy, and information) $Z = (M, E, \Phi)$. H is the entropy and establishes a recognition of the role of the second law of thermodynamics in the production activity with negative external effects. Potentially, however, there is extensive evidence in support of positive externalities of information, at least within

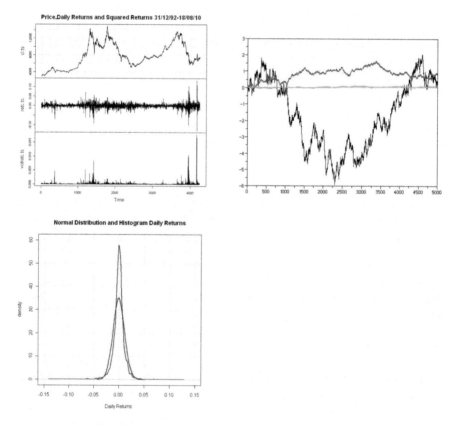

Fig. 9.2 Distribution of daily returns PSI20 (31/12/1992–12/8/2010). Source: Gil Martín and Rege (2010)

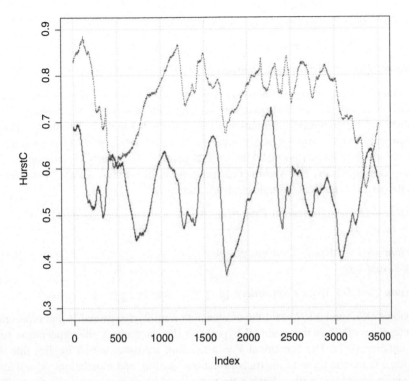

Fig. 9.3 Hurst exponent various durations 1-day moving window. Source: Gil Martín and Rege (2010)

the space of possible scenarios of sustainable growth. The right panel represents the flow of final goods and services. The process between production and consumption determines the dynamics of wealth distribution. The price system is bounded below and above by the ideal extreme cases of perfect competition (PC) and monopoly (mono). While facts support the view that production, the initial part of the economic system, is mostly dependent on flows of information rather than the intensive use of production factors, this does not seem to be the case when it comes to the final part of the flow, namely, consumption. This observation suggests that if it were not the case, some of the complexities observed in the economic realm would not have a considerable impact on allocation of resources, assets, and prices.

Appendix A The Expected Utility Theory: Axiomatization

Let $a, b, c \in \mathscr{A}$ convex and $\lambda \in [0, 1]$ and denote for simplicity the compounded act $(a, b)_\lambda := \lambda a + (1 - \lambda) b$. Assume a preference relation $\succsim$ on $\mathscr{A}$ such that

Axiom (ORD Partial Order). $\succsim$: *complete, nondegenerate, and transitive.*

Axiom (IND Independence). $a \succsim b \Leftrightarrow (a, c)_\lambda \succsim (b, c)_\lambda$ *for any* $0 \leq \lambda \leq 1$.

Axiom (IND* Certainty Independence). *IND restricted to the class* $c = \delta_x,\ x \in \mathscr{C}$.

Partial order allows to classify a set of lotteries in equivalence classes. Transitivity avoids cyclic decisions that may give rise, for instance, when customers and sellers hold asymmetric information about the product quality. The independence axiom allows the compounding of lotteries. To avoid unbounded functionals, a sort of continuity must be imposed which will depend on the underlying topology of the set of consequences. The Archimedean property is the weakest version of continuity as it does not require any topological structure on $\mathscr{C}$.

Axiom (ARC Archimedean Property). $a \succ b \succ c \Rightarrow \exists \lambda, \mu \in (0, 1) : (a, c)_\lambda \succ b \succ (a, c)_\mu$.

Axiom (MIX Mixed Continuity). $\{\lambda \in [0, 1] : (a, b)_\lambda \succsim c\}, \{\lambda \in [0, 1] : c \succsim (a, b)_\lambda\}$ *are closed sets.*

Axiom (w-CON Weak Continuity). $\{g : f \succsim g\}$ *and* $\{g : g \succsim f\}$ *are closed sets.*

Axioms ORD and ARC grant the existence of a utility functional V that represents preferences. IND is the essential hypothesis that guarantees the expectation representation (9.2). The functional V is a *cardinal* measure, which implies that the index u is unique up to affine transformations: scaling and translations do not alter the primitive order relation. Such is the meaning of uniqueness in results that follow.

Proposition 1 (VNM). *Let* $\succsim$ *denote a binary relation of preferences over a convex subset of a linear space* $\mathscr{C}$ *that satisfies the axioms ORD and IND:*

a. *ARC grants the existence of a unique linear functional V defined on* $\mathscr{A}$ *that represents* $\succsim$.
b. *In addition, w-CON holds if and only if* $V(c) = \int u(c)\, dP$.
c. *If* $\mathscr{C}$ *is a discrete set, MIX* $\Leftrightarrow$ $V(c) = \int u(c)\, dP = \sum \pi_s u(x_s)$, *where* $\pi_s = P(c = x_s)$.

Appendix B Axiomatic Formulation of Preferences Under Uncertainty

Assuming convexity conveys more generality than betweenness, as proven in experimental analysis.

Axiom (CONV Convexity). $c \sim c' \Rightarrow (c, g)_\lambda \succsim c,\quad 0 < \lambda < 1$.

Ghirardato et al. (2004) show that a necessary and sufficient condition for a unique linear representation reduces to the axiom of invariant biseparate preferences (IBP).

Consider a monotone sequence of measurable sets asymptotically null $B_n \downarrow \emptyset$, and $x, y, z \in \mathscr{C}$, with $x \gtrsim z$:

Axiom (CONm Monotonic Continuity). $\forall y \in X \exists n : x \gtrsim (y, z)_{B_n}$.

A nondegenerate class of preferences is said to be IBP if it obeys the axioms ORD, CONm, IND*,MON' $(f \geq g \Rightarrow f \gtrsim g)$.

Proposition 2. $\gtrsim \in$ IBP $\Leftrightarrow V(c) = \mathscr{L}u(c)$, where $\mathscr{L}$ is a linear functional.

The term biseparable highlights a separation principle between beliefs and tastes (IND*). It implies the binary relation preferencias y creencias (IND*) and conveys the following implication: if v is a capacity $v(A) \equiv \mathscr{L}(1_A)$, with $A \in \mathscr{F}$, then[17]

$$\mathscr{L}(x, y)_A = u(x) v(A) + u(y)(1 - v(A)), \quad \forall x, y \in \mathscr{C}.$$

This capacity resumes attitude toward ambiguity—our ignorance of the context. There exists evidence of different patterns of behavior in the face of uncertainty environments (i.e., ambiguity seekers) compatible with CEU. IND means a neutral attitude, while CONV displays *hedging*.

CEU assumes *comononotonic independencia* ICO, which is a condition restricted to the class of actions whose underlying preferences are preserved by Ω. Following Ghirardato et al. (2004), we can state the following result:

Proposition 3. *Let* $\gtrsim \in$ IBP

a. CEU $\gtrsim$ obeys ICO if and only if $V(c) = \int u(c) dv$, where the integral is of Choquet (Schmeidler 1989).
b. Minmax $\gtrsim$ obeys CONV if and only if there exists a unique weak* compact and convex set of $\mathscr{P}$ such that $V(c) = \min\{\int udP : P \in \mathscr{P}\}$. (Gilboa and Schmeidler 1992)

Bewley (1986) introduced the maximal closed set under the axiom of independence. This set defines from the primitive relation a monotonic preorder monotonic $\gtrsim^*$ that obeys the sure-thing principle. Let $\mathscr{P}$ the class of probability measures in $(S, \mathscr{F})$, such that $\int u(c) dP$ represents $\gtrsim^*$. $\mathscr{P}$ can be proven to be weak* compact and convex. Let A^* maximal set of acts such that $\gtrsim^*$ is a complete order. Given two preference relations over $\mathscr{A}$, we say that $\gtrsim_1^*$ *reveals more ambiguity than* $\gtrsim_2^*$ if $\gtrsim_1^* \subset \gtrsim_2^*$. Therefore,[18] $A_1^* \subset A_2^*$ and $\mathscr{P}_1 \supset \mathscr{P}_2$. Hold when there exist $c_1, c_2 \in \mathscr{C}, x_1, x_2 \in \mathscr{C}; \lambda, \lambda' \in [0, 1]$ such that $(f, x)_\lambda \sim^* (g, x)_{\lambda'}$. Accordingly, an equivalence relation $f \asymp g$ in $\mathscr{A}$. Let $[c]$ be such class associated to c. By Lemma 8 in Ghirardato et al. (2004), $\mathscr{C} \subset [x] =: K$. This set is made up of the so-called *crisp* acts; it is the maximal element which satisfies the independence axiom[19]: its

---

[17]The only difference between a probability measure and a capacity is that the latter does not require additivity.

[18]Ghirardato et al. (2004, Propositions 5 and 6).

[19]Ghirardato et al. (2004, Proposition 10).

elements, not necessarily constant, cannot be combined to hedge. In addition, the following representation holds[20]:

Proposition 4. $\succsim \in$ IBP *if and only if there exist a function* $u : X \to R$ *and a function* $a : A| \rightthreetimes \to [0, 1]$ *such that the utility functional takes the form*

$$V (c) = a ([c]) \min_{P \in \mathscr{C}} \int u (c) \, dP + (1 - a ([c])) \max_{P \in \mathscr{C}} \int u (c) \, dP.$$

Note that the maxmin theories α-MEU and SEU are special cases of this functional, with $a ([f])$ being naturally interpreted as an index of ambiguity aversion.

Appendix C The Core of Preferences

In Sect. 9.3.2 we have observed a relation between $V = \mathscr{V}u$ and P (tastes and beliefs) through the *Gateaux* derivative. The equality $DV = P$ admits an interpretation in terms of the shadow value $P (ds)$ that measures the welfare gains associated to such state. However, there is no unique P when the derivative does not exist. Concavity of V ensures that the superdifferential is well defined, though. The core can be obtained by evaluating the superdifferential at 0, that is, Core $(V (c)) =$ Core $(P \circ c^{-1}) = \partial V (0)$. The local Lipschitzian property satisfied by V allows one to utilize an alternative notion of derivative.[21] Let $DV (z, \theta)$ stand for the directional derivative of V along the admissible (measurable) direction $\Theta \ni \theta : \Omega \to R$. The *Clarke derivative* at c is defined to be the set

$$\partial V (c) = \left\{ P : \int \theta dP \geq DV (c, \theta), \forall \theta \in \Theta \right\},$$

The Clarke derivative is deeply connected to the notion of core of a given functional $\mathscr{V}$:

$$\text{Core} (\mathscr{V}) = \left\{ P : \int x dP \geq \mathscr{V} (x) \, \forall x \in u (\mathscr{C}) \right\}.$$

CEU representation coincides with the core of v in Proposition 2.a, namely, Core $(\mathscr{V}) =$ Core (v). Proposition 15 in Ghirardato et al. (2004) shows that $\succsim \in$ IBP implies $\partial \mathscr{V} (0) = \mathscr{C}$ (see Proposition 2.b). Let the anticore be defined as

---

[20]Ghirardato et al. (2004, Proposition 11).

[21]A function F is *Lipschitz* when $|F (x) - F (y)| \leq \beta \|x - y\|$ with $\beta > 0$. Here $\|\cdot\|$ is the sup-norm or an equivalent one. A function is *locally Lipschitz* when for each point of its domain, there exists a neighborhood which satisfies the Lipschitz condition. It can be (easily) proven that, under SEU, $\mathscr{L}$ is Lipschitz with $\beta = 1$.

Anti $(\mathcal{V})$ = Core $(-\mathcal{V})$: the main properties of the core are stated in the following proposition[22]:

Proposition 5 (Properties of the Core).

a. *If $\mathcal{V}$ is linear,* Core $(\mathcal{V}) \cup$ Anti $(\mathcal{V}) \subset \partial H(0)$
b. *$\mathcal{V}$ is linear iff* Core $(\mathcal{V}) \vee$ Anti $(\mathcal{V}) = \emptyset$
c. *$\mathcal{V}$ is concave iff* Core $(\mathcal{V}) = \partial H(0)$

If the number of states is finite (n), then it is possible to write $\partial H(0) = \overline{\text{co}}\{\nabla H(z) : z \in \Phi\}$, where $\phi \subset u(A) \subset \mathbb{R}^n$ is the domain of differentiability of $\mathcal{V}$. In this case, there exists a numerable systems of cones of nonempty closure Γ_s containing $u(\mathscr{A})$. The functionals known in the literature are linear when restricted to each Γ_s. Therefore, we can define for each Γ_s a probability P_s. It follows $\partial H(0) = \overline{\text{co}}\{P_s\}$.

Acknowledgements The author wishes to thank Robert Waldmann, Roger Farmer, Ramon Marimon, and Costas Azariadis for the unvaluable insights. I would also like to thank the editors Elvio Accinelli and Carlos Hervés, one anonymous referee, the organizing committee, and the assistants of the II Escuela Potosina taken place at the Universidad Autónoma of San Luis Potosí (México) in April 2015.

References

Allais, M.: Le comportement de l'homme rationnel devant le risque: Critique des postulats et axiomes de l'ecole americaine. Econometrica **21**(4), 503–546 (1953). http://www.jstor.org/stable/1907921

Anscombe, F.J., Aumann, R.J.: A definition of subjective probability. Ann. Math. Stat. **34**(1), 199–205 (1963). http://www.jstor.org/stable/2991295

Backus, D.K., Routledge, B.R., Zin, S.E.: Exotic preferences for macroeconomists. In: NBER Macroeconomics Annual, vol. 19 (2004)

Bewley, T.F.: Knightian decision theory: Part I. Cowles Foundation Discussion Papers 807, Cowles Foundation for Research in Economics, Yale (1986)

Black, F., Scholes, M.: The pricing of options and corporate liabilities. J. Polit. Econ. **81**(3), 637–654 (1973). http://www.jstor.org/stable/1831029

Boi, J., Philppon, T., Savov, A.: Have financial markets become more informative? Tech. rep., Federal Reserve of New York, Staff Report Number 578 (2012)

Brunnermeier, M.K.: Asset Pricing Under Asymmetric Information: Bubbles, Crashes, Technical Analysis, and Herding. Oxford University Press, Oxford (2001)

Chew, S.H.: A generalization of the quasilinear mean with applications to the measurement of income inequality and decision theory resolving the Allais paradox. Econometrica **51**(4), 1065–1092 (1983). http://www.jstor.org/stable/1912052

Chew, S.H.: Axiomatic utility theories with the betweenness property. Ann. Oper. Res. **19**, 273–298 (1989)

Common, M., Stagl, S.: Ecological Economics: An Introduction. Cambridge University Press, Cambridge (2005)

---

[22]Ghirardato et al. (2004, p. 151).

Duffie, D., Epstein, L.G., Appendix C by Skiadis, C.: Stochastic differential utility. Econometrica **60**(2), 353–394 (1992)

Easley, D., O'Hara, M.: Microstructure and asset pricing . In: Handbook of Economics and Finance. Elsevier, Amsterdam (2003)

Ellsberg, D.: Risk, ambiguity, and the savage axioms. Quart. J. Econ. **75**(4), 643–669 (1961). http://www.jstor.org/stable/1884324

Epstein, L.G., Zin, S.E.: Substitution, risk aversion, and the temporal behavior of consumption and asset returns: a theoretical framework. Econometrica **57**(4), 937–969 (1989). http://www.jstor.org/stable/1913778

Fishburn, P.C.: Non-transitive measurable utility for decision under uncertainty. J. Math. Econ. **18**, 187–207 (1989)

Geoffard, P.Y.: Discounting and optimizing: capital accumulation problems as variational minimax problems. J. Econ. Theor. **69**, 53–70 (1996)

Ghirardato, P., Maccheroni, F., Marinacci, M.: Differentiating ambiguity and ambiguity attitude. J. Econ. Theor. **118**, 113–173 (2004)

Gilboa, I., Schmeidler, D.: Maximin expected utility with a non-unique prior. J. Math. Econ. **18**, 141–153 (1989)

Gil Martın, S., Rege, S.: Portuguese stock markets: a long-memory process? Bus. Theory Pract. **12**(1), 75–84 (2011)

Guidolin, M., Rinaldi, F.: Ambiguity in asset pricing and portfolio choice: a review of the literature. Theor. Decis. **74**(2), 183–217 (2013)

Hansen, L.P.: Dynamic valuation decomposition within stochastic economies. Econometrica **80**(3), 911–967 (2012). http://www.jstor.org/stable/41493841

Heath, C., Tversky, A.: Preference and belief: ambiguity and competence in choice under uncertainty. J. Risk Uncertainty **4**, 5–28 (1991)

Jacod, J., Shiryaev, A.N.: Limit Theorems for Stochastic Processes. Springer, New York (2003)

Kahneman, D., Tversky, A.: Advances in prospect theory: cumulative representation of uncertainty. J. Risk Uncertainty **5**, 297–323 (1992)

Karni, E., Schmeidler, D.: A temporal dynamic consistency and expected utility theory. J. Econ. Theor. **54**, 401–408 (1991)

Keynes, J.M.: A Treatise on Probability. MacMillan and Co., London (1921)

Klibanoff, P., Marinacci, M., Mukerji, S.: A smooth model of decision making under ambiguity. Econometrica **73**(6), 1849–1892 (2005). http://www.jstor.org/stable/3598753

Knight, F.H.: Risk, Uncertainty and Profit. Hart, Schaffner & Marx/Houghton Mifflin Co., Boston (1921)

Koopmans, T.: Stationary ordinary utility and impatience. Econometrica **28**(2), 287–309 (1960)

Kreps, D.M., Porteus, E.L.: Temporal resolution of uncertainty and dynamic choice theory. Econometrica **46**(1), 185–200 (1978). http://www.jstor.org/stable/1913656

Maccheroni, F., Marinacci, M., Rustichini, A.: Ambiguity aversion, robustness, and the variational representation of preferences. Econometrica **74**(6), 1447–1498 (2006). http://www.jstor.org/stable/4123081

Machina, M.J.: "Expected utility" analysis without the independence axiom. Econometrica **50**(2), 277–323 (1982). http://www.jstor.org/stable/1912631

Mehra, R., Prescott, E.C.: J. Monet. Econ. **15**, 145–161 (1985)

Øksendal, B.K.: Stochastic Differential Equations: An Introduction with Applications. Springer, Berlin (2003)

Maddison, A.: Contours of the World Economy 1–2030 AD: Essays in Macro-Economic History. OUP Catalogue. Oxford University Press, Oxford (2001)

Quiggin, J.: A theory of anticipated utility. J. Econ. Behav. Organ. **3**, 323–343 (1982)

Samorodintsky, G., Taqqu, M.S.: Stable Non-Gaussian Random Prcessses: Stochastic Models with Infinite Variance. Chapman and Hall, New York (1994)

Savage, L.J.:The Foundations of Statistics. Wiley, New York (1954)

Schmeidler, D.: Subjective probability and expected utility without additivity. Econometrica **57**(3), 571–587 (1989). http://www.jstor.org/stable/1911053

Thimme, J. Völkert, C.: High Order Smooth Ambiguity Preferences and Asset Prices (2012). Available at SSRN: http://ssrn.com/abstract=2021815
Von Neumann, J., Morgenstern, O.: Theory of Games and Economic Behavior. Princeton University Press, Princeton (1944)

Chapter 10
Additional Properties of the Owen Value

Oliver Juarez-Romero, William Olvera-Lopez, and Francisco Sanchez-Sanchez

Abstract In this work we present two properties of coalitional values for games with coalitional structure. The main goal of the paper is to prove that the Owen value satisfies these properties, which are related to a gain game and a lost game. The satisfaction of these properties provides a greater stability for this value because it is immune to a possible manipulation given by these games.

Keywords Cooperative games • Coalitional structures • Owen value • Shapley value • Stability • Manipulability • Gain and lost games

10.1 Introduction

The organization of agents in coalitional structures is an important fact in many real-world contexts, such as the formation of cartels, trading blocks among nation states, research joint ventures, and political parties. These situations can be modeled through transferable utility games, in which the players are partitioned into coalitions for the purpose of bargaining. All players in the same coalition agree before the game that any cooperation with other players will only by carried out collectively. Given a coalitional structure, the bargaining occurs between coalitions and between players in the same coalition. The main idea is that coalitions play among themselves as individual players in a game among coalitions and then, the profit obtained by each coalition is distributed among its members. Owen (1977) studied the allocation that arises from applying the Shapley value (Shapley (1953)) twice: first, in a game among coalitions and then in a game inside each coalition.

O. Juarez-Romero (✉) • F. Sanchez-Sanchez
UASLP, School of Economics, Av. Pintores S/N, Burócratas del Estado, C.P. 78213, San Luis Potosı, SLP, México
e-mail: ojuarez@cimat.mx; william.olvera@uaslp.mx

W. Olvera-Lopez
CIMAT, Jalisco S/N, Valencia, C. P. 36240 Guanajuato, Guanajuato, Gto, México
San Luis Potosí, SLP, Mexico
e-mail: sanfco@cimat.mx

© Springer International Publishing Switzerland 2016
A.A. Pinto et al. (eds.), *Trends in Mathematical Economics*,
DOI 10.1007/978-3-319-32543-9_10

In this paper we study a kind of stability of the Owen value. Given a game with coalitional structure and a coalitional value, we build two games called the gain and the lost games by association, respectively. These games model the changes in the payoffs when new coalitional structures arise by manipulation of the coalitional value. The main result establishes that if the coalitional value is the Owen value, manipulating does not help: the payoffs that the players receive in the gain and the lost games by association are the same as in the original game. A similar property was studied by Hamiache (2001).

This article is organized as follows: Sect. 10.2 provides the notation that we use in the rest of the document. On Sect. 10.3, we show that the Owen value is immune to the manipulation given in the gain and lost games by association.

10.2 Notation and Preliminaries

Let $N = \{1, \ldots, n\}$ be a finite set of agents. A *cooperative game* with transferable utility (TU-game) is a pair (N, υ) where $\upsilon : 2^N \to \mathbb{R}$ is a *characteristic function* defined on the power set of N satisfying $\upsilon(\emptyset) = 0$. An element i of N is called a *player*, every nonempty subset S of N a *coalition*, and N the *grand coalition*. The real number $\upsilon(S)$ is called the *worth* of coalition S. $\upsilon(S)$ is interpreted as the total payoff that the coalition S can obtain for its members. Let G^N be the set of all cooperative TU-games with players set N. A *payoff vector* $x \in \mathbb{R}^n$ of $\upsilon \in G^N$ is a vector giving a payoff x_i to any player $i \in N$. A solution on G^N is a map φ that assigns to each $\upsilon \in G^N$ a payoff vector $\varphi(\upsilon) \in \mathbb{R}^n$. A player $i \in N$ is a *null player* in $(N, \upsilon) \in G^N$ if for every $S \subseteq N$ such that $S \ni i$ we have that $\upsilon(S) = \upsilon(S \setminus \{i\})$. Two players $i, j \in N$ are *symmetric* in $(N, \upsilon) \in G^N$ if for every $S \subseteq N$ such that $i \in S$ we have that $\upsilon(S) = \pi \upsilon(\pi S)$, where $\pi(S) = \pi^{-1}(S)$, $\pi \upsilon(T) = \upsilon(\pi(T))$ for all $T \subseteq N$ and $\pi \in S_n$ [1] is a permutation such that $\pi(i) = j$.

A very well-known solution on G^N is the *Shapley value*. The Shapley value can be defined by orders as follows:

$$\mathrm{Sh}_i(N, \upsilon) = \frac{1}{|S_n|} \sum_{\pi \in S_n} \left[\upsilon(P_i^\pi \cup \{i\}) - \upsilon(P_i^\pi) \right], \quad \text{for all } i \in N,$$

where P_i^π denotes the set of all predecessors of i in π, that is, $P_i^\pi = \{j \in N : \pi(j) < \pi(i)\}$. For further information about the Shapley value, please refer Shapley (1953).

For all finite set N, a *coalitional structure* over N is a partition of N, i.e., $B = \{B_1, \ldots B_m\}$ is a coalitional structure if it satisfies that $\bigcup_{1 \leq k \leq m} B_k = N$ and $B_k \cap B_l = \emptyset$ when $k \neq l$. We assume that $B_k \neq \emptyset$ for all k. The sets $B_k \in B$ are called unions or blocks. There are two trivial coalitional structures. The first one, which we denote by B^N, where only the grand coalition is formed, that is, $B^N = \{N\}$,

---

[1] S_n is the group of permutations of N.

and the second one is the coalitional structure where each union is a singleton and it is denoted by B^n, that is, $B^n = \{\{1\}, \{2\}, \ldots, \{n\}\}$. We denote by $\mathcal{B}(N)$ the set of all coalitional structures over N. A TU-game (N, υ) with coalitional structure $B \in \mathcal{B}(N)$, is denoted by (B, υ). Let CSG^N denote the family of all TU-games with coalitional structure with player set N.

For every game $(B, \upsilon) \in CSG^N$, with $B = \{B_1, \ldots B_m\}$, the *quotient game* is the TU-game $(M, \upsilon_B) \in G^M$ where $M = \{1, 2, \ldots, m\}$ and $\upsilon_B(T) = \upsilon(\bigcup_{i \in T} B_i)$ for all $T \subseteq M$. (M, υ_B) is the TU-game induced by (B, υ) considering the coalitions of B as players. Notice that for the trivial coalitional structure B^n, we have $(M, \upsilon_B) \equiv (N, \upsilon)$. For all $\{k, l\} \subseteq M$, we say that B_k and B_l are *symmetric* in (B, υ) if k and l are symmetric in the TU-game (M, υ_B). For all $k \in M$, we say that B_k is a *null coalition* if k is a null player in the quotient game (M, υ_B).

Given $B \in \mathcal{B}(N)$, for all $k \in M$ and all $S \subseteq B_k$, we denote by $B \mid_S$ the new coalitional structure defined on $(\bigcup_{j \neq k} B_j) \cup S$ which appears when the complementary of S in B_k leaves the game, that is,

$$B \mid_S = \{B_1, \ldots, B_{k-1}, S, B_{k+1}, \ldots, B_m\}.$$

A *coalitional value* is a function Φ that assigns to each $(B, \upsilon) \in CSG^N$ a payoff vector $\Phi(B, \upsilon) \in \mathbb{R}^n$. One of the most important coalitional values is the *Owen value* (Owen (1977)). Given $(B, \upsilon) \in CSG^N$ and $k \in M$, for all $S \subseteq B_k$, we denote $\bar{S} = B_k \backslash S$. Owen (1977) defined a game $(M, \upsilon_{B \mid_S})$ that describes what would happen in the quotient game if union B_k was replaced by S, i.e.,

$$\upsilon_{B \mid_S}(T) = \upsilon(\bigcup_{j \in T} B_j \backslash \bar{S}), \qquad \text{for all } T \subseteq M.$$

Next, he defines an internal game (B_k, υ_k) by setting $\upsilon_k(S) = \text{Sh}_k(M, \upsilon_{B \mid_S})$ for all $S \subseteq B_k$. Thus, $\upsilon_k(S)$ is the payoff to S in $\upsilon_{B \mid_S}$. The *Owen value* of the game (B, υ) is the payoff vector $\text{Ow}(B, \upsilon) \in \mathbb{R}^n$ defined by

$$\text{Ow}_i(B, \upsilon) = \text{Sh}_i(B_k, \upsilon_k), \qquad \text{for all } k \in M \text{ and for all } i \in B_k.$$

This procedure has the next interpretation: First, the union k plays the quotient game (M, υ_B) among the unions; then, the payoff obtained is shared among its members by playing the internal game (B_k, υ_k). In both levels of bargaining, the payoffs are obtained using the Shapley value.

The Owen value satisfies the *quotient game property*: that is,

$$\sum_{i \in B_k} \text{Ow}_i(B, \upsilon) = \text{Sh}_k(M, \upsilon_B), \qquad \text{for all } k \in M.$$

Notice that for the trivial coalition structures B^N and B^n,

$$\text{Ow}(B^N, \upsilon) = \text{Ow}(B^n, \upsilon) = Sh(N, \upsilon).$$

The Owen value can be defined by orders. Let B be a coalitional structure over N and $\pi \in S_n$. We say that π is *admissible respect to* B if for all $\{i, j, k\} \subseteq N$ and $l \in M$ such that $\{i, k\} \subseteq B_l$, if $\pi(i) < \pi(j) < \pi(k)$, then $j \in B_l$. In other words, π is admissible respect to B if players in the same group of B appear successively in π. We denote by $\mathcal{A}(B, N)$ the set of all admissible orders (on N) respect to B. The Owen value is given by the formula

$$\mathrm{Ow}_i(B, \upsilon) = \frac{1}{\mid \mathcal{A}(B, N) \mid} \sum_{\pi \in \mathcal{A}(B, N)} \left[\upsilon(P_i^\pi \cup \{i\}) - \upsilon(P_i^\pi) \right], \quad \text{for all } i \in N,$$

(10.1)

where P_i^π denotes the set of all predecessors of i in π, that is, $P_i^\pi = \{j \in N : \pi(j) < \pi(i)\}$.

Now, we present the axioms that characterize the Owen value on CSG^N. Let Φ be a coalitional value. We define

$$\Phi(B, \upsilon)[S] = \sum_{i \in S} \Phi_i(B, \upsilon), \quad \text{for all } S \subseteq N.$$

(1) *Efficiency*: For all $(B, \upsilon) \in CSG^N$, $\Phi(B, \upsilon)[N] = \upsilon(N)$.
(2) *Additivity*: For all $(B, \upsilon), (B', \omega) \in CSG^N$, with $B = B'$, $\Phi(B, \upsilon + \omega) = \Phi(B, \upsilon) + \Phi(B, \omega)$.
(3) *Intracoalitional symmetry*: For all $(B, \upsilon) \in CSG^N$, all $k \in M$ and every $\{i, j\} \subseteq B_k$, if i and j are symmetric players in (N, υ), then $\Phi_i(B, \upsilon) = \Phi_j(B, \upsilon)$.
(4) *Coalitional symmetry*: For all $(B, \upsilon) \in CSG^N$ and all $\{k, l\} \subseteq M$, if B_k and B_l are symmetric players in (B, υ), then $\Phi(B, \upsilon)[B_k] = \Phi(B, \upsilon)[B_l]$.
(5) *Null player axiom*: For all $(B, \upsilon) \in CSG^N$ and all $i \in N$, if i is a null player in (N, υ), then $\Phi_i(B, \upsilon) = 0$.

Theorem 1 (Owen 1977). *A value Φ on CSG^N satisfies efficiency, additivity, intracoalitional symmetry, coalitional symmetry, and null player axiom if and only if Φ is the Owen value.*

For more details regarding the Owen value, see Owen (1977).

10.3   Additional Properties

In this section, we consider that (B, υ) and Φ are given and fixed. Given $S \subseteq N$, we assume that the members of S decide to play jointly and to leave the coalitions that they belong. Let $B'(S)$ be the *coalitional structure induced* by this behavior of the members of S, i.e.,

$$B'(S) = \{B_1', B_2', \cdots, B_m', S\}, \quad \text{where } B_k' = B_k \backslash S, \text{ for all } B_k \in B.$$

We define (N, v_Φ^*) where the characteristic function v_Φ^* is given by

$$v_\Phi^*(S) = \Phi(B'(S), v)[S], \quad \text{for all } S \subseteq N.$$

This game is called the *gain game by association* because for all $S \subseteq N$, $v_\Phi^*(S)$ represent the amount obtain by S when the members of S decide to form one collaborative unit, leaving their original coalitions.

Now, we define the TU-game $(N, v_\Phi^\diamond)$, where $v_\Phi^\diamond$ is given by

$$v_\Phi^\diamond(S) = \sum_{B_k : B_k \cap S \neq \emptyset} \left[\Phi(B, v)[B_k] - \Phi(B'(S), v)[B_k'] \right], \quad \text{for all } S \subseteq N.$$

So, $v_\Phi^\diamond(S)$ represents the change in the payoffs of the groups B_k such that $B_k \cap S \neq \emptyset$ given some event that caused the formation of the new coalitional structure $B'(S)$. By this reason, $(N, v_\varphi^\diamond)$ is called the *lost game by association*.

Lemma 1. *Let $(B, v) \in CSG^N$ be a game with coalitional structure B. Then*

$$Ow(B, v_{Ow}^\diamond)[B_k] = Ow(B, v)[B_k] = Ow(B, v_{Ow}^*)[B_k], \quad \text{for all } k \in M.$$

Proof. We will prove the left identity because for proving the right identity, we can use similar arguments. From the quotient game property, we have that $Ow(B, v)[B_k] = Sh_k(M, v_B)$. Then, calculating the worth of each $B_k \in B$ in the game $(N, v_{Ow}^\diamond)$, we can see that

$$v_{Ow}^\diamond(B_k) = Ow(B, v)[B_k].$$

We define (M, v_{Ow}^B), where

$$v_{Ow}^B(T) = \sum_{t \in T} v_{Ow}^\diamond(B_t), \quad \text{for all } T \subseteq M.$$

Because the game (M, v_{Ow}^B) is additive, we have that

$$Ow(B, v_{Ow}^\diamond)[B_k] = Sh_k(M, v_{Ow}^B) = Ow(B, v)[B_k].$$

Now, we are going to present the main result of the paper:

Theorem 2. *For all TU-game (N, v) with coalitional structure B and for all $i \in N$, we have that*

$$Ow_i(B, v_{Ow}^\diamond) = Ow_i(B, v) = Ow_i(B, v_{Ow}^*).$$

Proof. Again, we will prove the left identity, because for proving the right identity, we can use similar arguments. We will prove that the result is valid for games in which $B = B^n$. For doing that, we will use the expression of the Owen value given in (10.1).

So, the Owen value for $i \in N$ in the game $(B, w_\varphi^\diamond)$ is given by

$$Ow_i(B, \upsilon_{Ow}^\diamond) = \frac{1}{|\mathcal{A}(B,N)|} \sum_{\pi \in \mathcal{A}(B,N)} \left[\upsilon_{Ow}^\diamond(P_i^\pi \cup \{i\}) - \upsilon_{Ow}^\diamond(P_i^\pi) \right].$$

It is easy to check that

$$\upsilon_{Ow}^\diamond(P_i^\pi \cup \{i\}) = \sum_{j \in P_i^\pi \cup \{i\}} Ow_j(B, \upsilon) \quad \text{and} \quad \upsilon_{Ow}^\diamond(P_i^\pi) = \sum_{j \in P_i^\pi} Ow_j(B, \upsilon).$$

Thus,

$$\upsilon_{Ow}^\diamond(P_i^\pi \cup \{i\}) - \upsilon_{Ow}^\diamond(P_i^\pi) = Ow_i(B, \upsilon),$$

and

$$Ow_i(B, \upsilon_{Ow}^\diamond) = Ow_i(B, \upsilon).$$

Now we will prove that the result is valid when $B = B^N$. In this case we can see that

$$\upsilon_{Ow}^\diamond(P_i^\pi \cup \{i\}) = \upsilon(N) - Ow(B'', \upsilon)[B_r \backslash \{i\}],$$

and

$$\upsilon_{Ow}^\diamond(P_i^\pi) = \upsilon(N) - Ow(B', \upsilon)[B_r],$$

where $B'' = \{\{P_i^\pi \cup \{i\}\}, \{B_r \backslash \{i\}\}\}$ and $B' = \{\{P_i^\pi\}, \{B_r\}\}$ with $B_r = N \backslash P_i^\pi$.
Given that

$$Ow(B'', \upsilon)[B_r \backslash \{i\}] = \upsilon(B_r \backslash \{i\}) + \frac{\upsilon(N) - \upsilon(B_r \backslash \{i\}) - \upsilon(P_i^\pi \cup \{i\})}{2},$$

$$Ow(B', \upsilon)[B_r] = \upsilon(B_r) + \frac{\upsilon(N) - \upsilon(B_r) - \upsilon(P_i^\pi)}{2},$$

then

$$\upsilon_{Ow}^\diamond(P_i^\pi \cup \{i\}) - \upsilon_{Ow}^\diamond(P_i^\pi) = \frac{\upsilon(B_r) - \upsilon(B_r \backslash \{i\}) + \upsilon(P_i^\pi \cup \{i\}) - \upsilon(P_i^\pi)}{2}.$$

In this case it is easy to check that there is $\pi' \in \mathcal{A}(B,N)$ such that $P_i^{\pi'} \cup \{i\} = N \backslash \{P_i^\pi\} = B_r$ and $P_i^\pi = N \backslash \{P_i^{\pi'} \cup \{i\}\} = B_r \backslash \{i\}$. Thus, we conclude that

$$\text{Ow}_i(B, v_{\text{Ow}}^\diamond) = \frac{1}{|\mathcal{A}(B,N)|} \sum_{\mathcal{A}(B,N)} \left[\frac{v(B_r) - v(B_r \setminus \{i\})}{2} + \frac{v(P_i^\pi \cup \{i\}) - v(P_i^\pi)}{2} \right]$$

$$= \text{Ow}_i(B, v).$$

In general $m \geq 2$ and, when there is at least a coalition $B_k \in B$ such that $|B_k| \geq 2$, we will use the Lemma 1.

It is well known that the Owen value satisfies the axiom of balanced contributions within coalitions (Calvo et al. (1996)). That is, for any $i, j \in B_k \in B$, we have that

$$\text{Ow}_i(B, v) - \text{Ow}_i(B \setminus \{j\}, v) = \text{Ow}_j(B, v) - \text{Ow}_j(B \setminus \{i\}, v).$$

Adding over all players $j \in B_k \setminus \{i\}$ in the last expression, we have that

$$|B_k - 1| \text{Ow}_i(B, v) = \sum_{j \in B_k \setminus \{i\}} \text{Ow}_j(B, v) - \sum_{j \in B_k \setminus \{i\}} \text{Ow}_j(B \setminus \{i\}, v)$$

$$+ \sum_{j \in B_k \setminus \{i\}} \text{Ow}_i(B \setminus \{j\}, v),$$

which can be written as follows:

$$|B_k| \text{Ow}_i(B, v) = \text{Ow}(B, v)[B_k] - \text{Ow}(B \setminus \{i\}, v)[B_k \setminus \{i\}] + \sum_{j \in B_k \setminus \{i\}} \text{Ow}_i(B \setminus \{j\}, v).$$

$$(10.5)$$

In the rest of the proof, we will use mathematical induction. Consider the case where $|B_k| = 2$. Then, we have that $B_k = \{i, j\}$ and thus

$$B \setminus \{j\} = \{B_1, \dots, B_{k-1}, B_k^*, B_{k+1}, \dots, B_m\}$$

where $B_k^* = \{i\}$. By using the Lemma 1, in Eq. (10.5) we have

$$2 \, \text{Ow}_i(B, v) = \text{Ow}(B, v)[B_k] - \text{Ow}(B \setminus \{i\}, v)[B_k \setminus \{i\}] + \text{Ow}(B \setminus \{j\}, v)[B_k^*],$$

$$= \text{Ow}(B, v_{\text{Ow}}^\diamond)[B_k] - \text{Ow}(B \setminus \{i\}, v_{\text{Ow}}^\diamond)[B_k \setminus \{i\}] + \text{Ow}(B \setminus \{j\}, v_{\text{Ow}}^\diamond)[B_k^*],$$

$$= 2 \, \text{Ow}_i(B, v_{\text{Ow}}^\diamond),$$

thus

$$\text{Ow}_i(B, v) = \text{Ow}_i(B, v_{\text{Ow}}^\diamond).$$

Suppose that the theorem is valid for coalitions $B_k \in B$ such that $|B_k| = s$. We will prove that it is valid when $|B_k| = s + 1$. For doing that, we rewrite Eq. (10.5) where $|B_k| = s + 1$. So

$$(s+1) \, \text{Ow}_i(B, v) = \text{Ow}(B, v)[B_k] - \text{Ow}(B \setminus \{i\}, v)[B_k \setminus \{i\}] + \sum_{j \in B_k \setminus \{i\}} \text{Ow}_i(B \setminus \{j\}, v).$$

The sum of the above equation involves the Owen value for $i \in B_k$. By induction hypothesis, it is equal to the Owen value in the TU-game $(N, w_\varphi^\diamond)$, and applying Lemma 1, we have

$$(s + 1) \, \mathrm{Ow}_i(B, \upsilon) = \mathrm{Ow}(B, \upsilon)[B_k] - \mathrm{Ow}(B \setminus \{i\}, \upsilon)[B_k \setminus \{i\}] + \mathrm{Ow}(B \setminus \{j\}, \upsilon)[B_k^*],$$
$$= (s + 1) \, \mathrm{Ow}_i(B, \upsilon_{\mathrm{Ow}}^\diamond).$$

Then, we conclude that

$$\mathrm{Ow}_i(B, \upsilon) = \mathrm{Ow}_i(B, \upsilon_{\mathrm{Ow}}^\diamond).$$

The fact that the Owen value holds Theorem 2 can be seen as stability, because this value is immune to the manipulation that the games $\upsilon_{\mathrm{Ow}}^\diamond$ and υ_{Ow}^* mean.

Acknowledgements The authors acknowledge support from CONACyT grants 167924 and 240229.

References

Calvo, E., Lasaga, J.J., Winter, E.: The principle of balanced contributions and hierarchies of cooperation. Math. Soc. Sci. **31**(3), 171–182 (1996)

Hamiache, G.: The Owen value values friendship. Int. J. Game Theor. **29**, 517–532 (2001)

Owen, G.: Values of games with a priori unions. In: Henn, R., Moeschlin, O. (eds.) Essays in Mathematical Economics and Game Theory, pp. 76–88. Springer, New York (1977)

Shapley, L.S.: A value for n-person games. In: Kuhn, H.W., Tucker, A.W. (eds.) Contributions to the Theory of Games II. Annals of Mathematics Studies, vol. 28, pp. 307–317. Princeton University Press, Princeton (1953)

Chapter 11
The Gödelian Foundations of Self-Reference, the *Liar* and Incompleteness: Arms Race in Complex Strategic Innovation

Sheri Markose

Abstract Self-referential calculations of oppositional or contrarian structures and the necessity to innovate to outsmart hostile agents in an arms race are ubiquitous in socio-economic systems, immunology and evolutionary biology. However, such phenomena with strategic innovation, which entails novel actions beyond listable sets, are outside the ambit of extant game theory. How can strategic innovation with novel actions be a Nash equilibrium of a game? Based on the only known Gödel-Turing-Post (GTP) axiomatic framework on meta-analyses of offline simulations that involve recursive operations on encoded information, we show that mutually mentalising agents capable of such offline simulations can "think outside the box" and embark on an arms race in novelty or surprises. A key logical ingredient of this is the self-referential encoding of a proposition on mutual negation or opposition often referred to as the Gödel sentence. The only recursive best response function of a two-person game with an oppositional structure that can implement strategic innovation in a lock-step formation of an arms race is the productive function of the Emil set theoretic proof of the Gödel incompleteness result.

Keywords Gödel incompleteness • Self-reference • Contrarian • Offline simulation • Strategic innovation • Novelty • Surprises • Red Queen-type arms race • Creative and productive sets • Productive function • Surprise Nash equilibrium

JEL Classification: A12, C70, C79, B40.

S. Markose (✉)
Economics Department, University of Essex, Wivenhoe Park, Colchester CO4 3SQ, UK
e-mail: scher@essex.ac.uk

© Springer International Publishing Switzerland 2016
A.A. Pinto et al. (eds.), *Trends in Mathematical Economics*,
DOI 10.1007/978-3-319-32543-9_11

11.1   Introduction

The 2007 Global Financial Crisis has prompted calls to re-examine the foundations of traditional economic models, and many funding organisations[2] have encouraged that economic interactions be analysed using nonclassical economic modelling involving reflexivity and complexity. Reflexivity refers to the capacity to make self-referential mappings, and it is understood that interactions in social systems pose problems of epistemic undecidability and those of radical uncertainty regarding the complex nature of collective outcomes. The foremost feature of complexity of socio-economic interactions that has made it hard for economics science to make progress arises from the fact that large swathes of observed phenomena, viz., socio-economic proteanism in which myriad novel objects and technologies are produced in what appears to be an unceasing strategic arms race, is outside the ambit of extant formalised mathematical models in economics.

The foundational problems in economics stem from at least three sources. The first of these was noted by Binmore (1987), who seminally raised the 'spectre of Gödel' (Binmore 1987) in the context of game theory which attempts to restrict the scope of strategic behaviour to a system that is logically closed and complete. In game theory, there are strategy mappings to a *fixed* action set and indeterminism extends only to randomisations between given actions. Binmore states that in outlawing strategic indeterminism or novelty production, the question that is pertinent here is the centre piece of Gödel (1931): *'what of the Liar?'*

The pragmatic import of the question, *'what of the Liar?'*, for game theory is as follows: when faced by a hostile agent who will falsify or negate one's actions if he could deduce what they are, can one rationally play an action that is known or can be formally deduced, both of which will be called 'transparent', or does one innovate and 'surprise' the opposition? Secondly, compounding the first omission noted by Binmore (1987) on the assumption of a completeness paradigm in which game theory is couched, and what has been reiterated by many, is the impossibility to produce novelty or surprises in a Nash equilibrium, let alone the structure of an arms race in strategic innovation. Bhatt and Camerer (2005) succinctly state this: "in a Nash equilibrium nobody is surprised about what others actually do, or what others believe, because strategies and beliefs are synchronised, presumably due to introspection, communication or learning". What is missing in this statement is the category of mutual belief and expectation of surprise and the characterisation of a Nash equilibrium in which players logically expect that they will need to surprise

---

[2]These include the EC FP 7 calls, the Horizon 20–20 Global Systems Science initiatives and the UK Economic and Social Research Council (ESRC) endeavours at the Oxford Symposium of 2012 and the 2014 Essex Diversity in Macroeconomics conference. The latter aimed at breaking up the monoculture in mainstream economics by bringing together developments from at least three new branches of economics. These include, agent-based computational, complexity and behavioural economics. The aim of the highly interdisciplinary studies of computational and digital technologies, complexity sciences and neurophysiology of the brain is to address erstwhile gaps and controversies in the micro- and macroscopic aspects of socio-economic interactions.

and be surprised. As will be shown, there is nothing inherent to a Nash equilibrium in which the strategic necessity of a surprise cannot be formulated, indeed novelty and surprise become a logical necessity to avoid inconsistency.

Finally, the consequence of the extant mathematical paradigm of economics is that it cannot formally model the category of radical uncertainty in terms of Gödel undecidability, incompleteness and non-computability arising endogenously from strategic innovations from mutually mentalising agents of a very tall order of computational intelligence. Following the provenance of Gödel (1931), in what is called the Wolfram–Chomsky schema (see Wolfram 1984; Casti 1994; Albin 1998; Markose 2005), the sine qua non of a complex adaptive system (CAS) is identified with Type IV structure changing innovation based undecidable dynamics associated with the interaction of agents having computational capabilities first identified in Gödel (1931).

The remarkable significance of the Gödel incompleteness theorem is that to date there is only one mathematical exit route from known listable sets in order for agents to construct novel objects that fall outside of these sets. Hence, I will argue it is important, both in the discussions of the neurophysiology of social coordination and proteanism and also for the missing formalism in economics for Type IV dynamics, to delineate the key ingredients of the Gödel paradigm. Type IV dynamics take the co-evolutionary form of a Red Queen[3]-type arms race in innovation. Regulator-regulatee arms race (no different from a parasite host dynamics in immunology) involves monitoring and production of countervailing new measures (comparable to the production of antibodies) by authorities in response to regulatee deviations from rules due to outright opposition or perverse incentives in place. However, while there has been much disquiet on foundational issues in economics, the need for the Gödel paradigm has not yet been signposted in any substantive way.

As noted by Markose (2005), almost all accounts of dynamics in economic interactions, even by those who espouse the 'complexity vision' (Colander 2000), eschew the Gödel provenance of proteanism with Type IV arms races in novelty production as the hallmark of complex adaptive systems.[4] Despite the long-standing legacy of Schumpeter (1942), Baumol (2002, 2004) who has extensively discussed and documented the role of the relentless Red Queen-type strategic arms race in innovation by firms of products and processes in capitalism claims

---

[3] The Red Queen, the character in Lewis Carol's *Alice Through the Looking Glass*, who signifies the need "to run faster and faster to stay in the same square" has become the emblematic of the outcome of competitive co-evolution for evolutionary biologists in that no competitor gains absolute ground; see Markose (2005).

[4] Even though Durlauf (2012) notes that economics as a science is still evolving, he does not think it needs any substantive 'paradigm' shifts or new foundational studies to understand and model economic systems as complex adaptive systems. Of the issues on non-linearity, heterogeneity and interconnectedness that Durlauf (2012) considers, the latter two, along with the use of agent-based simulation models, are relatively new approaches in the literature (see Tesfatsion and Judd 2006). In the UK, there are no more than a dozen economists who pursue these nonmainstream approaches.

		Other(s)	
		Known	**Unknown**
Self	**Known**	(Known, Known) Common Knowledge	(Known, Unknown) Private Information Asymmetric Information
	Unknown	(Unknown, Known) Private Ignorance Asymmetric Information	(Unknown, Unknown) Radical Mutual Uncertainty with surprises and novelty production

Fig. 11.1 Taxonomy of uncertainty as asymmetry of information and radical uncertainty which proceeds from novelty production

that this is not addressed in mainstream economics. Recently, renewed efforts are being made to come to terms with arms races in novelty production, especially in the monetary and financial system (see Haldane 2012). These have wrought structural transformations[5] that many regard to be exogenous to macroeconomics and hence as noted by Axelrod (2003) widespread system failure has followed from having ignored competitive co-evolution. The practice of modelling innovation as exogenous white noise, especially in the area of macroeconomics, has led (Goodhart 1999) in his Keynes Lecture to state that: *One of the central problems is that uncertainty is far more insidious and pervasive than represented by additive error terms in standard models.*

In the schema given in Fig. 11.1 below of how uncertainty is modelled in mainstream economics, useful delineations start from what is mutually known by self and others yielding different classes of asymmetric information. However, radical uncertainty which can be modelled as Gödelian points of departure from known listable sets of actions is not what is considered by those who purport to model Knightian uncertainty.

The objective of this paper is to give the developments in mathematical logic pioneered by Gödel (1931), Turing (1937) and Post (1941) (GTP logic from here on) that underpin a complex adaptive system and to demonstrate how this can inform a two-person game in which Type IV novelty and surprises will emerge as in an arms race. There are four key formal key components outlined below, which will be dealt with in the paper.

(i) Agents/players have the capacity to operate on encoded information. The GTP theory of computation, also called recursion function theory, provides the only known formalism for operations on encoded information represented by integers, $n \in \aleph$, also known as Gödel numbers. These operations are

---

[5] One of the more prescient of macroeconomists, William White, has recently stated "it seems to me that nobody on the regulatory (and macroeconomics) side has really got to grips with the reality of this constant innovation" (Financial Times, June 25, 2014).

strictly syntactic ones that entail instructions utilising strings of symbols to achieve encoded outputs from inputs in a finite number of steps in terms of an algorithm or program. The power of recursion to reuse code in concatenations and recombinations is considered by many (see Beinhocker (2011))[6] to be a key building block of evolution and human technological progress.[7]

(ii) Agents utilise a framework well known as Gödel metamathematics (see Rogers 1967) which implements a 1-1 mapping between internal *offline meta-analysis* made by agents and their respective external machine executions. The offline operations, referred to as *simulations*, is based on encoded information of online external machine executions, both of which can be done on 'mechanisms' involving any substrata ranging from intracellular biology to silicon chips. This capacity of meta-representation without which CAS properties do not emerge yields the notion of a universal Turing machine (UTM) which can take encoded information of other machines and replicate them. Remarkably, UTMs can run codes involving themselves, which is the basis of self-reference. If codes of functions are not already given, then successful simulations require discovering fixed points of executable action functions. This involves the use of the Second Recursion Theorem. The significance of this for human cognition lies in the recent discovery of the mirror neuron system by the Parma group (Gallese et al. 1996; Rizzolatti et al. 1996, which can be regarded as one of the most important scientific discoveries of the late twentieth century. Gallese and Sinigaglia (2011) have characterised the MNS as a neuronal platform for conducting *offline embodied simulations* for action prediction in the other based on a parallel set of neurons that fire during actual action execution by one-self.[8] While Gallese and others use this as an analogy, it will be shown that the two-place Gödel substitution function in Gödel metamathematics can provide the mathematical wherewithal for processing action prediction regarding the other as an offline exercise. Nash equilibria will be found on the diagonal array of the fixed point mappings of action execution functions (strategy functions) using the Second Recursion Theorem.

(iii) It is well known that the *Liar* qua agent in Gödel logic can be viewed as the hostile agent or the contrarian who will falsify or negate one's actions if he can deduce what they are. Also the *Liar*, following the provenance of the Cretan Liar,[9] refers to a self-referential statement of negation which results in

---

[6]Arthur (1993) argues for the need to take a computational perspective to model complexity.

[7]Sayama (2008) suggests that routine programs called quines appended to the end of other programs to read, copy and 'print' are important building blocks in self-replication algorithms.

[8]The neurons that fire with actual action execution are called *canonical neurons*, Arbib and Fagg (1998), and represent online machine executions in the G-T-P language.

[9]Since antiquity, it has been known that self-refuting statements generate paradoxes as in the Cretan Liar proposition: *this is false*. Gödel's analogue of the Liar proposition is the undecidable proposition. The latter, denoted as A, has the following structure: $A \leftrightarrow \sim |-(A)$. That is, A says of itself that it is not provable ($\sim |-$). However, unlike the Cretan Liar, there is no paradox in Gödel's undecidable proposition as it can be proved that this is so. Any attempt to prove the proposition A

an undecidable truth predicate. In the Gödel framework, which is known to have transformed the problem from that of truth to one of computation, we have an encoding of a fixed point of a mapping entailing the negation of a self-regarding code, often referred to as the Gödel sentence. In the context of a game, as will be shown, the Gödel sentence encodes a non-computable fixed point involving a contrarian/Liar strategy and hence is a recording of a mutual recording of hostility or opposition. Ben-Jacob (1998) has suggested a similar interpretation of the Gödel sentence as a record of threats to a code at the level of the genome.[10] Again, in a remarkable set of experiments by Scott Kelso and his group (see Tognoli et al. (2007)), recording of neurophysiological markers of anti-coordination that arises from actions that need to be different/opposite from what is predicted of another indicates that these are also part of the human mirror neuron system. This is an important piece of evidence for the Gödel framework to be relevant as an analogue of cognitive incompleteness.[11]

(iv) The logical consequences of (i)–(iii) can be represented by so-called *creative* and *productive* sets using the set theoretic proof by Post (1944) of Gödel incompleteness (see also Smullyan 1961; Cutland 1980). The Post (1944) productive function implements points at which agents exit a given listable set, referred to as a recursively enumerable set that can be enumerated by any Turing machine including another negating or oppositional UTM, within a structure of an arms race in novelty production or 'surprises'. To rigorously define the set of novel objects, it is useful to consider the set of all potential technologies as the outputs of a *total* computable function that always stops and yields some output on any encoded input. This set denoted by $\Re$ is countably infinite, and there is no systematic way of 'searching' or listing this set. Some finite subset of this set entails known technologies and can represent a given listable action set A of traditional game theory. A novelty or a surprise is an encoded object in the set $(\Re - A)$, i.e. outside of set A that is already known to exist. The remarkable achievement of GTP mathematical logic is that there is only one exit route from a set like A, viz., with the incorporation

---

results in a contradiction with both A and $\sim A$, its negation being provable in the system. Simmons (1993, p. 29) has noted how with the Cantor diagonal lemma (which was used to prove that the power set of a set has greater cardinality than a set) we begin to have so-called 'good' uses of self-refuting structures that result in theorems rather than paradoxes.

[10]It is beyond the scope of this paper to discuss certain anti-machine views of Ben-Jacob (1998) which makes claims for semantic knowledge that goes beyond sub-personal syntactic expression of the Gödel sentence.

[11]F.A. Hayek is the first economist to have discussed the implications for economics that arise from the problems of non-computability that he called *the limits of constructive reason* and on the possibility that the brain manifests Gödel incompleteness (Hayek 1952, 1967). Hayek seminally redirected the discussion on the limits of deductive inference from Humean scepticism to the Gödel logic of incompleteness (Markose 2002, 2005) and indeed brought this to my attention by instructing me to read his book on the *Sensory Order* (1952). However, Hayek's own account of this did not go beyond the Cantor diagonal lemma (see footnote 9), which led him to a view on cognitive incompleteness that there is much knowledge that cannot be formally enumerated.

of the Liar or contrarian strategy function, denoted by $f^\neg$, which has a non-computable fixed point encapsulated in the Gödel sentence. Thereafter, any non-trivial computable function of the Gödel sentence is the so-called Post productive function Post (1941) that determines the logical necessity for surprise mappings into the set $(\mathfrak{R} - A)$. As the productive function in logic implements novelty and surprise, it will be denoted as $f^!$.

The significance of the mathematical logic of GTP is that any system incorporating such capacity for offline simulation will imply incompleteness or the capacity to produce new objects outside of a given set that can be enumerated by *any*, including the negating or oppositional UTM. The first intriguing corollary that this framework signifies is that (syntactic) recognition of hostile agents requires the highest level of computational intelligence (which Steven Wolfram (2002) claims is already ubiquitous even in the humble virus) and consequently, in the absence of contrarian agents or oppositional structures, there is no logical need to innovate or surprise. Finally, neither the formalism regarding the *Liar* nor the arms race in novelty is familiar to economists as fixed points of recursive functions in diagonal constructions such as in the Second Recursion Theorem. It should be noted, once an oppositional structure to an encoded action arises, the Liar can lead to the failure of the other party only *out of* (Nash) *equilibrium* when the identity of the Liar is not known or the formal structure of the game involving the Liar is not acknowledged. The arms race in novelty follows in a Nash equilibrium when both agents coexist. Hence, interestingly once opposition is in place, if innovation is not adopted, then expect to be 'negated' by the hostile agent, viz., innovate or die.

Though important, it is beyond the scope of this paper to give a fuller account of human and biological proteanism[12] and to give more evidence from recent discoveries such as that of the mirror neuron system which gives cellular neurophysiological evidence for common coding and offline simulation for action prediction among conspecifics. I will confine my literature review to papers that have used recursion function theory in game theory.

A number of game theory papers such as Albin (1982), Anderlini (1990), Anderlini and Sabourian (1995), Canning (1992), Nachbar and Zame (1996), which use recursion function theory, focus exclusively on defining the problem of indeterminacy associated with self-refuting decision structures. It is interesting to note that these game theory papers discuss neither the significance nor the possibility of innovation and surprise strategies arranged in a structure of an arms race. The problems here arise for two main reasons. These papers appear not to utilise the major methodological triumph of Gödel (1931) which is the meta-analysis that produces fully definable meta-propositions, in an ever extendable sequence, in terms of the Post (1944) productive functions that provide recursive mappings from self-refuting fixed points to the outside of given recursively enumerable sets in order to avoid logical inconsistency, hence, the aphorism that sufficiently rich formal sys-

---

[12]Byrne and Whiten (1999) have presented extensive evidence on the development of the Machiavellian Brain.

tems cannot be both consistent and complete. Secondly, the characterisation of Nash equilibria as fixed points of recursive functions seems not to be specified as such. It was in the seminal paper of Spear (1989) that the Second Recursion Theorem was introduced to formalise the problem of computing rational expectations equilibria as fixed points. Though Spear (1989) does not explicitly depict a game, it will be shown that without the proper formalism of defining fixed points of recursive functions, the Nash equilibria of a game which requires the identification of the meta-representations of mutual best response functions in a two-place diagonal alignment, one could be forced into different 'resolutions' of classic oppositional problems.[13]

The rest of the paper is organised as follows. Section 11.2 sets up the mathematical preliminaries for a recursive function approach to analyse a two-person finite game. Computability constraints are imposed on decision procedures and implementable action rules. This leads to a framework well known as Gödel metamathematics (see Rogers 1967) which implements a 1-1 mapping between internal offline recordings made by players with their respective external online calculations.

In Sect. 11.3, this enables us to exploit a variant of the Second Recursion Theorem used by Spear (1989) to establish the fixed points of recursive functions which are Nash equilibria strategy functions of a two-person game. In Sect. 11.4, this framework for determining fixed points is used to reformulate the original Binmore (1987) notion of surprise strategies in a Nash equilibrium in terms of the computability or not of fixed points of best response functions of the players. Here, the structure of opposition arising from the Liar strategy is also analysed. The first significant point is that the Liar can win only *out of equilibrium* when the identity of the Liar is not known or the formal structure of the game involving the Liar is not acknowledged. The second and more famous result is that when there is mutual or common knowledge of the Liar, we are at Gödel's non-computable fixed point where there is the recursive indeterminacy of the action-reaction functions of the agents. In the final section, we prove that the only best response function in the

---

[13]Koppl and Rosser (2002) attempt to characterise the Nash equilibrium of the zero sum game that depicts the machinations of a well-known oppositional game involving Holmes and Moriarty using recursive function theory. Moriarty, who seeks the demise of Holmes, has to be in proximity with him, while Holmes needs to elude Moriarty. They conclude as follows: 'We can see that there are best-reply functions, $f(x)$, such that $f(x) \neq x$, $\forall$ x. That is, there are best-reply functions without a fixed point. (A fixed point is defined by the condition that $f(x) = x$.)'. It will be shown that Gödel meta-representational system has no problem 'referring' to the fixed point of the best response function that seeks to negate or deceive as in the Holmes-Moriarty game. The important point here, therefore, is not that one or the other player has to find a best response function that does not have a fixed point, but that the fixed point of an important class of best response functions is not computable, and this is fully deducible from within the meta-representational system of the players.

Nash equilibrium with the Liar is the productive function implementing the surprise strategy or an innovation outside extant action sets. The productive set displays a structure of an arms race in innovation and hence grows in a nonanticipating way.

11.2 Recursion Function Theory and Gödel Incompleteness: An Introduction

Gödel (1931) pioneered the framework of analysis called *metamathematics* pertinent to self-referential structures where he obtains epochal results on the sort of statements an internal observer can make as a meta-theorist if he is constrained to be very precise in what he can know and how he can make inferences. Thus, as highlighted by Binmore (1987), the significance of using the Gödel meta-analysis to establish incompleteness or non-computability results for formalised decision problems and game theory stems precisely because this can be proven to arise not from incorrect or inconsistent reasoning or calculation but rather to avoid strategic irrationality and logical contradiction. To this end instrumentally rational players are accorded the full powers of an idealised computing machine in the calculation of what are effectively self-referential mappings for the determination of Nash equilibrium strategies. Following from the Church-Turing thesis, the computability constraint on decision procedures implies that these are computable functions that can only entail finitely describable set of instructions in the computation. Likewise, all information is in codifiable form.

Again by a method introduced by Gödel (1931) called Gödel numbering, all objects of a formalisable system describable on the basis of a countable alphabet are put into 1-1 mapping with the set of natural numbers referred to as their Gödel numbers (g.ns, for short). Thus, computable functions can be indexed by the g.n of their finitely describable program. Impossibility results on computation, therefore, become the only constraints on what rational/optimising players cannot calculate given the same information on the encoded primitives on the game.

11.2.1 Some Preliminaries on Computable Functions

By the Church-Turing thesis, computable functions are number theoretic functions, $f : \aleph \rightarrow \aleph$, where $\aleph$ is the set of all integers.[14] Each computable function is identified by the index or g.n of the program that computes it when operating on an

---

[14]The first limitative result on functions computable by T.Ms is that at most there can only be a countable number of these with the cardinality of $\aleph$ being denoted by $\aleph_0$, while from Cantor we know that the set of all number theoretic functions have cardinality of $2^{\aleph_0}$. Hence, not all number theoretic functions are computable (see Cutland 1980).

input and producing an output if the function is defined or the calculation terminates at this point. Following a well-known notational convention, we state this for a single valued computable function as follows

$$f(x) \cong \phi_a(x) = q. \tag{11.1}$$

That is, the value of a computable function $f(x)$ when computed using the program/TM with index a is equal to an integer $\phi_a(x) = q$, if $\phi_a(x)$ is defined or halts (denoted as $\phi_a(x) \downarrow$) or the function $f(x)$ is undefined ($\sim$) when $\phi_a(x)$ does not halt (denoted as $\phi_a(x) \uparrow$). The domain of the function $f(x)$ denoted by Dom ϕ_a or W_a is such that,

$$\text{Dom } \phi_a = W_a = \{x|\phi_a(x) \downarrow: \text{TM}_a(x) \text{ halts}\}. \tag{11.2}$$

Note, the range of the function $f(x)$ is denoted by E_a.

Definition 1. Computable functions that are defined on the full domain of $\aleph$ are called total computable functions. Partial computable functions are those functions that are defined only on some subset of $\aleph$.

Related to (11.2) is the notion of sets whose members can be enumerated by an algorithm or a TM.

Definition 2. A set which is the null set or the domain or the range of a recursive/computable function is a recursively enumerable set. Sets that cannot be enumerated by T.Ms are not recursive enumerable.

The one feature of computation theory that is crucial to game theory where players have to simulate the decision procedure of other players is the notion of the universal Turing machine (UTM).

Definition 3. The UTM is a partial computable function, defined as $\Psi(a, x)$, which uses the index a of the TM whose behaviour it has to simulate. By what is called the Parameter or Iteration Theorem, there is a total computable function u(a) which determines the index of the UTM such that

$$\Psi(a, x) = \phi_{u(a)}(x) \cong \phi_a(x). \tag{11.3}$$

Equation (11.3) says that the UTM, on the left-hand side of (11.3) on input x, will halt and output what the TM_a on the right-hand side does when the latter halts and otherwise both are undefined.

Of particular significance are Turing machines that use their own code/g.n as inputs in their calculation. We will refer to these as self-referential calculations.

Definition 4. The set denoted by C is the set of g.ns of all TMs that halt when operating on their own g.ns or alternatively C contains the g.ns of those recursively enumerable sets that contain their own codes

(see Cutland 1980, p. 123; Rogers 1967, p. 62).

$$C = \{x|\phi_x(x) \downarrow: TM_x(x) \text{ halts}; x \in W_x\}. \tag{11.4}$$

$$C^{\sim} = \{x | \phi_x(x) \uparrow : TM_x(x) \text{ does not halt}; \ x \notin W_x\}.$$

Theorem 1. *The set $C^{\sim}$ is not recursively enumerable.*

In the proof that $C^{\sim}$ is not recursively enumerable, viz., there is no computable function that will enumerate it, Cantor's diagonalisation method is used.[15]

11.2.2   Post (1944) Set Theoretic Representation of Gödel Incompleteness

As indicated in the introduction, we will now state the formal character of systems capable of the endogenous production of novelty or surprises in terms of the notion of creative and productive sets first defined by Emil Post (1944).

Definition 5. A *creative* set Q is a recursively enumerable set whose compliment, $Q^{\sim}$, is a *productive* set. The set $Q^{\sim}$ is productive if there exists a recursively enumerable set W_x disjoint from Q (viz., $W_x \subset Q^{\sim}$) and there is a total computable function $f(x)$ which belongs to $Q^{\sim} - W_x$. $f(x) \in Q^{\sim} - W_x$ is referred to as the productive function and is a 'witness' to the fact that $Q^{\sim}$ is not recursively enumerable. Any effective enumeration of $Q^{\sim}$ will fail to list $f(x)$, Cutland (1980, pp. 134–136).

Lemma 1. *Set C in (11.4) is a simple example of a creative set. The productive function $f(i) = i$ is the identity function.*
By the definition of C if any number $i \in C \leftrightarrow i \in W_i$ by the definition of C. Hence, for $f(i) = i$ if $f(i) \in C \leftrightarrow i \in W_i$. If W_i is disjoint from C, then $f(i) \notin C \cup W_i$. If $i \in W_i$, then $i \in C$ and W_i will not be disjoint from C.

Smullyan (1961, p. 58) claims that the construction of two recursively insep-arable disjoint subsets such as that for C and $C^{\sim}$ with W_i disjoint from any subset of C plays a fundamental role in modern approaches to incompleteness and undecidability. The inseparability of two recursively enumerable disjoint number sets, A and B ($A \subset C$ and $B \subset C^{\sim}$), arises from the property that any recursive reductions of these sets, respectively, denoted as A' and B', will imply that the Gödel number for the productive function for $A'^{\sim}$ lies outside of both A' and B' and will be a constructive 'witness' for incompleteness. The following Lemma 2 will be used to engineer a recursive reduction between sets such as A and B and the respective sets

---

[15] Assume that there is a computable function $f = \phi_y$, whose domain $W_y = C^{\sim}$. Now, if $y \in W_y$, then $y \in C^{\sim}$ as we have assumed $C^{\sim} = W_y$. But by the definition of $C^{\sim}$ in (11.5) if $y \in W_y$, then $y \in C$ and not to $C^{\sim}$. Alternatively, if $y \notin W_y$, $y \notin C^{\sim}$, given the assumption that $C^{\sim} = W_y$. Then, again we have a contradiction, as since from (11.5) when $y \notin W_y$, $y \in C^{\sim}$. Thus, we have to reject the assumption that for some computable function $f = \phi_y$ and that its domain $W_y = C^{\sim}$.

A' and B' to prove that the Nash equilibrium surprise strategy is none other than the productive function, as in the case for the set $A'^{\sim}$.

Lemma 2. *Consider two recursively enumerable disjoint number sets, A and B, with $A \subset C$ and $B \subset C^{\sim}$ and let $B = W_{\sigma(n^{\rightarrow})}$ of Lemma 1 with index $\sigma(n^{\rightarrow})$. Let the recursive function $h(i)$ define the following many-1 reduction of A to A': If $i \in A$, then $h(i) \in A'$ and $j \notin i$, if $j \in B$, then $h(j) \in B'$. Hence, $A = h^{-1}(A')$ and $B = h^{-1}(B') = W_{\sigma(n^{\rightarrow})}$. As $W_{\sigma(n^{\rightarrow})} \subset C^{\sim}$ and $f(\sigma(n^{\rightarrow}))$ is a productive function of $C^{\sim}$ with g.n $\sigma(n^{\rightarrow}) \notin C \cup W_{(n^{\rightarrow})}$, it also serves as the productive function for $A^{\sim}$. Likewise, $g(h(f((n^{\rightarrow}))$ is the productive function for $A'^{\sim}$ and with $B' = W_{\sigma'(n^{\rightarrow})}$ the g.n$(g(h(f(\sigma(n^{\rightarrow})))) \notin A' \cup W_{\sigma'(n^{\rightarrow})}$.*

The proof is as follows.[16] As A and B are disjoint and $B \subset C^{\sim}$ with f being an identity function, $\sigma(n^{\rightarrow})$ cannot belong to either A or B as this will imply that $\sigma(n^{\rightarrow}) \in W_{\sigma(n^{\rightarrow})}$. If $\sigma(n^{\rightarrow}) \in A$, then as $A \subset C$, $\sigma(n^{\rightarrow}) \in W_{\sigma(n^{\rightarrow})}$ and $W_{\sigma(n^{\rightarrow})} \subset C$, which entails a contradiction. Hence, $\sigma(n^{\rightarrow}) \notin A$ and $\sigma(n^{\rightarrow}) \notin A \cup W_{\sigma(n^{\rightarrow})}$. This also implies that $A^{\sim}$ is not recursively enumerable and that $A^{\sim}$ is productive. The recursive reduction function h will guarantee that as $A^{\sim}$ is productive, so is $A'^{\sim}$. The non-trivial productive function $g(h(f(\sigma(n^{\rightarrow}))$ for $A'^{\sim}$ in Lemma 2 which is not an identity function as in Lemma 1 will also be shown to arise from the surprise best response function in the Nash equilibrium of the two-person oppositional game.

Figure 11.2 illustrates the Post (1944) set theoretic representation of Gödel Incompleteness Result. The prototypical creative set is the set C in (11.4) which contains self-referential calculations that converge. They will be shown to correspond to computable fixed points. Contrarian propositions of the latter, on account of consistency of the system, belong to a set disjoint from C and hence though a subset of the complement of C, viz., $C^{\sim}$ in (11.5), its membership can be enumerated and also shown to be a set on which Turing machines can be logically deduced to be incapable of halting.[17] Thus, there is a recursively enumerable subset of $C^{\sim}$,

---

[16]This is analogous to the proof in Smullyan (1961, p. 96, Chap. V, Proposition 2).

[17]If sentences in a formal system, denoted as FS, are provable and have the status of being theorems (proof being defined as the operation of a Turing machine that halts), then their negations are refutable in that it is known that they belong to the domain on which Turing machines will not halt when attempting their proofs. If a *FS* is complete, then the set of all sentences satisfies the condition that $FS = TUR$, where T and R, respectively, are the set of provable and refutable sentences. *FS* is consistent if T and R are disjoint. The subset of R that is recursively enumerable are negations of those propositions that are known to be provable. The set *FS* is said to be incomplete if $TUR \subset FS$. The Gödel (1931) incompleteness result and the set theoretic proof of this by Post (1944) provide a constructive proof of a sentence denoted as u such that $u \in FS$ and u does not belong to TUR. The sentence u is undecidable and is the 'witness' that *FS* is incomplete.

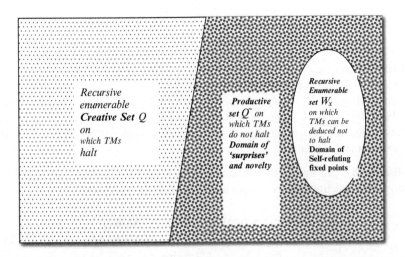

Fig. 11.2 Post (1944) set theoretic representation of Gödel incompleteness in the domain outside disjoint recursively enumerable sets (see Definition 5)

and it represents self-refuting fixed points that are fully deducible in the system. Any total computable function of these which defines the productive function can only map into the domain for novelty[18] in Fig. 11.2 outside both these recursively enumerable disjoint sets. We propose to show that $f(x)$ the productive function in Definition 5 which provides a 'witness' for the incompleteness of the formal system also corresponds to the best response surprise function in the Nash equilibrium of a game.

11.2.3   Meta-Representational System (MTS) and Simulation Theory for a Two-Person Game

This section sets out how a MTS organises encoded information involving the self and other. This interactive situation is best characterised by a two-person game. The primitives of the game, can be interpreted as one in which both cooperation and opposition arise such as in a regulatory/policy game or a parasite-host game, is codified as follows.

$$G = \{(p, g), (A_p, A_g), s \in S\}. \tag{11.6}$$

---

[18]Markose (2004) identifies this to be the domain of complex dynamics and the emergence of novelty proposed by Wolfram (1984), Casti (1994) and Langton (1992).

Here, (p, g) denote the respective g.ns of the objective functions, to be specified, of players, p, the private sector/regulatee and g, government/regulator. The action sets denoted by A_i are finite and countable with $a_{il} \in_i$, $i \in (p, g)$ being the g.n of an action of player i and $l = 0, 1, 2, \ldots, L$. An element $s \in S$ denotes a finite vector of state variables and other archival information, and S is a finite and countable set. The action set $A = A_p \cup A_g$ represents the known technologies. In order to highlight the fundamental recursive nature of actions as technologies and the potential for new technologies, the class of best response strategy functions will be defined as a set of total computable functions.

Definition 6. The best response strategy functions f_i, $i \in (p, g)$ that are total computable functions can belong to one of the following classes:

$$f_i = \begin{bmatrix} f_i^1 = 1(\text{unitFunction}) & - & \text{Rule} & \text{Abiding} \\ f_i^{\neg} & - & \text{Rule} & \text{Breaking/Liar} \\ f_i^! & - & \text{Surprise} \end{bmatrix} \qquad (11.7)$$

such that the g.ns of f_i are contained in set $\Re$,

$$\Re = \{m \mid f_i = \phi_m, \phi_m \text{ is total computable}\}. \qquad (11.8)$$

The set $\Re$ which is the set of all total computable functions is not recursively enumerable. The proof of this is standard; see Cutland (1980). The total computability of best response functions $f_i = \phi_m$, $m \in \Re$ in (11.7)–(11.8) yields the notion of constructible/effective action rules such that a finitely codifiable description of some (institutional) procedure which is defined for all mutually exclusive states of the world is obtained.

As will be clear, (11.8) draws attention to issues on how innovative actions/institutions can be constructed from existing action sets. The remarkable nature of the set $\Re$ is that potentially there is countable infinite number of ways in which 'new' institutions can be constructed from extant action set A. The task is to show the conditions under which it is mutually deducible that the best response function f_i, $i \in (p, g)$ satisfies Post's productive function and is a surprise strategy, $f_i = f_i^! = \phi_m$, such that $m \in \Re - A$. Only such innovations will be accorded with the status of strategic innovations. The trigger for the surprise strategy involves the identification of the negation or Liar strategy $f_i^{\neg}$. These total computable strategy functions with the exception of the identity function in (11.7) will be shown to generate dynamics in the system.

A major implication of imposing computability constraints on all aspects of the game is that all meta-information with regard to the outcomes of the game for any given set of state variables, $s \in S$, can be effectively organised by the so-called prediction function $\phi_{\sigma(x,y)}(s)$ in an infinite matrix Ξ of the enumeration of all partial computable functions. This is given in Fig. 11.2 (see Cutland 1980, p. 208). The tuple (x, y) identifies the row and column of this matrix Ξ whose rows are denoted as Ξ_j, $i = 0, 1, 2, \ldots$.

$$
\begin{array}{ccccccc}
\Xi_0 & \phi_{\sigma(0,0)} & \phi_{\sigma(0,1)} & \phi_{\sigma(0,2)} & \phi_{\sigma(0,3)} & \cdots & \phi_{\sigma(0,y)} & \cdots \\
\Xi_1 & \phi_{\sigma(1,0)} & \phi_{\sigma(1,1)} & \phi_{\sigma(1,2)} & \phi_{\sigma(1,3)} & \cdots & \phi_{\sigma(1,y)} & \cdots \\
\Xi_2 & \phi_{\sigma(2,0)} & \phi_{\sigma(2,1)} & \phi_{\sigma(2,2)} & \phi_{\sigma(2,3)} & \cdots & \phi_{\sigma(2,y)} & \cdots \\
\vdots & \vdots & \vdots & \vdots & \vdots & \vdots & \vdots \\
\Xi_x & \phi_{\sigma(x,0)} & \phi_{\sigma(x,1)} & \phi_{\sigma(x,2)} & \phi_{\sigma(x,3)} & \cdots & \phi_{\sigma(x,x)} & \cdots
\end{array}
$$

Fig. 11.3 Meta-information on outcomes and dynamics for two-person games : matrix Ξ

The function $\phi_{\sigma(x,y)}(s)$ if defined at a given state s and $\sigma(x,y)$ yields

$$\phi_{\sigma(x,y)}(s) = q. \tag{11.9}$$

Here, q in some code, determines the outcome of the decision problem of the game and $q \in E_{\sigma_x}$. Note, $\sigma(x,y)$ is the index of the program for this function ϕ that produces the outputs of the strategic decision problem of the two-person game. The tuple also identifies a point on the matrix Ξ in Fig. 11.2. The conditions under which the output of the prediction function for each (x,y) point in the above matrix is defined are given in the following Theorem (Fig. 11.3).

Theorem 2. *The representational system is a 1-1 mapping between meta-information in matrix Ξ in Fig. 11.3 and executable computations such that the conditions under which the prediction function which determines the output of the game for each (x,y) point is defined as follows:*

$$\phi_{\sigma(x,y)}(s) \cong \phi_{\phi_x(y)}(s) = q, \ \textit{iff} \ \phi_x(y) \downarrow . \tag{11.10}$$

Here, the total computable function $\sigma(x,y)$ modelled along the lines of Gödel's two-place substitution function[19] (see Rogers 1967, pp. 202–204) has the feature that it names or 'signifies' in the meta-system Ξ the points in the game that correspond to the different executed calculations on the right-hand side of (11.10) as we substitute different values for (x,y) for a given state s. The g.ns representing $\sigma(x,y)$ can always be obtained whether or not the partial recursive function on the right-hand side of (11.10) which executes programs halts or not.

Proof. See Rogers (1967).

By the necessary condition in (11.10) if the function $\phi_x(y)$ on the right-hand side (RHS) executing the internal calculation is defined, we say the prediction function $\phi_{\sigma(x,y)}$ in the meta-system on the (LHS) producing the output of the game is computable and the outcome q of the game at that point is predictable. Likewise,

---

[19]This approach economises on formalism and enables us to highlight and exploit the Fixed Point Theorems of recursive function theory to determine Nash equilibrium outcomes more readily than has been the case in, for instance, in Anderlini (1990) and Canning (1992).

the 'only if' condition in (11.10) implies that meta-statements that are valid on the predictability of the outcomes of the game at any (x, y) must give the correct inference on whether program executions on the right-hand side terminate. In view of the discovery of the mirror neuron system, as indicated in the introduction and discussed further in Markose (2015), the set-up in (11.10) formalises the relationship between a mirror and a meta-system on the LHS of (11.10) which records all 'successful' machine executions on the RHS of (11.10). The latter relates to the canonical system involving online activity, while the LHS of (11.10) denotes the *offline* simulations and the synchrony implied in (11.10) can integrate self and the other as actor and observer.

Definition 7. The two-place notation of the meta-system $\sigma(x, y)$ can be used to define second-order encodings of the following kind:

(a) When player i has to determine her own best response function, the first place entry x in $\sigma(x, y)$ refers to what the player i does (viz., the g.n of best response function f_i) given that player j plays a strategy that is consistent with player i's belief denoted by y of what player j believes player i has done. Note this is the self-referential second-order belief in Bhatt and Camerer (2005) that is linked with player i's choice of action.

(b) All Nash equilibria and other relevant fixed points of the game satisfying what has been referred to as consistent alignment of beliefs (CAB, for short, Osborne and Rubinstein 1994) have to be elements, $\sigma(x, x)$, along the diagonal array of this matrix. Note, $\sigma(x, y)$ which are off-diagonal entries in matrix Ξ violate the CAB condition.

Note, $\sigma(x, x)$ diagonal points in the meta-system Ξ assume perfect mutual mirroring. GTP meta-analyses are operations on Gödel numbers yielding simulated or virtual experience of the actual phenomena but in principle bypasses the online executions which involve canonical or motor activity. In other words, all permissible inferences are obtained in short hand from encoded information. Likewise, on account of the 'only if' condition in *Theorem 2*, many interesting aspects of the Nash equilibria of computable games can be established only with reference to the meta-analyses and information in the matrix Ξ in Fig. 11.3, with no explicit reference to physical executions of programs.

An out-of-equilibrium belief state will be defined here which represent off-diagonal terms in matrix Ξ. This corresponds to the case (a) in Definition 7 when a false state of belief is attributed regarding either player's best response function f_i, $i \in (p, g)$.

Definition 8 Deceit and False Belief. Denoting by $x^\neg$ the negation of x brought about by best response function $f_i^\neg$ or $f_j^\neg$ defined in (11.7), we have $\sigma(x^\neg, x)$ in the two-place meta-representation of the game by say p. In case (a) in Definition 7, this represents the case when player p knows that he has negated action with g.n x and attributes to player g the false belief that player p is playing x.

Both logically and neurophysiologically (see Markose 2015), this out-of-equilibrium situation involving false beliefs has great significance.

It will be shown how total computable functions for the best response function f_i, $i = p, g$ in a two-person game when applied to the diagonal array of the matrix Ξ can dynamically move it to a specific row in matrix Ξ. The Fixed Point or Second Recursion Theorem states that there exists an index n of a program/set of instructions that computes $f(n)$ and then *applies* $f(n)$ so that both n and $f(n)$ are instructions for the computation of the same recursive function, f, and if the fixed point is computable, the same outcome q is predicted by the operation of the two programs.

Theorem 3. *Fixed Point or Second Recursion Theorem (Cutland 1980, p. 200) Let f be a total unary computable function, then there exists a number n such that*

$$\phi_{f(n)} = \phi_n. \tag{11.11}$$

Note, $f(n) \neq n$ being codes for different programs, but they identify the same function and both sides of equation will yield an identical output if f has a computable fixed point.

From a perspective of the dynamics implied by (11.11), the property that any computable function f has an identifiable fixed point follows from the fact that a function representing an encoded set of instructions when applied to the diagonal array of matrix Ξ in Fig. 11.3 belongs to some row of the matrix Ξ, say v, such that the $v + 1$th element in the vth row, $\phi_{f(\sigma(v,v))}$, and the $v + 1$th element in the diagonal array of Ξ coincide, yielding

$$\phi_{f(\sigma(v,v))} = \phi_{\sigma(v,v)}. \tag{11.12}$$

This is demonstrated as follows by the $v + 1$th element of the vth row of matrix Ξ:

$$\Xi_v \; \phi_{f(\sigma(0,0))} \; \phi_{f(\sigma(1,1))} \; \phi_{f(\sigma(2,2))} \cdots \boxed{\phi_{f(\sigma(v,v))} = \phi_{\sigma(v,v)}} \cdots \phi_{f(\sigma(x,x))} \cdots \tag{11.13}$$

A major advantage of this framework is that the determination of Nash equilibrium strategies involves the use of total computable best response functions (f_p, f_g) which can be shown to operate directly on points such as $\sigma(x, x)$ to effect computable transformations of the system from one row to another of matrix Ξ with special reference to its diagonal array; see Fig. 11.3. *Theorem 3* is used in the determination of the fixed points for the total computable functions best response function f_i, $i = p, g$. When one player applies his best response f_i, Nash equilibria require that both players identify the same prediction function as producing the output of the game under conditions of consistent alignment of beliefs (that avoids false beliefs given in Definition 8).

$$\phi_{f_i\sigma(v,v)}(s) = \phi_{\sigma(v,v)}(s), \quad i \in (p, g). \tag{11.14}$$

11.3   Nash Equilibria : When Does One Surprise
the Opposition?

11.3.1   Total Computable Best Response Functions
and Optimal Strategy Functions

The optimisation algorithms entailed in achieving best responses in the game arise
from the objective functions of players.

Definition 8. The objective functions of players are computable functions Π_i,
$i \in (p, g)$ defined over the partial recursive payoff/outcome functions specified
in (11.10) and the strategy functions specified in (11.7)

$$\text{Arg} \underset{b_i \in B_i}{\text{Max}}\ \Pi_i(\phi_{\sigma(b_i, b_i)}(s)), \quad i \in (p, g). \tag{11.15}$$

The choice set B_i contains the g.ns of strategy functions. The Nash equilibrium
strategies (β_g^E, β_p^E) with g.ns denoted by (b_p^E, b_g^E) entail up to two subroutines
or iterations, to be specified below. In principle, the strategy functions (β_g, β_p)
are universal Turing machines that simulate optimal strategies of the players that
satisfy (11.15) and involve the total computable best response functions (f_p, f_g)
which incorporate elements from the respective action sets $A = (A_p, A_g)$ and given
mutual second-order self-referential beliefs of one another's optimal strategy. In
the Nash equilibrium best response calculus, the first subroutine denoted by g.n
b^1 simulates the other player's optimisation calculus to determine mutual optimal
actions. The problem is that actions can in general be implemented by any total
computable best response function, $f_i = \phi_m$, $m \in \Re$, $i \in (p, g)$ in (11.8).

In standard rational choice models of game theory, the optimisation calculus
(with Godel number z) in the choice of best response restricts choice to given actions
sets. Hence, starting from some point $\sigma(x, x)$, the strategy functions map from a
relevant tuple that encodes meta-information of the game into given action sets

$$\beta_i(f_i \sigma(x, x), z, s, A) \rightarrow A_i \text{ and } f_i = \phi_m, \ m \in A, \ i \in (p, g). \tag{11.16}$$

Unless this is the case, as the set $\Re$ is not recursively enumerable, there is in
general no computable decision procedure that enables a player to determine the
other player's best response function. However, in principle, a strategic decision
procedure (β_g, β_p) for choice of best response, $f_i = \phi_m$, $m \in \Re$, $i \in (p, g)$, can map
into $\Re - A$, implying that an innovative action not previously in given action sets is
used.

$$\beta_i(f_i \sigma(x, x), z, s, A) \rightarrow \Re - A \text{ and } f_i = f_i^! = \phi_m, \ m \in \Re - A, \ i \in (p, g). \tag{11.17}$$

The question is which fixed point $\sigma(x, x)$, fully deducible in the metamathematics, will trigger such Nash equilibrium surprise strategies, $(\beta_g^{E!}, \beta_p^{E!})$, with g.ns

denoted by $(b_p^{E!}, b_g^{E!})$? It has been noted in passing by Anderlini and Sabourian (1995, p. 1351), based on the work of Holland (1975), that heterogeneity in forms does not arise primarily by random mutation but by algorithmic recombinations that operate on existing patterns. However, a number of preconceptions from traditional game theory such as the "givenness" of actions sets prevent Anderlini and Sabourian (1995) from positing that players who as in (11.17), equipped with the wherewithal for algorithmic recombinations of existing actions, do indeed innovate from strategic necessity rather than by random mutation. Indeed, it is the very function of the Gödel meta-framework to ensure that no move in the game made by rational and calculating players can entail an unpredictable/surprise response function from set $\Re - A$ unless players can mutually infer by strictly codifiable deductive means from $\sigma(x, x)$ that (11.17) is a logical implication of the optimal strategy at the point in the game. In other words, the necessity of an innovative/surprise strategy as a best response and that an algorithmic decision procedure is impossible at this point are fully codifiable propositions in the meta-analysis of the game. While it will be shown what specific structure of opposition logically and strategically necessitates surprise strategies in the Nash equilibrium of the game, in keeping with the set theoretic formulation of novelty production in Fig. 11.2, the corresponding creative and productive disjoint subsets of the strategy sets have also to be developed.

11.3.2   Fixed Point/Second Recursion Theorem: The Base Point

The meta-analysis in the determination of Nash equilibrium strategies (β_p^E, β_g^E) with g.ns (b_p^E, b_g^E) will be undertaken here. In the classic matching pennies game format, the optimal outcomes for the government/regulator arise when the regulatee/private sector is rule abiding or coordinating. Calculations start at this so-called base point which is the fixed point of f_g which has to be arrived at by player p in (11.18):

$$\phi_{f_g \sigma(b_a, b_a)}(s) = \phi_{\sigma(b_a, b_a)}(s) = q. \tag{11.18}$$

Here, b_a is the g.n of the strategy f_g that selects the optimal action a from set A in (11.16) when g is put in for the index i. In the two-place notation in $\sigma(b_a, b_a)$ on the RHS of (11.18), the first b_a is the code of the program from (11.16) as adopted by p to simulate the optimal policy rule a, and the second place b_a denotes that p believes that g believes and acts on the basis that the p is rule abiding and has left the policy rule a unchanged. The prediction functions in (11.18) $\phi_{\sigma(b_a, b_a)}(s)$ are computable, and outcomes of the policy rule a are predictable and q is the desired outcome that g wants in state variables when applying this policy rule a. It is convenient to assume that policy rule a is optimal for g if the private sector is rule abiding. By rule abiding is meant that p will leave the system unchanged in terms of the row b_a of matrix Ξ in Fig. 11.3. However, the predictable outcome q may involve asset prices or quantity positions, which may attract profitable arbitrage in the form of the Liar strategy against it furnishes the conditions under which a transparent/predictable rule will fail to be a Nash equilibrium strategy.

11.3.3 The Liar/Rule Breaker Strategy: The Logic of Opposition

For player p, for the given (a, s) it may be optimal for p to apply the Liar strategy, $f_p^{\neg} \sigma(b_a, b_a)$, with code $b_a^{\neg}$. Formally, the Liar strategy has the following generic structure. For any state s when the rule a applies,

$$\phi_{f_p^{\neg} \sigma(b_a, b_a)}(s) = q^{\sim}, \quad q^{\sim} \notin E_{\sigma_{b_a}} \leftrightarrow \phi_{\sigma(b_a, b_a)}(s) = q, \quad q \in E_{\sigma_{b_a}}. \tag{11.19}$$

For all s when policy rule a does not apply,

$$f_p^{\neg} = 0 : \text{Do Nothing}. \tag{11.20}$$

The Liar can successfully subvert with certainty in [LHS of (11.19)] if and only if ($\leftrightarrow$) the policy rule a has predictable outcomes [RHS of (11.19)] and $f_p^{\neg}$ itself is total computable. Thus, $f_p^{\neg} = \phi_m$, $m \in A_p$, must include a codified description of an action rule if undertaken by the Liar can subvert the predictable outcomes of the policy rule a. Formally, if q is predicted, then the application of $f_p^{\neg}$ to $\sigma(b_a, b_a)$ is equivalent to the condition of deliberate deceit in Definition 7(a), and the g.n of this strategy is $\sigma(b_a^{\neg}, b_a)$. That is, p has negated b_a and he knows that g harbours a false belief about him, that p is rule abiding with b_a. This out-of-equilibrium $\sigma(b_a^{\neg}, b_a)$ point in the game is off diagonal in terms of the matrix Ξ and will bring about an outcome $q^{\sim} \notin E_{\sigma_{b_a}}$ which belongs to a set disjoint from the set that contains the desired output of rule a for all s for which rule a applies, viz., $E_{\sigma_{b_a}} \cap E_{\sigma_{b_a^{\neg}}} = \emptyset$. The outcomes $(q^{\sim}, q)$ can be zero sum, but in general we refer to property $q^{\sim} \notin E_{\sigma_{b_a}}$ in (11.19) as being oppositional or subversive.

Thus, we come to the point as to why agents who precipitate the Wolfram–Chomsky Type IV dynamics with innovation have to have powers of self-referential calculation. As discussed, in a Nash equilibrium, p has to remove any attribution of false belief to g about p's identity as the Liar.

Theorem 4 Non-computable Fixed Point. *The prediction function indexed by the fixed point of the Liar/rule breaker best response function $f_p^{\neg}$ in (11.21) is not computable and corresponds to the famous Gödel sentence for the game.*

$$\phi_{f_p^{\neg} \sigma(b_a^{\neg}, b_a^{\neg})}(s) = \phi_{\sigma(b_a^{\neg}, b_a^{\neg})}(s). \tag{11.21}$$

The non-computability of the fixed point follows from the property that the output of the game at this point cannot be predicted in the meta-system. Proof is standard and rests on premise that formal system is consistent. Assume that the fixed point of the recursive function $f_p^{\frown}$ in (11.21) is computable, then if the R.H.S of (11.21) produces the output q and the L.H.S by the definition of the Liar strategy produces output $q^{\sim}$. Hence, if (11.21) is computable, we have $q = q^{\sim}$ which is a contradiction. Though the conditions of the out-of-equilibrium success of the Liar are spelt out in (11.19) and are computable, in many fast moving co-evolutionary systems, predictable strategies such as b_a or $b_a^{\frown}$ may not be observed, and instead only the arms race in novelty given in the next section is what persists such that both players coexist.[20]

11.3.4   Surprise Nash Equilibria

There is no paradox in stating that as both players can prove the non-computability of (11.21), they will be able to mutually deduce that the only Nash equilibrium strategies for both players that is consistent with meta-information in the fixed point in (11.21) is one that involves strategies that elude prediction from within the system. On substituting the fixed point $\sigma(b_a^{\frown}, b_a^{\frown})$ in (11.21) for $\sigma(x, x)$ in (11.17), g's Nash equilibrium strategy β_q^E with g.n b_q^E implemented by an appropriate total computable function such as in (11.18) must be such that

$$\beta_g^E(f_g\sigma(b_a^{\frown}, b_a^{\frown}), z, s, A) \to \Re - A \text{ and } f_g = f_g^{E!} = \phi_m, \ m \in \Re - A. \quad (11.22)$$

That is, $f_g^{E!}$ implements an innovation and $b_g^E!$ is the g.n of the surprise strategy function in (11.22), hence $\sigma(b_g^E!, b_g^E!)$ is the fixed point of $f_g^{E!}$.

Likewise, for player p, $f_p^{E!}$ implements an innovation in (11.23) and $b_p^E!$ is the g.n of the surprise strategy function, viz., $\sigma(b_p^E!, b_p^E!)$ the fixed point of $f_p^{E!}$. Thus,

$$\beta_p^E(f_p\sigma(b_a^{\frown}, b_a^{\frown}), z, s, A) \to \Re - A \text{ and } f_p = f_p^{E!} = \phi_m, \ m \in \Re - A. \quad (11.23)$$

The intuition here is that from the non-computable fixed point with the Liar, the total computable best response function implementing the Nash equilibrium strategies can only map as above into domains of the action and strategy sets of the players that cannot be algorithmically enumerated in advance.

Using *Theorem 4*, Definition 5 and Lemma 2, we will now prove the incompleteness results for the strategy sets of the players from the Liar/rule breaking

---

[20]In Markose (2015), it is suggested that using environments suitable for neurophysiological experiments of such a game, it is interesting to identify the juncture at which a player p knows his best payoffs come from the out-of-equilibrium configuration wherein the other player g has to be kept in a state of false belief $\sigma(b_a^{\frown}, b_a)$ given in Definition 7.

strategy. Analysis will be done for p's strategy set B_p as the strategy functions β_p and β_g, respectively, can be shown to implement a reduction, as in Lemma 2, of the prototypical creative set C in (11.4).

Corresponding to those (a_{gl}, s) tuples, $a_{gl} \in A_g$ of g's base point optimal strategy for which p's best response f_p is to be rule abiding, viz., $f_p = 1$, the g.ns of these optimal strategies for p, $b_p^1 \in B_p$ result in computable fixed points. Here, in the case when p is rule abiding, b^1 indicates the subroutine 1, which is sufficient for the determination of the Nash equilibrium strategy. This set denoted by β_p^+ which contains all of g's actions for which p is rule abiding can be generated by recursive methodology. Thus,

$$\beta_p^1 = \{b_p^1 | \phi_{b_p1}(b_p^1) \downarrow \text{ for all } (a_{gl}, s), a_{gl} \in A_g, f_p = 1\}. \tag{11.24}$$

Using logic in (11.19), (11.20), a set $\beta_p^{\rightarrow}$ can be recursively generated so that it contains the g.ns of p's strategies for when it is optimal for p to use the Liar best response function $f_p^{\rightarrow}$ to those (a_{gl}, s) tuples, $a_{gl} \in A_g$ of g's action set. By *Theorem 4*, this is a set of p's strategies that can be proven to result in non-computable fixed points. Hence,

$$\beta_p^{\rightarrow} = \{b_p^1 | \phi_{b_p1}(b_p^1) \uparrow \text{ for all } (a_{gl}, s), a_{gl} \in A_g, f_p = f_p^{\rightarrow}\}. \tag{11.25}$$

For the same (a_{gl}, s) tuple, $a_{gl} \in A_g$ constituting g's base point optimal strategy, p's optimal strategy b_p^1 cannot belong to both β_p^+ and $\beta_p^{\rightarrow}$. Thus, logical consistency of the meta-analysis requires $\beta_p^+ \cap \beta_p^{\rightarrow} = \emptyset$, and these are disjoint sets as required by Lemma 2. It is convenient to index them as $\beta_p^1 = W_{\sigma_l^1}$ and $\beta_p^{\rightarrow} = W_{\sigma_n^{\rightarrow}}$ where n denotes a sequence of elements in the sets with the nth element of $W_{\sigma_n^{\rightarrow}}$ referring to $\sigma(b_a^{\rightarrow}, b_a^{\rightarrow})$ in (11.21). As the set $\beta_p^{\rightarrow}$ contains known strategies b_p^1 that imply the negation of g's objectives, denoted generically as q in (11.19), for p and g to achieve their, respective, desired objectives of $q^{\rightarrow}$ and q, in a Nash equilibrium, these can only be done by implementing surprise strategies that map outside of set A. The second iteration or subroutine of the Nash equilibrium strategy denoted as b^2 implements a consistent recursive reduction of the sets β_p^1 to β_p^+ and as in *Lemma 2*, $\beta_p^{\rightarrow}$ is mapped into the complement set of β_p^+ which is denoted as β_p^{+c}:

$$\beta_p^{+c} = \{x \mid \phi_x(x) \uparrow, x \in B_p\}. \tag{11.26}$$

Hence, the incompleteness of p's Nash equilibrium strategy set B_p that arises from the agency of the Liar strategy requires the proof that β_p^{+c} is productive as in *Definition 5* and *Lemma 2*.

Following the steps in Lemma 2, $b_p^{E!} = g.n(b^2(f_p(\sigma_n^{\rightarrow})) = b_n^1$ is the g.n for the productive function defined by the Nash equilibrium surprise strategy function $f_p^{E!}$ implementing an innovation in (11.23). Hence, the subroutine denoted by b^2 implements the recursive reduction from $\beta_p^{\rightarrow} = W_{\sigma_n^{\rightarrow}}$ to a recursively enumerable

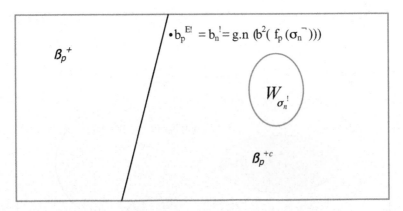

Fig. 11.4 The incompleteness of p's strategy set B_p: *surprise strategy (g.n)*

set $W_{\sigma_n^!}$ indexed by $\sigma_n^!$. The incompleteness of p's Nash equilibrium strategy set in given in Fig. 11.4.

Theorem 5: Productive Function and Surprise Nash Equilibrium. *(i) By construction of the recursive reduction using b^2 on $\beta_p^- = W_{\sigma_n^-}$, $b_p^{E!} = b_n^! = g.n(b^2(f_p(\sigma_n^-)))$ is g.n of the Nash equilibrium total computable best response function $f_p^{E!}$, which implements an innovation as in (11.23) and is productive function of the set β_p^{+c} such that $b_n^! \in \beta_p^{+c} - W_{\sigma_n^!}$ and $b_n^! \notin \beta_p^+ \cup W_{\sigma_n^!}$ as shown in Fig. 11.4. Here, $W_{\sigma_n^!}$ is a recursively enumerable subset of β_p^{+c}, and p's Nash equilibrium strategy set B_p is incomplete.*
(ii) Once the surprise Nash equilibrium strategy has been implemented by p with g.n $b_n^!$, the growth of the strategy set can be proven to take the following nonanticipating form and is shown in Fig. 11.5:

$$W_{\sigma_{n+1}^!} = W_{\sigma_n^!} \cup \{b_n^! = g.n(b^2(f_p(\sigma_n^-)))\} \tag{11.27}$$

Proof. The proof of (i) follows from *Lemma 2*. Proof of (ii) requires showing that surprise strategy functions have g.ns $b_n^!$ that can only be added on to the extant productive set $W_{\sigma_n^!}$ and cannot belong to $W_{\sigma_n^!}$ itself. As given in *Lemma 2*, $\beta_p^- = W_{\sigma_n^-} = (b^2)^{-1}(W_{\sigma_n^!})$, implying that $\sigma_n^! = g.n(b^2(\sigma_n^-))$. Hence, for any n, if $b_n^! \in W_{\sigma_n^!}$ will imply that $\sigma_n^- \in W_{\sigma_n^-}$ and hence lead to a conclusion that the fixed point $\sigma(b_a^-, b_a^-)$ in (11.21) is computable.

The significance of *Theorem 5* is that the surprise strategy is fully definable as a meta-proposition and is paradox free as the surprise strategy is indeed a pure innovation in the strategy set B_p and outside of sets $\beta_p^+ \cup W_{\sigma_n^!}$ that can be enumerated by recursive calculation and information in G; see Fig. 11.5. It is precisely the absence of logical inconsistency and strategic irrationality in the meta-proposition

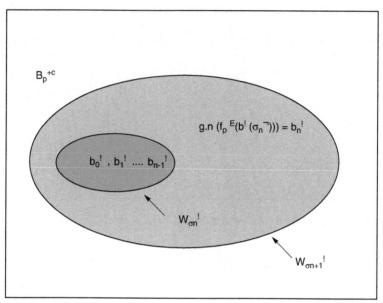

g.n: Godel Number

Fig. 11.5 Arms race in surprises/innovations: growth of the productive strategy set

on the surprise strategy that sustains the consistent alignment of beliefs condition of a Nash equilibrium with surprises. Thus, as already observed, for human players utilising ideal reasoning provided by Gödel meta-analysis, the set $\mathfrak{R}$ of best response functions in (11.8) should provide an inexhaustible source of surprise or innovative strategies. However, by the same token, by *Theorem 5*, there is no algorithmic way by which the prediction function with the index $\sigma(b_p^E!, b_p^E!)$ at the surprise equilibrium can produce an output q though both players can mutually identify that $\sigma(b_p^E!, b_p^E!)$ is the fixed point of the surprise Nash equilibrium best response function f_p^E. Indeed, $\sigma(b_p^E!, b_p^E!)$ says that this is so self-referentially. In a nutshell 'innovate or die' describes this Nash equilibrium in which neither party can unilaterally deviate without drastically impairing their prospects. The Gödel numbers $b_0^!, b_1^!, b_3^!, \ldots, b_n^!$ in Fig. 11.5 can be interpreted to be the g.ns of the 'antibodies' produced in the oppositional encounters.

   Theorem 5 and Fig. 11.5 on the surprise strategy in a Nash equilibrium of a game formally correspond to the set theoretic proof of Gödel's undecidable proposition in miniature, Cutland (1980). We have succeeded in showing the formal equivalence between the Nash equilibrium with surprise or novelty in Fig. 11.4 and the phase transition in dynamical systems theory that characterises the endogenous production of novelty as in Fig. 11.2. Following from part (ii) of *Theorem 5* and as seen in Fig. 11.5, once players are locked in an oppositional structure, the strategy set of each player will grow utilising the formalism of an arms race in novelty.

11.4   Concluding Remarks

The paper has given a unifying framework using recursion function theory and the vehicle of the Second Recursion Theorem to explain how the Gödelian foundations of non-computability that arises from self-refuting structures have far-reaching implications for some of the fundamental issues involving complex strategic behaviour in economic systems. The reaction function that systematically falsifies forecasts or subverts predictable outcomes of rules is the Liar strategy. Logicians and some decision theorists have long known of the problem of epistemic indeterminacy posed by the Liar (see Koons 1992; Simmons 1993). Brian Arthur (1994) has intuitively outlined how unique computable forecast rules cannot exist when games have contrarian- or minority-type payoff functions as in a stock market game when players get the best payoffs when they are contrarian vis-á-vis what the majority do.

For any rule that satisfies the formal conditions on the right-hand side of (11.19), especially if it involves predictable outcomes for asset prices or quantity positions, a Liar strategy against it furnishes us with conditions under which a transparent/predictable rule will fail to be a Nash equilibrium strategy. The agent applying the transparent/predictable rule can have his desired objectives contravened by the Liar. No rational agent can be assumed to operate such a predictable rule, and no institution based on it will survive unless one of the following occurs: the rule is attenuated, the Liar is eliminated or the agents resort to an arms race in innovations.[21]

This paper has indicated that Gödel metamathematics that underpin offline simulations and incompleteness are at the heart of both social coordination and human proteanism. The discussions suggest that the capacity for syntactical encoding of a mutual state of hostility, negation or deceit, viz., the embedding of the Gödel sentence, is what is logically necessary for novelty-producing behaviours. The deeply contextual points of exit and innovation, which follow in lock step, have been demonstrated in *Theorem 5*. The exit routes are guided by the encoded information on specific hostile interactions. The radical uncertainty associated with the Gödel exit points, as noted in Markose (2004, 2005), does not correspond to white noise that economists typically use to model innovations and surprise. Thus, the explication of the logical foundations of novelty production in a strategic setting suggests many rich lines of investigation. These include empirical neurophysiological experiments for evidence of the capacity of the mirror neuron system

---

[21] It is beyond the scope of this paper to discuss why a complexity perspective on radical uncertainty has a bearing on the long tradition of liberal jurisprudence associated with Kant (1781) and Hayek (1960) that universalisable rules of just society are end neutral and have no predictable consequences that are person, place or time specific. It can be conjectured, along the lines of the famous dictum of Hirschman Albert (1991) that certain rules that aim to achieve specific ends in society may lead to futility, perversity and jeopardy that culminates in system failure as a result of gaming and protean agents. Likewise, absenting predictability of outcomes is the basis of a no arbitrage equilibrium in many economic models.

to do offline simulations that can produce social coordination, anti-coordination and innovation. The urgency for studies on the logical and neurophysiological foundations of complex Type IV dynamics in the Wolfram–Chomsky schema arises from the fact that extant mathematical models in economics of strategic behaviour cannot account for protean behaviour which is ubiquitous in socio-economic and biological systems. These foundational studies are necessary to meet the renewed call to arms for a complexity perspective in the aftermath of the 2007 financial crisis as the continued black box approach to innovation and protean behaviours, that can both enhance economies and destabilise them by regulatory arbitrage, proves to be a stumbling block in the modelling and management of such complex systems.

Acknowledgements I'm grateful for the interest at the 2015 AUEB 12th Annual Summer School in ideas relating to computability and complexity for economics. Discussions with AUEB Summer School participants in July 2015 and encouragement from Thanos Yannacouplos have contributed to this paper. Over the years, there have been discussions with Steve Spear, Peyton Young, Aldo Rustichini, Ken Binmore, Arthur Robson, Kevin McCabe, Steven Durlauf, Shyam Sunder and Vela Velupillai. The 2014 ESRC-funded Diversity in Macroeconomics Conference where I had the chance to assemble Vittorio Gallese, Scot Kelso and Eshel Ben Jacob has helped me to take this field to a new frontier. I have also benefitted from feedback from participants at invited Graduate lectures at the following institutions/workshops: 2012 Kiel Institute of the World Economy Summer School, 2010 Ruhr Bochum Economics Department invited Graduate lectures, 2009 Institute for Advanced Studies at Glasgow Workshop, 2002–2009 Lectures at the Centre for Computational Finance and Economic Agents and presently the new MSc Computational Economics, Financial Markets and Policy at the University of Essex.

References

Albin, P.: The metalogic of economic predictions, calculations and propositions. Math. Soc. Sci. **3**, 329–358 (1982)

Albin, P.: In: Duncan F. (ed.) Barriers and Bounds to Rationality, Essays on Economic Complexity and Dynamics in Interactive Systems. Princeton University Press, Princeton (1998)

Anderlini, L.: Some notes on Church's thesis and the theory of games. Theory Decis. **29**, 19–52 (1990)

Anderlini, L., Sabourian, H.: Cooperation and effective computability. Econometrica **63**, 1337–1369 (1995)

Arthur, W.B.: On the evolution of complexity. Working Paper, 93-11-070. Santa Fe Institute, Santa Fe, NM (1993)

Arthur, W.B.: Inductive reasoning and bounded rationality. Am. Econ. Rev. (Pap. Proc.) **84**, 406–411 (1994)

Arbib, M., Fagg, A.: Modeling parietal-premotor interactions in primate control of grasping. Neural Netw. **11**, 1277–1303 (1998)

Axelrod, R.: Risk in Networked Information Systems, Mimeo. Gerald R. Ford School of Public Policy, University of Michigan. (2003); Baumol, W.: The Free Market Innovation Machine. Princeton University Press, Princeton (2002)

Baumol, W.: The Free Market Innovation Machine. Princeton University Press, Princeton (2002)

Baumol, W.: Red Queen games: arms races, rule of law and market economies. J. Evol. Econ. **14**(2), 237–247 (2004)

Beinhocker, E.: Evolution as computation: integrating self-organization with generalized. J. Inst. Econ. **7**(3), 393–423 (2011)

Ben-Jacob, E.: Bacterial wisdom, Gödel's theorem and creative genomic webs. Phys. A Stat. Mech. Appl. **248**(1), 57–76 (1998)

Bhatt, M., Camerer, C.: Self-referential thinking and equilibrium as states of mind in games: FMRI evidence. Games Econ. Behav. **52**, 424–459 (2005)

Binmore, K.: Modelling rational players: Part 1. J. Econ. Philos. **3**, 179–214 (1987)

Byrne, R., Whiten, A.: (1999) Machiavellian Intelligence: Social Expertise and the Evolution of Intellect in Monkeys, Apes, and Humans. Oxford University Press, Oxford (1988)

Canning, D.: Rationality, Computability and Nash equilibrium. Econometrica **60**, 877–895 (1992)

Casti, J.: Complexification: Explaining a Paradoxical World Through the Science of Surprises. London Harper Collins, London (1994)

Colander, D.: The Complexity Vision and the Teaching of Economics, Cheltenham (2000)

Cutland, N.J.: Computability: an introduction to recursive function theory, Cambridge University Press. expectations, learning and convergence to rational expectations equilibrium. Am. Econ. Rev. **72**, 652–68 (1980)

Durlauf, S.: Complexity, economics and public policy. Pol. Philos. Econ. **11**(1), 45–75 (2012)

Gallese, V., Sinigaglia, C.: What is so special about embodied simulation? Trends Cogn. Sci. **15**(11), 515 (2011)

Gallese, V., Fadiga, L., Fogassi, L., Rizzolatti, G.: Action recognition in the premotor cortex. Brain **119**(2), 593–609 (1996). doi: 10.1093/brain/119.2.593. PMID 8800951

Gödel, K.: On Formally Undecidable Propositions of Principia Mathematica and Related Systems [Translation in English in Gödel's Theorem in Focus, ed. by S.G Shanker, 1988, Croom Helm] (1931) Goodhart C.: Central Bankers and Uncertainty, Keynes Lecture to the Royal Academy. Reprinted in the Bank of England Quarterly Bulletin, Feb, pp. 102–114 (1999)

Goodhart, C.: Central Bankers and Uncertainty, Keynes Lecture to the Royal Academy. Reprinted in the Bank of England Quarterly Bulletin, Feb, pp. 102–114 (1999)

Haldane, A.: Financial arms races, speech delivered at the institute for new economic thinking, Berlin, 14 April 2012

Hayek, F.A.: The Sensory Order. The University of Chicago Press, Chicago (1952)

Hayek, F.A.: The Constitution of Liberty. University of Chicago Press, Chicago (1960)

Hayek, F.A.: The Theory of Complex Phenomena. Studies in Philosophy, Politics, and Economics. The University of Chicago Press, Chicago (1967)

Hirschman, A.O.: The Rhetoric of Reaction: Perversity, Futility, Jeopardy. The Belknap Press of Harvard University Press, Cambridge (1991)

Holland, J.: Adaptation in Natural and Artificial Systems. MIT Press, Cambridge (1975)

Kant, I.: Critique of Pure Reason (1965). Translated by Norman Kemp Smith. St. Martin's, New York (1781)

Koons, R.: Paradoxes of Beliefs and Strategic Rationality. Cambridge University Press, Cambridge (1992)

Koppl, R., Rosser B.: Everything i might say will already have passed through your mind. Metroeconomica **53**(4), 339–360 (2002)

Langton, C.: Life at the edge of chaos. In: Langton, G.C., Taylor, C., Doyne Farmer, J., Rasmussen, S. (eds.) Artificial Life II (Sante Fe Institute Studies in the Sciences of Complexity, vol. 10). Addison-Wesley, Reading (1992)

Markose, S.M.: The new evolutionary computational paradigm of complex adaptive systems: challenges and prospects for economics and finance. In: Chen, S.-H. (ed.) Genetic Algorithms and Genetic Programming in Computational Finance, pp. 443–484. Kluwer Academic Publishers, Boston (2002). Also Essex University Economics Department DP no. 552, July 2001

Markose, S.M.: Novelty in complex adaptive systems (CAS): a computational theory of actor innovation. Phys. A Stat. Mech. Appl. **344**, 41–49 (2004). Fuller details in University of Essex, Economics Department. Discussion Paper No. 575, January 2004

Markose, S.M.: Computability and evolutionary complexity: markets as complex adaptive systems (CAS). Econ. J. **115**, F159–F192 (2005)

Markose, S.M.: Mirroring, Offline Simulation and Complex Strategic Interactions: Coordination, Anti-Coordination and Innovation. University of Essex Mimeo, Essex (2015)

Nachbar, J.H., Zame,W.R.: Non-computable strategies and discounted repeated games. Econ. Theory **8**, 111–121 (1996)

Osborne, M., Rubinstein, A.: A Course in Game Theory. MIT Press, Cambridge (1994)

Post, E.: Recursively enumerable sets of positive integers and their decision problems. Bull. Am. Math. Soc. **50**, 284–316 (1944)

Rizzolatti, G., Fadiga, L., Gallese, V., Fogassi, L.: Premotor cortex and the recognition of motor actions. Cogn. Brain Res. **3**, 131–141 (1996)

Rogers, H.: Theory of Recursive Functions and Effective Computability. Mc Graw-Hill, New York (1967)

Sayama, H.: Construction theory, self-replication, and the halting problem. Complexity (2008). doi:10.1002/cplx

Schumpeter, J.A.: Capitalism, Socialism and Democracy, p. 139. Routledge, London (1994). ISBN 978-0-415-10762-4. (1942)

Simmons, K.: Universality of the Liar, Cambridge University Press, Cambridge (1993)

Smullyan, R.: Theory of Formal Systems. Princeton University Press, Princeton (1961)

Spear, S.: Learning rational expectations under computability constraints. Econometrica **57**, 889–910 (1989)

Tesfatsion, L., Judd, K.L. (eds.): Handbook of Computational Economics: Volume2, Agent-Based Computational Economics. Handbooks in Economics Series, 904pp. Elsevier, North-Holland Imprint, Amsterdam (2006)

Tognoli, E., Lagarde, J., de Guzman, G.C., Kelso, J.A.S.: The phi complex as a neuromarker of human social coordination. Proc. Natl. Acad. Sci. USA **104**(19), 8190–8195 (2007)

Turing, A.M.: On computable numbers, with an application to the Entscheidungsproblem. Proc. Lond. Math. Soc. 2 **42**, 230–265 (1937)

Wolfram, S.: Cellular automata as models of complexity. Physica **10D**, 1–35 (1984)

Wolfram, S: A New Kind of Science. Wolfram Media, Inc., Champaign (2002)

Chapter 12
Revenue Sharing in European Football Leagues: A Theoretical Analysis

Bodil Olai Hansen and Mich Tvede

Abstract In the present chapter, a general model of competition between clubs in sports leagues with flexible supply of inputs is studied. There are externalities between clubs because it takes more than one club to produce games and tournaments. It is assumed that the externalities take the form of complementarities. Firstly, it is shown that revenue sharing leads to lower overall quality of sports leagues. Secondly, it is shown that the optimal quality for the league is lower (higher) than the quality in a league without revenue sharing in case of negative (positive) externalities between clubs. Thirdly an example is used to illustrate the findings.

Keywords Complementarity • Revenue sharing • Sports leagues • Supermodularity • Competition in sports leagues • European vs. American sports leagues • Level of talent • Profit maximization

JEL Classification: C72, D21, L83.

12.1 Introduction

In the present chapter, a model of competition between clubs in a European football league is presented. Externalities are important in sports leagues because it takes more than one club to produce as explained in Rottenberg (1956) and Neale (1964). Indeed two clubs are needed for a game and several clubs are needed for a tournament. The competitive balance is related to the relative distribution of quality

B.O. Hansen
Department of Economics, CBS, Porcelaenshaven 16A, 2000 Frederiksberg, Denmark
e-mail: boh.eco@cbs.dk

M. Tvede (✉)
Newcastle University, 5 Barrack Road, Newcastle upon Tyne, Tyne and Wear,
Newcastle NE1 4SE, UK
e-mail: mich.tvede@ncl.ac.uk

© Springer International Publishing Switzerland 2016
A.A. Pinto et al. (eds.), *Trends in Mathematical Economics*,
DOI 10.1007/978-3-319-32543-9_12

across clubs and the aggregate level to the aggregate sum of talent in the sports league. The value of a sports league depends on the competitive balance in as well as the aggregate quality of the sports league: if the quality of clubs is balanced and high, then the outcome of the competition in the sports league is more unpredictable and games are more exciting to watch. However, since clubs are concerned with their own performance rather than the performance of the sports league, there is no reason to expect the distribution of quality between clubs to be sufficiently balanced or the overall quality of the sports league to be sufficiently high to be welfare maximizing.

The presence of externalities in sports leagues implies that there are potential gains from internalization. Suppose that the quality of a club is measured by the level of talent in that club. On the one hand, if one additional unit of talent is distributed between clubs such that the competitive balance is unchanged, then the value of the sports league should increase because the competitive balance is unchanged and the aggregate level of talent is increased. On the other hand, if one additional unit of talent is distributed to the club with the highest level of talent, then the change in the value of the league should be ambiguous, because the sports league is less balanced but the aggregate level of talent is increased. Hence, there are complementarities in value creation because the competitive balance matters. Several instruments including different forms of revenue sharing of both matchday revenues and revenues from sale of broadcast rights, regulation of wages including salary caps and luxury taxes, and more recently Financial Fair Play introduced by UEFA for European football can be seen as attempts to improve the competitive balance.

There are several differences between American sports leagues such as MLB, NBA, and NFL and European football leagues such as the Premier League in UK, Primera Division in Spain, Serie A in Italy, and Bundesliga in Germany. One difference is that American sports leagues are closed in the sense that there is no relegation or promotion, while in European football leagues there is. Consequently clubs close to relegation or promotion face more uncertainty about future income than American clubs and other clubs in European football leagues. Another difference is that in American sports leagues, the market for talent is closed in the sense that the supply of talent is more or less fixed, while it is open in European football leagues in the sense that the supply of talent in every league is flexible. On the one hand, American sports leagues face little or no international competition since there are no other leagues of compatible quality for these sports. Therefore players are left with no alternatives to the American sports leagues. Consequently, the supply of talent is essentially fixed. On the other hand, European football leagues face fierce competition since the quality of the Premier League, Primera Division, Serie A, and the Bundesliga is compatible. Therefore players can move between different leagues. Consequently, the different leagues compete over talent making the supply of talent variable in every league.

There has been quite some interest in analyzing the effects of revenue sharing, salary caps, and luxury taxes. The focus has been on the competitive balance rather than the overall quality of leagues. One reason being that the overall quality of American sports leagues is fixed because the market for talent is closed. Another reason being that in a large part of the literature, revenues of clubs have been modeled as depending on competitive balance and not on overall quality. See Atkinson et al. (1988), Burg and Prinz (2005), Dietl and Lang (2008), Dietl et al. (2010), Dobson and Goddard (2014), El-Hodiri and Quirk (1971), Fort and Quirk (1995), Kesenne (2000a,b, 2007), Marburger (1997), Szymanski (2003, 2004, 2007), Szymanski and Kesenne (2004), Vrooman (1995, 2007), and Whitney (2005). However, the overall quality is not fixed in European football leagues because talent can be exported to and imported from other football leagues. Consequently leaving out the overall quality of leagues in the modeling of revenues of clubs makes the analysis less relevant for European football leagues. A couple of recent exceptions are Dietl et al. (2009) and Dobson and Goddard (2014) where revenues depend on both the competitive balance and the overall quality of the league.

In the present we focus on the relation between revenue sharing and the overall quality in European football leagues. See Atkinson et al. (1988), Burg and Prinz (2005), Dietl and Lang (2008), Dietl et al. (2010), Dobson and Goddard (2014), El-Hodiri and Quirk (1971), Fort and Quirk (1995), Kesenne (2000a), Marburger (1997), Szymanski (2003, 2004), Szymanski (2004), Vrooman (1995, 2007), and Whitney (2005) for more on revenue sharing. See Burg and Prinz (2005), Dietl et al. (2009), Fort and Quirk (1995), and Kesenne (2000b) for more on salary caps. See Marburger (1997) and Szymanski (2003, 2004) for more on luxury taxes.

We present a model of competition between clubs in European football leagues. In the model: (1) the quality of a club is described by the level of talent in the club; (2) the revenue of every club depends on the distribution of talent in the sports league; (3) the cost of every club depends on its level of talent; and (4) clubs maximize profits. Moreover, there are complementarities in revenues. The analysis is based on complementarity and supermodularity; see Milgrom and Shannon (1994) and Topkis (1998) for more on these notions.

Firstly the distribution of talent in sports leagues without revenue sharing is compared with the distribution of talent in sports leagues with revenue sharing. It is found that revenue sharing lowers the aggregate level of talent. Secondly the distributions of talent in sports leagues without and with revenue sharing are compared with the distribution of talent that maximizes the aggregate profit of the sports leagues. It is found that if there are positive (negative) externalities between every club and the rest of the league, then the level of talent in sports leagues without redistribution is too low (high). Thirdly an example is used to illustrate the findings. It may clarify to consider the example while reading the rest of the chapter.

12.2   Setup and Assumptions

Consider a sports league with a finite number n of clubs $j \in \mathcal{N} = \{1, \ldots, n\}$. Clubs hire talent and compete. Every club is characterized by a revenue function that depends on the level of talent in the club as well as the level of talent in the rest of the league $R_j : \mathbb{R}_+ \times \mathbb{R}_+^{n-1} \to \mathbb{R}$ and a cost function that depends on the level of talent in the club $C_j : \mathbb{R}_+ \to \mathbb{R}$. For a level of talent t_j in club j and a list of levels of talent $t_{-j} = (t_1, \ldots, t_{j-1}, t_{j+1}, \ldots, t_n)$ in the rest of the league, the profit of club j is $R_j(t_j, t_{-j}) - C_j(t_j)$. With some abuse of notation, the revenue in club k is sometimes written $R_k(t_j, t_{-j})$ rather than $R_k(t_k, t_{-k})$.

The following assumptions are supposed to be satisfied:

(A.1) $C_j \in C(\mathbb{R}_+, \mathbb{R})$.

(A.2) $R_j \in C(\mathbb{R}_+^n, \mathbb{R})$.

(A.3)   There is $\alpha_j > 0$ such that $R_j(t_j, t_{-j}) - C_j(t_j) \leq 0$ for all $t_j \geq \alpha_j$ and $t_{-j} \in \mathbb{R}_+^{n-1}$.

(A.4)   There is $\beta_j > 0$ such that $\sum_k R_k(t_j, t_{-j}) - C_j(t_j) \leq 0$ for all $t_j \geq \beta_j$ and $t_{-j} \in \mathbb{R}_+^{n-1}$.

Assumptions (A.1) and (A.2) are natural. Assumption (A.3) implies that for every club there is an upper limit on the level of talent for which its profit is positive. Assumption (A.4) implies that for every club there is an upper limit on the level of talent for which the aggregate revenue minus cost of the club is positive. Revenues being bounded from above and costs tending to infinity as levels of talent tend to infinity imply (A.3) and (A.4) are satisfied. Clearly the advantage of imposing weak assumptions on revenue and cost functions is that results apply to all functions satisfying these assumptions.

There are externalities between clubs because the revenue in every club depends on the level of talent of the club as well as the level of talent in the rest of the league. The form of the externalities is discussed in Sects. 12.4 and 12.5.

The competitive balance is related to the relative distribution of talent across clubs $(t_1 / \sum_k t_k, \ldots, t_n / \sum_k t_k)$ and the aggregate level to the sum of talent in the league $\sum_k t_k$. Let $\triangle \subset \mathbb{R}_+^n$ be defined by

$$\triangle = \{ (w_1, \ldots, w_n) \in \mathbb{R}_+^{n-1} \mid \sum_k w_k = 1 \}$$

be the set of possible relative distributions of talent. Then the map $F : \mathbb{R}_+^n \setminus \{0\} \to \triangle \times \mathbb{R}_{++}$ defined by

$$F(t_1, \ldots, t_n) = \left(\frac{t_1}{\sum_k t_k}, \ldots, \frac{t_n}{\sum_k t_k}, \sum_k t_k \right)$$

mapping distributions of talent of to relative distributions of talent across the first $n-1$ clubs and the aggregate level of talent is a diffeomorphism. Indeed the inverse of F is $G : \triangle \times \mathbb{R}_{++} \to \mathbb{R}_+^n \setminus \{0\}$ defined by

$$G(w_1, \ldots, w_n, T) = (w_1 T, \ldots, w_n T)$$

mapping relative distributions of talent across the first $n-1$ clubs and the aggregate level of talent to distributions of talent. Therefore, it does not matter whether revenue functions are considered to depend on distributions of talent $(t_1, \ldots, t_n)$ or relative distributions of talent and the sum of talent $(w_1, \ldots, w_n, T)$. However, information is lost in case revenue functions depend on relative distributions of talent and nothing else. In the present chapter revenue functions are considered to depend on distributions of talent rather than relative distributions of talent and the sum of talent making the assumptions on revenue functions simpler to interpret.

12.3 Equilibrium

The problem of club j is to choose a level of talent t_j to maximize its profit given the level of talent in the rest of the league t_{-j}.

$$\max_{t_j \geq 0} R_j(t_j, t_{-j}) - C_j(t_j).$$

In equilibrium every club maximizes its profit given the level of talent in the rest of the league. However, there is no market clearing for talent because as explained in the introduction a club in one European football league is able to attract talent from another club in another European football league.

Definition 1. An **equilibrium** is a list of levels of talents $\bar{t} = (\bar{t}_1, \ldots, \bar{t}_n)$ such that $\bar{t}_j$ is a solution to the problem of club j given $\bar{t}_{-j}$.

The assumptions on revenue and cost functions are too weak to ensure existence of equilibrium. Therefore consider the effect of increasing the level of talent in club j on revenue of club j. Suppose that the higher the level of talent in the rest of the league, the more the revenue of club j increases.

Increasing Differences in Revenue (IDR). *For all pairs of lists of levels of talent* (t, u), *where* $u_k \geq t_k$ *for every club* k,

$$R_j(u_j, u_{-j}) - R_j(t_j, u_{-j}) \geq R_j(u_j, t_{-j}) - R_j(t_j, t_{-j})$$

for every club j.

IDR implies that there are complementarities in the revenue of a club between its level of talent and the levels of talent in other clubs. Suppose that revenue functions are twice differentiable. Then

$$\frac{\partial^2 R_j(t_j, t_{-j})}{\partial t_j \partial t_k} \geq 0$$

for every j and $k \neq j$ implies IDR is satisfied.

Theorem 1. *Assume IDR is satisfied. Then there is an equilibrium.*

Proof. Firstly it is shown that the game $(\mathcal{N}, (T_j, \pi_j)_j)$, where $T_j = \mathbb{R}_+$ is the strategy set for choice of talent and $\pi_j : \mathbb{R}_+^n \to \mathbb{R}$ is defined by

$$\pi_j(t_j, t_{-j}) = R_j(t_j, t_{-j}) - C_j(t_j)$$

is the payoff function, is a supermodular game. Secondly Theorem 4.2.1. in Topkis (1998) on existence of equilibrium in supermodular games is applied.

The game $(\mathcal{N}, (T_j, \pi_j)_j)$ is supermodular: 1. T_j is a lattice; 2. π_j satisfies IDR because R_j satisfies IDR; and 3. π_j is supermodular because all functions from $\mathbb{R}_+$ to $\mathbb{R}_+$ are supermodular. For every t_{-j} if t_j is a solution to the problem of club j, then $t_j \leq \alpha_j$. Therefore the set of equilibria for the games $(\mathcal{N}, (T_j, \pi_j)_j)$ and $(\mathcal{N}, (S_j, \pi_j)_j)$, where $S_j = [0, \alpha_j]$, coincide. According to Theorem 4.2.1. in Topkis (1998), the game $(\mathcal{N}, (S_j, \pi_j)_j)$ has an equilibrium.   *Q.E.D.*

In Theorem 1 it is shown that there is an equilibrium, but it is not shown that the equilibrium is unique. However, as stated in Theorem 4.2.1 in Topkis (1998), the set of equilibria is a lattice with a greatest equilibrium $G \in \mathbb{R}_+^n$ and a least equilibrium $L \in \mathbb{R}_+^n$. Therefore if $\bar{t}$ is an equilibrium, then $G_j \geq \bar{t}_j \geq L_j$ for every j and if $\bar{t}$ and $\bar{t}'$ are equilibria, then $(\max\{\bar{t}_1, \bar{t}_1'\}, \ldots, \max\{\bar{t}_n, \bar{t}_n'\})$ and $(\min\{\bar{t}_1, \bar{t}_1'\}, \ldots, \min\{\bar{t}_n, \bar{t}_n'\})$ are equilibria too.

12.4 Revenue Sharing

In the present section, it is discussed how revenue sharing, where each club pays a tax proportional with its revenue and receives a subsidy proportional with total revenue of the league, influences equilibria.

Consider revenue sharing such that every club pays a share of its revenues and receives a share of the collected revenues. Let $\tau \in [0, 1]$ be the degree of revenue sharing, then the problem of club j is

$$\max_{t_j \geq 0} (1 - \tau) R_j(t_j, t_{-j}) + \frac{\tau}{n} \sum_k R_k(t_k, t_{-k}) - C_j(t_j).$$

If performance and revenue are positively correlated, then revenue sharing corresponds to making subsidies negatively correlated with performance.

IDR is not enough to ensure existence of equilibrium for every level of revenue sharing. Therefore consider the effect of increasing the level of talent in club j on aggregate revenue of the league. Suppose that the higher the level of talent is in the rest of the league, the more the aggregate revenue increases.

Increasing Differences in Aggregate Revenue (IDAR). *For all pairs of lists of levels of talent* (t, u), *where* $u_k \geq t_k$ *for every club k,*

$$\sum_k (R_k(u_j, u_{-j}) - R_k(t_j, u_{-j})) \geq \sum_k (R_k(u_j, t_{-j}) - R_k(t_j, t_{-j}))$$

for every club j.

IDAR implies that there are complementarities in the aggregate revenue between the levels of talent in the different clubs. Suppose revenue functions are twice differentiable. Then

$$\sum_j \frac{\partial^2 R_j(t_j, t_{-j})}{\partial t_k \partial t_\ell} \geq 0$$

for every k and $\ell \neq k$ implies IDAR is satisfied.

Theorem 2. *Assume IDR and IDAR are satisfied. Then for every $\tau \in [0, 1]$, there is an equilibrium.*

Proof. Firstly it is shown that the game $(\mathcal{N}, (T_j, \pi_j^\tau)_j)$, where $T_j = \mathbb{R}_+$ is the strategy set, $\pi_j^\tau : \mathbb{R}_+^n \to \mathbb{R}$ defined by

$$\pi_j^\tau(t_j, t_{-j}) = (1 - \tau)R_j(t_j, t_{-j}) + \frac{\tau}{n} \sum_k R_k(t_j, t_{-j}) - C_j(t_j)$$

is the payoff function and $\tau \in [0, 1]$, is a supermodular game. Secondly Theorem 4.2.1. in Topkis (1998) on existence of equilibrium in supermodular games is applied.

The game $(\mathcal{N}, (T_j, \pi_j^\tau)_j)$ is supermodular: 1. T_j is a lattice; 2. π_j^τ satisfies IDR because R_j satisfies IDR and $\sum_k R_k$ satisfies IDAR; and 3. π_j^τ is supermodular because all functions from $\mathbb{R}_+$ to $\mathbb{R}_+$ are supermodular. For every t_{-j} if t_j is a solution to the problem of club j, then $t_j \leq k_j$. Therefore the sets of equilibria for the two games $(\mathcal{N}, (T_j, \pi_j^\tau)_j)$ and $(\mathcal{N}, (S_j, \pi_j^\tau)_j)$, where $S_j = [0, \max\{\alpha_j, \beta_j\}]$, coincide. According to Theorem 4.2.1. in Topkis (1998), the game $(\mathcal{N}, (S_j, \pi_j^\tau)_j)$ has an equilibrium.   *Q.E.D.*

In order to address how the level of revenue sharing influences the level of talent, compare the effect of increasing the level of talent in club j on revenue of club j and the other clubs. Suppose that the change in revenue of club j is larger than the average change in the revenue of the rest of the league.

Dominant Differences in Revenues (DDR). *For all pairs of lists of levels of talent (t, u), where $u_k \geq t_k$ for every club k,*

$$R_j(u_j, t_{-j}) - R_j(t_j, t_{-j}) \geq \frac{1}{n - 1} \sum_{k \neq j} (R_k(u_j, t_{-j}) - R_k(t_j, t_{-j}))$$

for every j.

DDR implies that on average changes in levels of talent are more important for clubs that change their levels of talent than for other clubs. Suppose revenue functions are once differentiable. Then

$$\frac{\partial R_j(t_j, t_{-j})}{\partial t_j} \geq \frac{1}{n - 1} \sum_{k \neq j} \frac{\partial R_k(t_j, t_{-j})}{\partial t_j}$$

for every j implies DDR is satisfied.

Theorem 3. *Assume IDR, IDAR, and DDR are satisfied. Then for every $\tau \in [0,1]$ there exist a greatest equilibrium $G^\tau \in \mathbb{R}_+^n$ and a least equilibrium $L^\tau \in \mathbb{R}_+^n$ in the sense that $G_j^\tau \geq \bar{t}_j^\tau \geq L_j^\tau$ for all equilibria $\bar{t}^\tau$ and every j. Moreover, the greatest and the least equilibria are decreasing in τ.*

Proof. IDR and IDAR imply the profit function

$$\pi_j^\tau(t_j, t_{-j}) = (1 - \tau)\pi_j(t_j, t_{-j}) + \frac{\tau}{n}\sum_k \pi_k(t_j, t_{-j}) - C_j(t_j)$$

of club j has increasing differences in (t_j, t_{-j}) for every j and DDR implies that the profit function of club j has increasing differences in $(t_j, -\tau)$ for every j. Therefore Theorem 4.2.2. in Topkis (1998) on comparative statics of equilibria in supermodular games can be applied to obtain Theorem 3.   Q.E.D.

Theorem 3 implies the levels of talent are decreasing in the degree of revenue sharing.

12.5   Aggregate Profit

Equilibria are typically inefficient because there are externalities between clubs. In the present section the equilibrium outcome and the efficient outcome are compared.

The problem of the league is to choose a list of levels of talent such that the aggregate profit is maximized

$$\max_{t_1,\ldots,t_n \geq 0} \sum_j (R_j(t_j, t_{-j}) - C_j(t_j)).$$

Consider the pair of games $(\mathcal{N}, (T_j, \pi_j^\lambda)_j)_{\lambda \in \{0,1\}}$ where $T_j = \mathbb{R}_+$ and

$$\pi_j^\lambda(t_j, t_{-j}) = (1 - \lambda)R_j(t_j, t_{-j}) + \lambda \sum_k R_k(t_j, t_{-j}) - C_j(t_j).$$

For $\lambda = 0$ the game is a league with clubs maximizing their own profits. For $\lambda = 1$ if the aggregate profit $\sum_j (R_j(t) - C_j(t_j))$ is concave in the levels of talent t, then the game is a league with clubs maximizing the aggregate profit of the league.

In Theorem 4 equilibria are shown to exist for the pair of games though equilibria for the game with $\lambda = 0$ are already shown to exist in Theorem 1.

Theorem 4. *Assume that IDR and IDAR are satisfied. Consider the pair of games $(\mathcal{N}, (T_j, \pi_j^\lambda)_j)_{\lambda \in \{0,1\}}$ where $T_j = \mathbb{R}_+$ and*

$$\pi_j^\lambda(t_j, t_{-j}) = (1 - \lambda)R_j(t_j, t_{-j}) + \lambda \sum_k R_k(t_j, t_{-j}) - C_j(t_j).$$

Then both games have equilibria.

Proof. The proof is identical to the proof of Theorem 2.   *Q.E.D.*

The externality between club j and the rest of the league is positive (negative) if an increase in the level of talent in the club j leads to an increase (decrease) in the revenue of the rest of the league. Below in Theorem 5, the levels of talent in equilibria of the two games are compared. Indeed it is shown that if the externality between every club and the rest of the league is positive (negative), then the level of talent is too low (high) in the league without revenue sharing compared to the level of talent that maximizes the aggregate profit of the league.

Theorem 5. *Assume that IDR, IDAR, and DDR are satisfied. Consider the pair of games $(\mathcal{N}, (T_j, \pi_j^\lambda)_j)_{\lambda \in \{0,1\}}$ defined in Theorem 4. Then for both games, there exist a greatest equilibrium $G^\lambda \in \mathbb{R}_+^n$ and a least equilibrium $L^\lambda \in \mathbb{R}_+^n$ in the sense that $G_j^\lambda \geq \bar{t}_j^\lambda \geq L_j^\lambda$ for all equilibria $\bar{t}^\lambda$ and every j.*

- *Suppose the externality between every club and the rest of the league is positive: for every pair of lists of talent (t, u), where $u_k \geq t_k$ for every club k,*

$$\sum_{k \neq j} (R_k(u_j, t_{-j}) - R_k(t_j, t_{-j})) \geq 0$$

for every club j. Then the greatest and least equilibria are increasing in λ.
- *Suppose the externality between every club and the rest of the league is negative: for every pair of lists of talent (t, u), where $u_k \geq t_k$ for every club k,*

$$\sum_{k \neq j} (R_k(u_j, t_{-j}) - R_k(t_j, t_{-j})) \leq 0$$

for every club j. Then the greatest and least equilibria are decreasing in λ.

Proof. The proof is identical to the proof of Theorem 3.   *Q.E.D.*

12.6 Comparison of Equilibria

Theorem 3 and the first claim of Theorem 5 imply that if the externality between every club and the rest of the league is positive, then revenue sharing works in the wrong direction.

Corollary 1. *Assume that IDR, IDAR, and DDR are satisfied. Suppose the externality between every club and the rest of the league is positive.*

- *The level of talent in equilibria with revenue sharing is lower than the level of talent in equilibria without revenue sharing.*
- *The level of talent in equilibria without revenue sharing is lower than the level of talent that maximizes the aggregate profit.*

Proof. The claims follow from Theorems 3 and 5.   *Q.E.D.*

In case of a positive externality between every club and the rest of the league, no revenue sharing $\tau = 0$ or perhaps even negative revenue sharing $\tau < 0$ appears to be the right way to redistribute. If performance and revenue are correlated, then negative revenue sharing corresponds to making subsidies positively correlated with performance.

Theorem 3 and the second claim of Theorem 5 imply that if the externality between every club and the rest of the league is negative, then revenue sharing works in the right direction.

Corollary 2. *Assume that IDR, IDAR, and DDR are satisfied. Suppose that the externality between every club and the rest of the league is negative.*

- *The level of talent in equilibria with revenue sharing is lower than the level of talent in equilibria without revenue sharing.*
- *The level of talent in equilibria without revenue sharing is higher than the level of talent that maximizes the aggregate profit.*

Proof. The claims follow from Theorems 3 and 5.   *Q.E.D.*

In case of negative externality between every club and the rest of the league, a positive degree of revenue sharing $\tau > 0$ appears to be the right way to redistribute. However, complete revenue sharing ($\tau = 1$) results in too low levels of talent provided that an increase in the level of talent in any club leads to an increase of the aggregate revenue of the league.

Theorem 6. *Assume that IDAR is satisfied. Suppose that for all pairs of lists of levels of talent (t, u), where $u_k \geq t_k$ for every club k,*

$$\sum_k (R_k(u_j, t_{-j}) - R_k(t_j, t_{-j})) \geq 0.$$

for every club j. Then the level of talent in equilibria with complete revenue sharing ($\tau = 1$) is lower than the level of talent that maximizes the aggregate profit.

Proof. Consider the pair of games $(\mathcal{N}, (T_j, \pi_j^\mu)_j)_{\mu \in \{0,1\}}$ where $T_j = \mathbb{R}_+$ and

$$\pi_j^\mu(t_j, t_{-j}) = \left(1 - \frac{n-1}{n}\mu\right) \sum_k R_k(t_j, t_{-j}) - C_j(t_j).$$

For $\mu = 0$ the game is a game, where every club is concerned with the aggregate revenue of the league, and for $\mu = 1$ the game is the game of the league with complete revenue sharing ($\tau = 1$). IDAR implies that the payoff function of club j has increasing differences in (t_j, t_{-j}) for every j. The assumption that $\sum_k (R_k(u_j, t_{-j}) - R_k(t_j, t_{-j})) \geq 0$ for all lists of levels of talent t and every club j and level of talent u_j in club j with $u_j \geq t_j$ implies that the payoff function of club

j has increasing differences in $(t_j, -\mu)$. Therefore, Theorem 4.2.2. in Topkis (1998) on comparative statics of equilibria in supermodular games can be applied to obtain Theorem 6.   *Q.E.D.*

For every club if an increase of the level of talent leads to a decrease of the aggregate revenue of the rest of league and an increase of the aggregate revenue of the league, then the right degree of revenue sharing appears to be somewhere between no revenue sharing and complete revenue sharing. Indeed the level of talent is too high in a league without revenue sharing and too low in a league with complete revenue sharing compared with the level of talent that maximizes the aggregate profit of the league as shown in Theorems 3 and 6.

12.7 Sports Leagues with Fixed Supply of Talent

As explained in the introduction, the supply of talent is more or less fixed for American sports leagues such as MLB, NBA, and NFL because there are no other leagues of compatible quality. Therefore the markets for talent in American sports leagues should be modeled differently than the markets for talent in European football leagues. The market for talent should probably be characterized by a price $w > 0$ for talent and an equilibrium condition for talent $\sum_k t_k = T$, where T is the supply of talent, rather than cost functions. An equilibrium would be a list of levels of talent and a price of talent such that every club maximizes its profit given the level of talent in the rest of the league and the price of talent, and the market for talent clears. Since the aggregate level of talent is fixed in American sports leagues, only the competitive balance matters, while the aggregate level of talent is without importance.

The results of the present chapter can be applied to American sports leagues with some modifications. However, the qualitative results in Theorems 3, 5, and 6 and Corollaries 1 and 2 would relate the degree of revenue sharing and the price for talent rather than the degree of revenue sharing and the level of talent. As an example Theorem 4.2.2. in Topkis (1998) implies that the price for talent is decreasing in the degree of revenue sharing. In NFL there is a tradition for salary caps such that clubs maximize their profits subject to the constraint that their costs have to be equal to or lower than the salary cap. As another example Theorem 4.2.2. in Topkis (1998) implies that the price for talent is decreasing in the salary cap.

12.8 An Example

12.8.1 The Model

For a league with n clubs, the revenue R_j of club j has the form

$$R_j(t_j, t_{-j}) = a_j + \sum_k b_{jk} \ln(t_k) + c_j \ln(\textstyle\sum_k t_k)$$

$$= a_j + \sum_k b_{jk} \ln\left(\frac{t_k}{\sum_\ell t_\ell}\right) + \left(\sum_k b_{jk} + c_j\right) \ln(\textstyle\sum_k t_k)$$

where $a_j, b_{j1}, \ldots, b_{jn}, c_j$ are constants. Clearly, the revenue of club j depends on the competitive balance $(t_1/\sum_k t_k, \ldots, t_n/\sum_k t_k)$ and the aggregate level of talent $\sum_k t_k$.

The natural logarithm in revenue functions has the undesired implication that the revenues tend to minus infinity as the level of talent of one club tends to zero. However, in order to ensure that revenue of club j converges to zero as t_k converges to zero for every k, revenue functions could be modified to

$$R_j(t_j, t_{-j}) = a_j + \sum_k b_{jk} \ln(t_k + \varepsilon) + c_j \ln(\textstyle\sum_k t_k + n\varepsilon)$$

where $\varepsilon > 0$ and $a_j = -(\sum_k b_{jk} + c_j) \ln(\varepsilon) - c_j \ln(n)$.

Concerning properties of revenue functions, the derivative of the revenue of club j with respect to the level of talent in club k is

$$\frac{\partial R_j(t_j, t_{-j})}{\partial t_k} = \frac{b_{jk}}{t_k} + \frac{c_j}{\sum_\ell t_\ell}.$$

If $b_{jk} > 0$ ($b_{jk} < 0$), then R_j is increasing (decreasing) in t_k for t_k sufficiently small and $\sum_{\ell \neq k} t_\ell > 0$. It is assumed that $b_{jj} > 0$ for every j. If $b_{jk} + c_j > 0$ ($b_{jk} + c_j < 0$), then R_j is increasing (decreasing) in t_k for t_k sufficiently large. Since the derivative of the revenue $R_j(\alpha t_j, \alpha t_{-j})$ of club j with respect to α is

$$\frac{\partial R_j(\alpha t_j, \alpha t_{-j})}{\partial \alpha} = \frac{\sum_k b_{jk} + c_j}{\alpha},$$

because $R_j(\alpha t_j, \alpha t_{-j}) = R_j(t_j, t_{-j}) + (\sum_k b_{jk} + c_j) \ln(\alpha)$, the revenue is increasing (decreasing) in the level of talent for $\sum_k b_{jk} + c_j > 0$ ($\sum_k b_{jk} + c_j < 0$).

Concerning the axioms about properties of revenue functions, club j satisfies IDR if and only if

$$\frac{\partial^2 R_j(t_j, t_{-j})}{\partial t_j \partial t_k} = -\frac{c_j}{(\sum_k t_k)^2} \geq 0,$$

where $k \neq j$, so club j satisfies IDR provided $c_j \leq 0$. The league satisfies IDAR if and only if

$$\sum_j \frac{\partial^2 R_j(t_j, t_{-j})}{\partial t_k \partial t_\ell} = -\frac{\sum_j c_j}{(\sum_j t_j)^2} \geq 0,$$

where $\ell \neq k$, so the league satisfies IDAR provided $\sum_j c_j \leq 0$ so IDR implies IDAR. The league satisfies DDR if and only if

$$\frac{\partial R_j(t_j, t_{-j})}{\partial t_j} = \frac{b_{jj}}{t_j} + \frac{c_j}{\sum_k t_k}$$

$$\geq \frac{1}{n-1}\left(\frac{\sum_{k\neq j} b_{kj}}{t_j} + \frac{\sum_{k\neq j} c_k}{\sum_k t_k}\right)$$

$$= \frac{1}{n-1}\sum_{k\neq j}\frac{\partial R_k(t_j, t_{-j})}{\partial t_j}$$

so the league satisfies DDR provided $(n-1)b_{jj} \geq \sum_{k\neq j} b_{kj}$ and $(n-1)(b_{jj} + c_j) \geq \sum_{k\neq j}(b_{kj} + c_k)$ for every j. IDR, IDAR, and DDR are assumed to be satisfied in the sequel.

For the sign of the externalities

$$\sum_{k\neq j}(R_k(u_j, t_{-j}) - R_k(t_j, t_{-j}))$$

$$= \sum_{k\neq j} b_{kj}(\ln(u_j) - \ln(t_j)) + \sum_{k\neq j} c_k(\ln(u_j + \sum_{\ell\neq j}t_\ell) - \ln(t_j + \sum_{\ell\neq j}t_\ell)).$$

Therefore, the externalities are positive provided $\sum_{k\neq j} b_{kj} \geq 0$ and $\sum_{k\neq j}(b_{kj} + c_k) \geq 0$ for every j and negative provided $\sum_{k\neq j} b_{kj} \leq 0$ and $\sum_{k\neq j}(b_{kj} + c_k) \leq 0$ for every j. For the effect of an increase in the level of talent in any club on the aggregate revenue of the league considered in Theorem 6

$$\sum_k(R_k(u_j, t_{-j}) - R_k(t_j, t_{-j}))$$

$$= \sum_k b_{kj}(\ln(u_j) - \ln(t_j)) + \sum_k c_k(\ln(u_j + \sum_{\ell\neq j}t_\ell) - \ln(t_j + \sum_{\ell\neq j}t_\ell)).$$

Hence, an increase in the level of talent in any club leads to an increase in aggregate revenue of the league provided $\sum_k b_{kj} \geq 0$ and $\sum_k(b_{kj} + c_k) \geq 0$ for every j.

The cost function of club j is supposed to be $C_j(t_j) = wt_j$, where $w > 0$ is the price for one unit of talent. Hence, talent is bought and sold on a perfectly competitive market.

12.8.2 Equilibria

Equilibria can be characterized as solutions to n equations with n unknowns.

Observation 1. *For equilibria:*

- $\bar{t} = (\bar{t}_1, \ldots, \bar{t}_n)$ *is an equilibrium without revenue sharing if and only if $\bar{t}$ is a solution to*

$$\bar{t}_j = \frac{b_{jj}}{w\sum_k t_k - c_j} \sum_k t_k$$

for every j

- $\bar{t}^\tau$ *is an equilibrium with revenue sharing if and only if $\bar{t}^\tau$ is a solution to*

$$t_j^\tau = \frac{(1 - \tau)b_{jj} + \frac{\tau}{n}\sum_k b_{kj}}{w\sum_k t_k^\tau - (1 - \tau)c_j - \frac{\tau}{n}\sum_k c_k} \sum_k t_k^\tau$$

for every j.

- t^* *is a solution to the problem of the league if and only if t^* is a solution to*

$$t_j^* = \frac{\sum_k b_{kj}}{w\sum_k t_k - \sum_k c_k} \sum_k t_k$$

for every j.

Proof. The first-order condition to the problem of club j in a league without revenue sharing is

$$\frac{b_{jj}}{t_j} + \frac{c_j}{\sum_k t_k} - w = 0.$$

If t_j is isolated in the first-order condition, then it becomes

$$t_j = \frac{b_{jj}}{w\sum_k t_k - c_j} \sum_k t_k.$$

The two other expressions are obtained by isolation of t_j is isolated in first-order condition for the problem of club j in a league with revenue sharing and the problem of the league.   *Q.E.D.*

From a practical point of view, an equilibrium is found in two steps: firstly the aggregate level of talent is determined in

$$1 = \sum_j \frac{b_{jj}}{w\sum_k t_k - c_j}$$

for a league without revenue sharing, in

$$1 = \sum_j \frac{(1 - \tau)b_{jj} + \frac{\tau}{n}\sum_k b_{kj}}{w\sum_k t_k - (1 - \tau)c_j + \frac{\tau}{n}\sum_k c_k}$$

for a league with revenue sharing, and in

$$1 = \frac{\sum_{j,k} b_{jk}}{w \sum_k t_k - \sum_k c_k}$$

for the problem of the league and secondly the equations in Observation 1 are used to determine the level of talent in the clubs.

For a league without revenue sharing, there is an equilibrium if and only if either there is a club j such that $c_j \geq 0$ or for every j, $c_j < 0$ and

$$\sum_j \frac{b_{jj}}{c_j} < -1.$$

Suppose that $b_{jk} > 0$ for every j and k. Then the optimal distribution of talent for club j is proportional with $(b_{j1}, \ldots, b_{jn})$. For the problem of the league, there exists a solution if and only if either $\sum_j c_j \geq 0$ or $\sum_j c_j < 0$ and

$$\frac{\sum_{j,k} b_{jk}}{\sum_j c_j} < -1.$$

Suppose that $\sum_j b_{jk} > 0$ for every k. Then the optimal distribution of talent for the league is proportional to $(\sum_j b_{j1}, \ldots, \sum_j b_{jn})$. Nonexistence is caused by revenue functions not being defined for levels of talent being zero.

12.8.3   Equilibria in a League with Symmetrical Clubs

In order to focus on the relation between the degree of revenue sharing and the aggregate level of talent, suppose that the clubs are symmetrical so there exist parameters α, β_{own}, β_{other}, and γ such that $a_j = \alpha$, $b_{jj} = \beta_{\text{own}}$, $b_{jk} = \beta_{\text{other}}$ for $k \neq j$ and $c_j = \gamma$ for every j. Then the revenue function of club j is

$$R_j(t_j, t_{-j}) = \alpha + \beta_{\text{own}} \ln(t_j) + \beta_{\text{other}} \sum_{k \neq j} \ln(t_k) + \gamma \ln(\textstyle\sum_k t_k).$$

Clearly equilibria are symmetrical. Therefore by assumption, the competitive balance is perfect independently of the degree of redistribution. However, the overall quality of the league depends on the degree of redistribution.

Observation 2.   *For a league with symmetrical clubs:*

- $\bar{t}$ *is an equilibrium without revenue sharing if and only if*

$$\bar{t}_j = \frac{\beta_{\text{own}}}{w} + \frac{\gamma}{nw}$$

for every j.

- $\bar{t}^{\tau}$ is an equilibrium with revenue sharing if and only if

$$\bar{t}_j^{\tau} = \frac{(n - (n-1)\tau)\beta_{\text{own}} + (n-1)\tau\beta_{\text{other}}}{nw} + \frac{\gamma}{nw}$$

for every j.
- t^* is a solution to the problem of the league if and only if

$$t_j^* = \frac{\beta_{\text{own}} + (n-1)\beta_{\text{other}}}{w} + \frac{\gamma}{w}$$

for every j.

Proof. The claims of Observation 2 follow from Observation 1.   *Q.E.D.*

If $\beta_{\text{own}} > \beta_{\text{other}} > 0 > \gamma$, then IDR, IDAR, and DDR are satisfied. Moreover, if $\beta_{\text{own}} + (n-1)\beta_{\text{other}} + \gamma > 0$, then the levels of talent $\bar{t}_j$, $\bar{t}_j^{\tau}$, and t_j^* in Observation 2 are positive.

The level of talent $\bar{t}_j^{\tau}$ is decreasing in the degree of revenue sharing as shown in Theorem 3. The level of talent t_j^* in the solution to the problem of the league is higher or lower than the level of talent $\bar{t}$ in the league without revenue sharing depending on the externalities between clubs as shown in Corollaries 1 and 2. Indeed $t_j^* > \bar{t}_j$ if and only if $n\beta_{\text{other}} + \gamma > 0$ and $t_j^* < \bar{t}_j$ if and only if $n\beta_{\text{other}} + \gamma < 0$.

The level of talent t_j^* in the solution to the problem of the league is n times higher than the level of talent $\bar{t}_j^{\tau}$ in the league with complete revenue sharing. If $t_j^* \geq \bar{t}_j$ or equivalently $n\beta_{\text{other}} + \gamma \geq 0$, then the level of talent is too low in the equilibrium without revenue sharing, so revenue sharing works in the wrong direction as shown in Theorem 6. If $t_j^* < \bar{t}_j$ or equivalently $n\beta_{\text{other}} + \gamma < 0$, then the level of talent is too high in the equilibrium without revenue sharing, so revenue sharing works in the right direction as shown in Theorem 6. The optimal level of revenue sharing is found by solving the equation $\bar{t}_j^{\tau} = t_j^*$ with respect to τ. The optimal degree of revenue sharing is

$$\tau^* = \frac{n\beta_{\text{other}} + \gamma}{-\beta_{\text{own}} + \beta_{\text{other}}}$$

and for this degree of revenue sharing, the aggregate profit of the league is maximized. Clearly the optimal degree of revenue sharing τ^* is negative for $t_j^* > \bar{t}_j$ and positive for $t_j^* < \bar{t}_j$.

12.9   Conclusion

In the present chapter, we have presented and studied a general model of competition between clubs in sports leagues with flexible supply of inputs. The externalities between clubs are assumed to take the form of complementarities. The focus has been on the relation between revenue sharing and the overall quality of sports leagues. Our main findings are: the level of talent in every club is decreasing in the degree of revenue sharing, and, the optimal degree of revenue sharing is negative in case of positive externalities between every club and the rest of the league and less than complete revenue sharing optimal in case of negative externalities between every club and the rest of the league.

Our main findings rest on assumptions about properties of revenue functions, namely, complementarity assumptions and externality assumptions. Clearly these assumptions could be tested empirically to address the empirical relevancy of the model.

Acknowledgements   The authors wish to thank Elvio Accinelli and two anonymous referees for constructive suggestions.

References

Atkinson, S., Stanley, L., Tschirhart, S.: Revenue sharing as an incentive in an agency problem: an example from the National Football League. Rand J. Econ. **19**, 27–43 (1988)

Burg, T., Prinz, A.: Progressive taxation as a means for improving competitive balance. Scott. J. Econ. **52**, 65–74 (2005)

Dietl, H., Lang, M.: The effect of gate revenue sharing on social welfare. Contemp. Econ. Policy **26**, 448–459 (2008)

Dietl, H., Lang, M., Rathke, A.: The effect of salary caps in professional team sports on social welfare. BE J. Econ. Anal. Policy **9** (2009). Article 17

Dietl, H., Lang, M., Rathke, A.: The combined effect of salary restrictions and revenue sharing in sports leagues. Econ. Inq. **49**, 447–463 (2010)

Dobson, S., Goddard, J.: Revenue sharing in a sports league with an open market in playing talent. Theor. Econ. Lett. **4**, 410–414 (2014)

El-Hodiri, M., Quirk, J.: An economic model of a professional sports league. J. Polit. Econ. **79**, 1302–1319 (1971)

Fort, R., Quirk, J.: Cross-subsidization, incentives, and outcomes in professional team sport leagues. J. Econ. Lit. **33**, 1265–1299 (1995)

Kesenne, S.: Revenue sharing and competitive balance in professional team sports. J. Sports Econ. **2**, 204–228 (2000a)

Kesenne, S.: The impact of salary caps in professional team sports. Scott. J. Polit. Econ. **47**, 422–430 (2000b)

Kesenne, S.: The peculiar international economics of professional football in Europe. Scott. J. Polit. Econ. **54**, 388–399 (2007)

Marburger, D.R.: Gate revenue sharing and luxury taxes in professional sports. Contemp. Econ. Policy **15**, 114–123 (1997)

Milgrom, P., Shannon, C.: Monotone comparative statics. Econometrica **62**, 157–180 (1994)

Neale, W.: The peculiar economics of professional sports: a contribution to the theory of the firm in sporting competition and market competition. Q. J. Econ. **78**, 1–14 (1964)

Rottenberg, S.: The base ball players' labor market. J. Polit. Econ. **64**, 242–258 (1956)

Szymanski, S.: The economic design of sporting contests. J. Econ. Lit. **41**, 1137–1187 (2003)

Szymanski, S.: Professional team sports are only a game: the Walrasian fixed supply conjecture model, contest-Nash equilibrium and the invariance principle. J. Sports Econ. **5**, 111–126 (2004)

Szymanski, S.: The champions league and the Coase theorem. Scott. J. Polit. Econ. **54**, 355–373 (2007)

Szymanski, S., Kesenne, S.: Competitive balance and gate revenue sharing in team sports. J. Ind. Econ. **52**, 165–177 (2004)

Topkis, R.: Supermodularity and Complementarity. Princeton University Press, Princeton (1998)

Vrooman, J.: A general theory of professional sports leagues. South. J. Econ. **61**, 971–990 (1995)

Vrooman, J.: Theory of the beautiful game: the unification of European football. Scott. J. Polit. Econ. **54**, 314–354 (2007)

Whitney, J.: The peculiar externalities of professional team sports. Econ. Inq. **43**, 230–242 (2005)

Chapter 13
Weakened Transitive Rationality: Invariance of Numerical Representations of Preferences

Leobardo Plata

Abstract The ordinal invariance of utility functions representing the same preference is a fundamental issue for solving decision problems using mathematical techniques. As it is well known, this property holds when preferences are complete preorders. However, it is not necessarily verified for numerical representations of weaker concepts of preferences. Moreover, there are cases in which it is not possible to build order-preserving maps between two different numerical representations of the same preference. In this work, we characterize the classes of numerical representations that preserve the order introduced by a given preference in a set of alternatives.

Keywords Numerical representations of preferences • Invariance of numerical representations • Utility correspondences • Weakened transitivity rationality • Nontransitive preferences • Preference representation

JEL Classification: C65, D11, D19.

13.1 Introduction

The numerical representation of preferences is very useful tool for facing problems of decision between different alternatives. This is a topic widely discussed in the literature and a relevant issue. Once that a numerical representation of a preference is obtained, it is possible to consider different mathematical approaches to solve complex optimization programs that often are involved in the decision-making processes. Of no less importance, but much less considered in the literature, is the question to know when two different numerical representations of the same preference can be related by means of a map that preserves order.

L. Plata (✉)
Facultad de Economía, UASLP, Av. Pintores S/N, Colonia Burocratas del Estado,
CP, 78263 San Luis Potos, S.L.P., México
e-mail: lplata@uaslp.mx

© Springer International Publishing Switzerland 2016
A.A. Pinto et al. (eds.), *Trends in Mathematical Economics*,
DOI 10.1007/978-3-319-32543-9_13

The rational choice in a set of alternatives implies the existence of complete and transitive preferences, i.e, complete preorders. The simultaneous verification of these two conditions require transitivity of the strict preference and transitivity of the indifference relationship. Since Armstrong (1950), the simultaneous fulfillment of these two conditions has been questioned. Fundamentally, because many laboratory experiments deny the transitivity in choice, this gap, between theory and empirical evidence, has motivated the search of new axioms allowing for the weakening of rationality as well as new conditions for the existence of numerical representations of the weaker concepts of preferences now considered.

To overcome this gap, different authors have considered different approaches. Several of them have considered weakening the assumption of transitivity of the indifference, and others have avoided the assumption of completeness. For instance, the more general concepts of semiorders and interval orders were introduced and have been represented by means of pairs of numerical functions. See for instance, Luce (1956), Scott and Suppes (1958), Jaminson and Lau (XXXX), Fishburn (1970a), Fishburn (1970b), Fishburn (1973), Fishburn (1985), Suppes et al. (1989), Bridges (1985), and Pirlot (1990). For partial orders, Fishburn (1970a) introduces the concept of weak representation. The weakest condition required by a preference supporting a rational behavior and allowing its numerical representation is the acyclicity. The existence of a numerical representation for such preferences is introduced in Bridges (1983) where the concept of weak representations of preferences is considered. The utility correspondences were introduced in Herrero and Subiza (1991) and Subiza (1994). These correspondences associate different alternatives with different bounded sets of real numbers. These sets of numbers can have nonempty intersection, only when they are representing indifferent alternatives. In the case of strict preference, the intersection of the numerical sets corresponding with different alternatives is necessarily empty, and the greater supremum represents the most preferred alternative. More recent jobs are in Nakamura (2002), Oloriz et al. (1998), Pirlot and Vinche (1997), Rodriguez-Palmero (1997), Candeal et al. (2002), Fishburn (1999), Bridges and Mehta (1995), Fagin et al. (2006), and Plata (2004), among others. Almost all of these works show existence results for numerical preference representation. As we shall show, weakening the assumptions of transitivity and completeness of preferences can give place to the existence of different numerical representations not immediately linked.

The invariance of the numerical representations has been scarcely considered in the literature. In this job, we characterize the maps which allow to transform a numerical representation of a given preference in another of the same preference. This is known as invariance analysis. We do this analysis for each preference with weakened rationality. We show that ordinality measurement of numerical representations disappears when we use weaker concepts of preference than complete preorders. Moreover, there are cases in which it is impossible to build a function that transforms a given numerical representation of a preference in another representation of the same preference. We introduce the concept of *similarity* between two numerical representations of the same preference. This means that the partitions of the set of alternatives generated by the inverse image of this two different numerical representations are the same. The concept of similarity allows us to partition the set of numerical representations of a preference in classes, characterized by the existence of a function able to transform a representation in another inside the same class of representations. We say that two representations in the same case are analogous.

Although we are interested in the economic meaning of each preference with weakened rationality, in this paper, we restrict ourselves to consider sets with countable alternatives. This is done in order to avoid topological assumptions. To introduce such considerations is necessary when we look for problems of invariance and existence of numerical representations of preferences in economics. But doing it now will lead us beyond our present purpose. Certainly, the topological structure of the alternative set plays a relevant role in economics.

This paper is a first step in this direction, so we leave these considerations for future works.

Section 13.2 is dedicated to define weakened rationality assumptions. The third section presents the definitions of numerical representations and the existence results for each one. Finally, the fourth section presents and proves our invariance results.

13.2 Preferences Structures and Rationality Assumptions

Let X be a set of alternatives. The set X may have a multiplicity of structural properties, a set of probabilities, or a Cartesian product, or a set of consumption. A binary relation on X is a subset of $X \times X$.

A preference relationship R on X is a complete and transitive binary relationship defined on a set of alternatives X. In such case (X, R) is a preordered set, for $(x, y) \in R$, we use denotation xRy. We denote xIy if xRy and yRx.

The relation R could denote is as good as, or is no more probable than, or is no longer than, and so forth. The corresponding interpretation for I is indifferent to or equally probable. If I is an equivalence relation (reflexive, symmetric, transitive), then X/I is the set of classes of indifference determined by I in X.

The notation P to denote xRy, but not yRx,, is useful in order to denote the fact that x is strictly preferred to y. The binary relation P denote strictly preferred to, or strictly more probable, an so for. This relation establishes a partial order in the set of alternatives.

The notation $\neg xRy$ means that the relationship xRy does not hold.

Recall that P is:

- A **partial order** if P is transitive, i.e., xPz whenever xPy and yPz, and irreflexive i.e., whenever xPx. Then the pair (X, P) is a partially ordered set.
- A **weak order** if it is a partial order for which I is transitive.

The set of equivalence classes is ordered by the relationship aP^*b if and only if xPy for some $x \in a$ and $y \in b$.

A fundamental result for (X, P) says that if the set of alternatives X is countable, and P is a weak order on X, then there exists a utility function $u : X \to \mathbb{R}$ representing the partial order, meaning that

$$xPy \Leftrightarrow u(x) > u(y) \text{ and } xIy \Leftrightarrow u(x) = u(y). \tag{13.1}$$

But this result is no longer true when the set of alternatives is uncountable. The lexicographic preference over $\mathbb{R}^2$ is a classical example of the not existence of a numerical representation. To obtain a similar result when X/I is uncountable, it needs to be assumed also that in X/I, there is a countable subset that is order dense in X/I. By definition, $S \subset X$ is an order dense in (X, P) if whenever x and y in X/S, there is $s \in S$ such that $xPsPy$. Countable order denseness is often replaced in economic discussions by a sufficient but non-necessary topological assumption which implies that u can be defined to be continuous in the topology used for X.

The standard assumption is that R is complete and transitive in X. In this case, we say that the structure of preferences is a complete preorder. The relation R can be decomposed in two relationship, the strict part P and the indifference part I.

It follows that xPy if and only if xRy and $\neg yRx$. Given the assumed completeness of R, we obtain that xIy if and only if $\neg xPy$ and $\neg yPx$. If neither of the two alternative is considered strictly preferred to the other, then they must be considered as indifferent. This means that a pair of alternatives are considered indifferent in two separate cases. In the first one, each alternative is as good as another. In the second case, both alternatives are not comparable using the relationship P. In Plata (2004), this difference is explained, and a result of representativeness for acyclic preferences is proved. In the next definitions, we can find some of the most known preference structures.

Definition 1. The preference structure (X, R) is a **complete preorder** if it satisfies:

a) Completeness: $\forall x, \forall y \in X$: xRy or yRx.
b) Transitivity: $\forall x, y, z \in X$: If xRy and yRz, then xRz.

These two requirements mean that (1) the individual has the ability to compare any two alternatives and (2) the individual does not show cycles with respect to the strict preferences. If such were the case, it would involve the inability to decide.

This axiom is very important for the decisiveness. There can only be indifference cycles in I but not in the cycle relation P. Note that if an agent prefers x to y, y to z, and z to x, she has no better alternative among the three.

Definition 2. The preference structure (X, R) is P-**acyclic** if $\forall n \in \mathbb{N}$ *and* $\forall x_1, x_2, \ldots, x_n \in X$ and if $x_1 P x_2, x_2 P x_3, \ldots, x_{n-1} P x_n$, then not $x_n P x_1$.

This condition prevents the existence of cycles under P, but not under R because the condition $x_n P x_1$ does not avoid the possibility $x_n I x_1$.

It is very easy to see that in the case of finite X, the acyclicity is equivalent to the existence of maximal elements in X. A z element X is R-*maximal* if there is no another element of X that strictly preferred to z. In Sen (1970), A. Sen has shown the following theorem:

Theorem 1. *Let (X, R) be a preference structure with X, a finite set. There is a R-maximal for each $B \subseteq X$ if only if R is P-acyclic.*

This result can be extended in a natural way to compact set in a topological space when P satisfies a condition of continuity (see Fagin et al. 2006).

There are many structures where the transitivity of strict preference is preserved, but intransitivity of indifference is verified. Suppose that coffee with n grams of sugar and coffee with $n - 1$ grams of sugar are to me indifferent and so on. We conclude that I am indifferent between coffee with 1 gram of sugar and coffee without sugar. But this does not mean that I am indifferent between coffee with n grams of sugar and the coffee without sugar.

In general, this is the case with semiorders and interval orders; the first one is a special case of the second one.

Definition 3. We say that (X, R) is an **interval order** if R is a pseudotransitive binary relationship; this means that: $\forall x, y, z, w \in X: xPyRzPw \longrightarrow xPw$.

Note that if R is pseudotransitive, from xPy and yPz we cannot conclude that xPz. For such a conclusion, we need to verify that yRz. Neither from xIy and yIz, we can conclude xIz. Think in the case of coffee with sugar previously considered.

Definition 4. We say that (X, R) is a **semiorder** if it is an interval order satisfying additionally the following property $\forall x, y, z \in X : xPy$ and $yPz \longrightarrow \forall w \in X, xPw$ or wPz.

When we have an interval order, X/I is not necessarily a set of classes of indifference.

13.3 Numerical Representations

Here, we present numerical representation theorems for each one of the preference structures in Sect. 13.2. References are provided. Definitions of the numerical representation for each type of rationality is first provided.

A typical theorem for the existence of numerical representation for a given structure (X, R) with a rationality assumption given by *** has the form:

Proposition. **Let (X, R) be a preference structure. Then (X, R) satisfies *** if and only if there is a function u with domain X that represents the preference R.**

13.3.1 Numerical Representation for Complete Preorders

Definition 5. $u : X \mapsto \mathbb{R}$ is a **utility function** for the complete preorder (X, R) if

$$\forall x, y \in X : xRy \Leftrightarrow u(x) \geq u(y).$$

A utility function measures the preference for each alternative. Then if u represents R, then xPy if and only if $u(x) > u(y)$ and if x and y are indifferent if and only if $u(x) = u(y)$. This theorem formally establishes the classic measurement of ordinal utility.

In what follows, we assume X countable. So, we do not need topological axioms for the existence of numerical representations. Such is the claim of the following theorem:

Theorem 2 (Cantor 1955; Debreu 1954). *Let (X, P) be a preordered set where X is countable, and if P is a partial order on X, then there exists a utility function $u : X \to \mathbb{R}$ representing the partial order.*

In case where (X, R) is a preference structure, but R is a partially ordered binary relationship that is not necessarily a weak order, to obtain a representation theorem, analogous to (2), we need to replace u by another quantitative representation or to look for some additional properties for the preference structure.

13.3.2 Numerical Representation for Acyclic Preferences

The use of simple numbers appears insufficient for the representation of ordered sets having a non-transitive indifference relation. In (Subiza 1994), a numerical representation of preferences by means of correspondences (or set-valued real function) is presented. Such representation provide a complete characterization of acyclic preferences on countable sets.

Definition 6. $\mu : X \mapsto \mathbb{R}$ is a **utility correspondence** for (X, R) if

a) $\forall x \in X$: $\mu(x)$ is a bounded nonempty set of real numbers.
b) $xPy \leftrightarrow \mu(x) \cap \mu(y) = \emptyset$ and $\sup \mu(x) > \sup \mu(y)$

Each alternative $x \in X$ is associated with a bounded set of real numbers. These sets can be intersected when they are representing indifferent alternatives. In the case of strict preference, intersection is empty, and the greater supremum represents the most preferred alternative.

Theorem 3 (Subiza 1994). *Let (X, R) be a preference structure. Then (X, R) is P-acyclic if and only if there is $\mu : X \mapsto \mathbb{R}$, utility correspondence for (X, R).*

13.3.3 Weak Representations for Acyclic Preferences or Partial Orders

Every proposition claiming the existence of a representation for a preference structure (X, R) must assume that R is a weak order. It cannot hold when R is not P-acyclic or is a partial order that is not also a weak order (i.e., I is not transitive). In what follows, we shall discuss some alternatives for such assumption.

Definition 7. Let (X, R) be a preference structure. We say that $u : X \mapsto \mathbb{R}$ **weakly represents** (X, R) if $\forall x, y \in X : xPy \rightarrow u(x) > u(y)$.

Theorem 4. *Let (X, R) be a preference structure. Then (X, R) is P-acyclic if and only if there is $u : X \mapsto \mathbb{R}$ that weakly represents (X, R).*

13.3.4 Interval Orders and Semiorders

A small difference of evaluation between two alternatives can remain inadequate to affirm a strict preference between them. This leads to the introduction of a positive threshold number q in such a way that the alternative x is said to be preferred to the alternative y if and only if the evaluation of x is greater than the evaluation of y plus the threshold q. The most classical structure respecting such an idea is a semiorder.

Definition 8. Let (X, R) be a preference structure. The pair of functions $u, v : X \mapsto \mathbb{R}$ is an **interval representation** for (X, R) if

$$\begin{cases} \forall \, x, y \in X, xPy \Leftrightarrow u(x) > v(y) \\ \forall \, x \in X \qquad v(x) \geq u(x). \end{cases}$$

For instance, we can consider $v(x) = u(x) + q$ where $q > 0$, and so $xPy \Leftrightarrow u(x) > u(y) + q$.

Theorem 5. *Let (X, R) a preference structure. Then (X, R) is an interval order if and only if there is an interval representation $u, v : X \mapsto \mathbb{R}$ for (X, R).*

Roughly speaking, when we have an interval representation then we can associate for each $x \in X$ an interval $I(x) = [u(x), v(x)]$ such that xPy if and only if $I(x) \cap I(y) = \emptyset$ and $I(x)$ is on the right of $I(y)$.

13.4 Invariance of Numerical Representations of Preferences

The invariance analysis aims to characterize the set of numerical representations of the same preference structure (X, R). This numerical representation is defined in X. Each point in X has associated a numerical object. These objects can be real numbers, intervals of real numbers, sets of real numbers, and vectors of real numbers, among others. This of course depends on the type of rationality and the type of numerical representation chosen.

It is well known that in the case of complete preorders, the utility function is an ordinal measure of the preference. This means that if we have two utilities representing the same complete preorder, they can be transformed into each other by means of an increasing function from real numbers to real numbers. However, this is not always possible in the rationality weakening. Sometimes it is not even possible to find a function that transforms a numerical representation to another.

Example 1 (Untransformable Numerical Representations). Suppose we have a preference defined for a set with four alternatives; u and v are two numerical representations of preference with the following property:

$$i, j \in \{1, 2, 3, 4\}$$

$$u(x_i) \neq u(x_j) \quad \text{if} \quad i \neq j \quad \text{except} \quad i = 2, j = 3$$

$$v(x_i) \neq v(x_j) \quad \text{if} \quad i \neq j \quad \text{except} \quad i = 3, j = 4.$$

Under this situation, it is not possible to transform u into v using a function. See first that there is not a function ϕ such that

$$\phi(u(x_i)) = v(x_i) \text{ for } i = 1, \dots, 4.$$

The reason is as follows. We need $u(x_2) = u(x_3)$ because of the first requirement above. But the second requirement above implies $v(x_2) \neq v(x_3)$. So, $\phi(u(x_2))$ must be different from $\phi(u(x_3))$. But in such a case, ϕ cannot be a function. In the same way, we cannot transform v to u. The natural question is under what condition can we transform a numerical representation in another by means of a function?

We can associate a partition of X with each numerical representation u. Two alternatives of X are in the same class if they have the same image under u. Formally, the relation $\sim u$ is defined as $x_1 \sim u \ x_j$ if and only if $u(x_i) = u(x_j)$ is an equivalence relation in X.

Definition 9. Let u, v be two numerical representations for the structure (X, R). We say that u is **analogous** to v if the partition of X induced by u is the same partition that is induced by v. This means that $X/ \sim u = X/ \sim v$.

In the above example, representations u and v are not analogous.

Partition of X induced by u is $\{\{1\}, \{2, 3\}, \{4\}\}$.

Partition of X induced by v is $\{\{1\}, \{2\}, \{3, 4\}\}$.

Definition 10. Let u, v be two numerical representations for the structure (X, R). We say that u and v are **transformable** if and only if there are functions ϕ and φ such that $v = \phi \circ u$ y $u = \varphi \circ v$.

The next proposition show that the fact that u and v are analogous is equivalent to transformability between u and v.

Theorem 6. *Let (X, R) be a preference structure. Let u and v be two numerical representations for preference R. Then, u and v are transformable if and only if u is analogous to v.*

Proof. If u and v are not analogous, the partition of X induced by u is not the same as the partition of X induced v. We have two cases, and they are not necessarily mutually exclusive cases. (a) There exist x_i and x_j such that $u(x_i) = u(x_j)$ but $v(x_i) \neq$

$v(x_j)$, or (b) there are x_i and x_j such that $v(x_i) = v(x_j)$ but $u(x_i) \neq u(x_j)$. In the first case, there is no function transforming u in v. In the second case, there is no function transforming v in u. So u and v are not transformable.

Let us see the part of the proposition. In order to transform u in v, we need to define ϕ such that associates $u(x_i)$ to $v(x_i)$. This means that $\phi = \{\langle u(x_i), v(x_i)\rangle : x_i \in X\}$. We note that this ϕ is a function. Suppose that $\langle u(x_i), v(x_i)\rangle \in \phi, \langle u(x_j), v(x_j)\rangle \in \phi$ and that $u(x_i) = u(x_j)$, we want to obtain that $v(x_i) = v(x_j)$. Because $u(x_i) = u(x_j)$, we have that $x_i \sim ux_j$. Using the hypothesis of u is analogous to v we have that $x_i \sim v \ x_j$ too. So, we have that $v(x_i) = v(x_j)$. For transforming v in u, we proceed similarly. $\square$

In what follows, we present the invariance analysis of numerical representations. An invariance theorem for a preference structure (X, R) assumes some rational conditions on this structure and some mathematical conditions on the numerical representations and shows necessary and sufficient conditions for the two representations to be analogous. This claim is generally equivalent with the existence of a functional ϕ verifying that $v = \phi \circ u$.

13.4.1 Invariance of Utility Functions

Let $u : X \to \mathbb{R}$ be a utility function representing (X, R). We denote by

$$u(X) = \text{rank } u = \{y \in \mathbb{R} : \exists x \in X, u(x) = y\}$$

the image of X under u. As it is easy to see, the next proposition holds:

Proposition 1. *Let* $u : X \mapsto \mathbb{R}$ *be a utility function representing the complete preorder* (X, R). *Then* $v : X \mapsto \mathbb{R}$ *is a utility function for* (X, R) *if and only if there is a strictly increasing function* $\phi : u(X) \mapsto \mathbb{R}$ *such that* $v = \phi \circ u$.

In Krantz et al. (1971), the reader can find an interesting discussion about the reaction, measured in terms of utilities, of different agents to several stimulus.

13.4.2 Invariance of Utility Correspondences

We present first an example showing some utility correspondences for a very simple acyclic preference. We will see some problems to achieve the transformation of a utility correspondence in another.

Example 2 (Case of No Transformable Utility Correspondences). We consider the P-acyclic preference structure given by

$$X = \{x_1, x_2, x_3\} \quad y \quad R = \{\langle x_1, x_2\rangle, \langle x_1, x_3\rangle\}.$$

The following two utility correspondences represent the acyclic preference R,

$$\mu(x_1) = \{3\} \qquad \mu'(x_1) = \{3\}$$
$$\mu(x_2) = \{1/2, 2\} \qquad \mu'(x_2) = \{1/2, 1\}$$
$$\mu(x_3) = \{1/2, 1\} \qquad \mu'(x_3) = \{1/2, 2\}.$$

Both utility correspondences say that x_1 is strictly preferred to x_2 and that x_1 is strictly preferred to x_3. We can obtain μ' from μ using set transformations. The identity is the unique function sending the union of $\mathrm{rank}(\mu)$ in the union of $\mathrm{rank}(\mu')$. However, this function does not help us to preserve the corresponding suprema:

$$\sup \ \mu(x_2) > \sup \ \mu(x_3) \quad \text{but not} \quad (\sup \ \mu'(x_2) > \sup \ \mu'(x_3)).$$

Let us consider now the following pair of utility correspondences for (X, R)

$$\mu(x_1) = (0, 1/2) \cup (1/2, 1) \cup (1, 2) \cup \{3\} \qquad \mu'(x_1) = \{3\}$$
$$\mu(x_2) = \{1/2, 2\} \qquad \mu'(x_2) = (0, 1/2] \cup \{1\}$$
$$\mu(x_3) = \{1/2, 1\} \qquad \mu'(x_3) = \{1/2\} \cup (1, 2].$$

We note that there is not a "reasonable numerical transformation" indeed, even a one to one numerical transformation that obtains μ' from the union of the sets formed by each $\mu(x)$.

In order to obtain a utility correspondence from another, we need transformation of sets satisfying some special requirements. We need the preservation of empty intersections, when be needed. Additionally, there must be monotonicity among suprema comparisons. Finally, bounded nonempty sets should be sent to bounded nonempty sets.

Given a utility correspondence $\mu : X \mapsto \mathbb{R}$, the range or image of μ is the collection of all real subsets $\mu(x_i)$ for each $x_i \in X$ i.e.,

$$\mathrm{rank} \ \mu \subset 2^{\mathbb{R}}.$$
$$\mathrm{rank} \ \mu = \{\mu(x_i) | x_i \in X\}$$

The admissible transformations for μ, which conserve invariance of the representation, are transformations of sets with domain in $\mathrm{rank} \ \mu$. We can denote this class of set transformations by $\Psi^{uc}(\mu)$; the formal definition of this class is as follows.

Definition 11. Let ϕ a correspondence such that $\phi : \mathrm{rank}\mu \mapsto 2^R$.

We say that ϕ is the class $\Psi^{uc}(\mu)$ or $\phi \in \Psi^{uc}(\mu)$ if the next two conditions hold:

1. If A is a nonempty and bounded set of real numbers, then $\phi(A)$ is nonempty and bounded set of real numbers.

2. If $x_i P x_j$, $\mu(x_i) \cap \mu(x_j) = \emptyset$ and

$$\sup A_i > \sup A_j \leftrightarrow \phi(A_i) \cap \phi(A_j) = \emptyset \text{ y } \sup \phi(A_i) > \sup \phi(A_j)$$

This definition requires preservation of empty intersections and preservation of monotonicity of sups in the strict part of preference.

Theorem 7 (Invariance of Utility Correspondences). *Let (X, R) a P-acyclic preference structure and μ a utility correspondence for (X, R). Then, μ' is a utility correspondence for (X, R) and μ' analogous to μ if and only if there is a transformation $\phi : \text{rank}\mu \mapsto 2^R$ belonging to $\Psi^{uc}(\mu)$ and such that for each $x_i \in X$, $\mu'(x_i) = \phi(\mu(x_i))$.*

Proof. $\Longrightarrow$ Fix the utility correspondence μ. Now we consider any other analogous utility correspondences μ' for (X, R). Let $\phi : \text{rank}\mu \to 2^\mathbb{R}$ be a function such that $\phi(\mu(x_i)) = \mu'(x_i) \forall x_i \in X$.

We shall show that ϕ belongs to $\Psi^{uc}(\mu)$. It is clear that ϕ transforms nonempty and bounded sets in nonempty and bounded sets; this is by the definition of ϕ and the fact that μ and μ' are utility correspondences for (X, R).

In order to show the second condition, we suppose that $x_i P x_j$, $\mu(xi) = A_i$ and that $\mu(x_j) = A_j$. Because μ and μ' represent (X, R), we have that

$$x_i P x_j \leftrightarrow \mu(x_i) \cap \mu(x_j) = \emptyset \quad \text{and} \quad \sup\mu(x_i) > \sup\mu(x_j)$$

$$x_i P x_j \leftrightarrow \mu'(x_i) \cap \mu'(x_j) = \emptyset \quad \text{and} \quad \sup\mu'(x_i) > \sup\mu'(x_j)$$

then

$$\mu(x_i) \cap \mu(x_j) = \emptyset \quad \text{and} \quad \sup\mu(x_i) > \sup\mu(x_j)$$

$$\leftrightarrow \mu'(x_i) \cap \mu'(x_i) = \emptyset \quad \text{and} \quad \sup\mu(x_i) > \sup\mu'(x_j)$$

by definition of ϕ, this is equivalent to

$$\mu(x_i) \cap \mu(x_j) = \emptyset \quad \text{and} \quad \sup\mu(x_i) > \sup\mu(x_j)$$

$$\leftrightarrow \phi(\mu(x_i)) \cap \phi(\mu(x_j)) = \emptyset \quad \text{and} \quad \sup\phi(\mu(x_i)) > \sup\phi(\mu(x_j))$$

so, we arrive $\phi \in \Psi^{uc}(\mu)$.

$\Longleftarrow$ Let $\phi : \text{rank } \mu \mapsto 2^R$ a transformation of sets of the class $\phi \in \Psi^{uc}(\mu)$ and such that $\mu'(x_i) = \phi(\mu(x_i)) \forall x_i \in X$.

We shall show that μ' is a utility correspondence for (X, R).

First, we note that for each x_i, $\mu(x_i)$ is nonempty and bounded set, because μ is a utility correspondence for (X, R). Since $\phi \in \Psi^{uc}(\mu)$, we have that $\phi(\mu(x_i))$ is nonempty and bounded. So, $\mu'(x_i)$ is nonempty and bounded for each x_i belonging to X.

Consider now that $(x_i, x_j) \in X$ and suppose in addition that $x_i P x_j$.

Because μ is a utility correspondence for (X, R), then we have that $\mu(x_i) \cap \mu(x_j) = \emptyset$ and $\sup \mu(x_i) > \sup \mu(x_j)$. But ϕ belongs to the class $\Psi^{uc}(\mu)$; this assertion is equivalent to $\phi(\mu(x_i)) \cap \phi(\mu(x_j)) = \emptyset$ and sup $\phi(\mu(x_i)) >$ sup $\phi(\mu(x_j))$. But this is precisely $\mu'(x_i) \cap \mu'(x_j) = \emptyset$ y sup $\mu'(x_i) >$ sup $\mu'(x_j)$.

We have proved that μ' is a utility correspondence for (X, R) too. $\qquad\square$

13.4.3 Invariance for Weak Representations

The class of admissible transformations, for this case, is built from the next definition.

Definition 12. Let (X, R) be a preference structure, a function $u : X \mapsto \mathbb{R}$ and $\phi :$ rank $u \mapsto \mathbb{R}$. We say that ϕ is *P-increasing* in *rank u* if for each pair $x_i, x_j \in X$: If $u(x_i) > u(x_j)$ and $x_i P x_j$ then $\phi(u(x_i)) > \phi(u(x_j))$.

This definition says that ϕ is increasing only when P "says the same." The class of transformation *P-increasing* includes, as a subset, the class of all increasing transformations.

Example 3 (P-Increasing Transformations). Consider the same structure (X, R) with $X = \{x_1, x_2, x_3\}$ and $R = \{\langle x_1, x_2 \rangle, \langle x_1, x_3 \rangle\}$. Suppose $u(x_1) = 7$, $u(x_2) = 6$ y $u(x_3) = 5$, this is a weak representation for this structure. The following transformations are *P-increasing*:

$$\phi(u(x_1)) = 3 \qquad \phi(u(x_2)) = 2 \qquad \phi(u(x_3)) = 1$$
$$\psi(u(x_1)) = 3 \qquad \psi(u(x_2)) = 1 \qquad \psi(u(x_3)) = 2.$$

Both transformations are forming new weak numerical representations for (X, R). But it is impossible to obtain one from the other using an increasing transformation.

Theorem 8 (Invariance of Weak Representations). *Let (X, R) be a preference structure P-acyclic and $u : X \mapsto \mathbb{R}$ a weak representation for (X, R). Then, $v : X \mapsto \mathbb{R}$ is a weak representation (X, R), and v is analogous to u if and only if there is $\phi :$ rank $u \mapsto \mathbb{R}$ such that ϕ es P-increasing in rank u and such that $v = \phi \circ u$.*

Proof. $\implies$ Let u a weak representation for (X, R). Let v any other analogous weak representation for the structure (X, R). Using these two numerical representations, we define $\phi :$ rank $u \mapsto \mathbb{R}$ such that $\phi(u(x_i)) = v(x_i) \forall x_i \in X$.

We need to see that ϕ is *P-increasing* in *rank u*. We take x_i, x_j in X such that $x_i P x_j$ and $u(x_i) > u(x_j)$. Because v weakly represents the preference structure, we have that $v(x_i) > v(x_j)$), but this means that $\phi(u(x_i)) > \phi(u(x_j))$. We concluded that ϕ is *P-increasing* in *rank u*.

$\impliedby$ Now we consider a function $u : X \mapsto \mathbb{R}$ weakly representing (X, R) a transformation ϕ, *P-increasing* defined in *rank u* in such way that we build the new function $v(x_i) = \phi(u(x_i))$ with domain X. We shall show that v is a weak representation for (X, R) too. In order to see this, suppose that $x_i P x_j$. Because u weakly represents (X, R), we have that $u(x_i) > u(x_j)$. Because ϕ is *P-increasing* in *rank u*, we have then that $\phi(u(x_i)) > \phi(u(x_j))$. But this means precisely that $v(x_i) > v(x_j)$. This concludes the proof. $\qquad\square$

13.4.4 Invariance for Interval Orders and Semiorders

Finally, we consider now the invariance analysis for the interval order representations. We need to think this type of representations as a pair of functions (u, v). In order to transform one pair in another pair of functions, we use the following definition.

Definition 13 (The Class of Transformations $\Psi^{RI}(u, v)$).

$$(\phi, \xi) \in \Psi^{RI}(u, v) \quad \text{if and only if} \quad \phi : \text{rank } u \mapsto \mathbb{R}, \quad \xi : \text{rank } v \mapsto \mathbb{R},$$

$$\phi(u(x)) \le \xi(v(x)) \quad \forall x \in X,$$

$$\text{and if} \quad xPy, \quad \text{then} \quad u(x) > u(y) \Leftrightarrow \phi(u(x)) > \xi(v(y))$$

Theorem 9 (Invariance of Interval Representations). *Let (X, R) be an interval order structure of preferences. Let (u, v) be an interval order representation for (X, R). Then (u', v') is an interval order representation for (X, R), u is analogous to u', and v is analogous to v' if and only if there is a pair of transformations $(\phi, \xi) \in \Psi^{RI}(u, v)$ such that for each $x \in X$, $(u'(x), v'(x)) = (\phi(u(x)), \xi(v(x)))$.*

Proof. $\Longrightarrow$ Consider that (u, v) defines an interval representation for (X, R). Let (u', v') be any other interval representation for the same structure of preferences. We build $\phi : \text{rank } u \mapsto \mathbb{R}$ and $\xi : \text{rank } v \mapsto \mathbb{R}$ in such a way that for each $x \in X : \phi(u(x)) = u'(x)$ and $\xi(v(x)) = v'(x)$. We shall show that (ϕ, ξ) is an element of the class $\Psi^{RI}(u, v)$. Because (u', v') represents (X, R), we have that for each $x \in X$, $u'(x) \le v'(x)$. This means, by the construction of ϕ and ξ, that $\phi(u(x)) \le \xi(v(x))$ for each $x \in X$. In order to verify the second condition, we suppose xPy. By the representability of the pair (u, v), this means that $u(x) > v(y)$. By the representability of the pair (u', v'), xPy is equivalent to $u'(x) > v'(y)$ too. So $u(x) > v(y) \Leftrightarrow u'(x) > v'(y)$, but this means precisely that $u(x) > v(y) \Leftrightarrow \phi(u(x)) > \xi(v(y))$. We have then that $(\phi, \xi) \in \Psi^{RI}(u, v)$.

$\Longleftarrow$ Suppose that $(\phi, \xi) \in \Psi^{RI}(u, v)$ and that $u'(x) = \phi(u(x))$ and $v'(x) = \xi(v(x))$. We want to see that (u', v') is an interval representation for (X, R). Because $(\phi, \xi) \in \Psi^{RI}(u, v)$ we have that $\phi(u(x)) \le \xi(v(x))$, this means that $v'(x) > u'(x) \forall x \in X$. We suppose now that xPy. Because (u, v) is an interval representation, we have that xPy is equivalent to $u(x) > v(y)$. This last is equivalent to $\phi(u(x)) > \xi(v(y))$ because (ϕ, ξ) are admissible transformations for (u, v). The last inequality is equivalent to $u'(x) > v'(y)$; we have then that xPy is equivalent to $u'(x) > v'(y)$. $\qquad\square$

13.5 By Way of Conclusion

In this paper, in order to avoid topological considerations, we have considered only sets with a countable number of alternatives. Just to show that our work can be, even under these limitations, relevant in economic theory, we refer to Debreu

(1954) where it is shown that a continuous preference relation defined in a subset of a separable normed space (and, with more generality, in a perfectly separable space) has a utility representation. However, even in a convex subset of a non-separable normed space (or more generally in any non-separable metric space), there are continuous preference relation without utility representation (see Estévez and Hervés 1995). For the case of non-separable spaces, the result by Monteiro (1987) characterizes representability: a continuous preference relation in a path connected set of alternative X has a continuous numerical representation, if and only if it is countably bounded, i.e., there is some countable subset F of X such that for all $x \in X$, there exist y and z in F with $y \succeq x \succeq z$.

Acknowledgements I wish to thank Elvio Accinelli, Erubiel Ordaz, and two anonymous referees for their helpful comments.

References

Armstrong, W.E.: A note on the theory of consumers behavior. Oxf. Econ. Pap. **2**, 119–122 (1950)

Bridges, D.S.: Numerical representation of intransitive preferences on a countable set. J. Econ. Theory **30**, 213–217 (1983)

Bridges, D.S.: Representating interval - orders by a single real - valued function. J. Econ. Theory. **36**, 149–165 (1985)

Bridges, D.S., Mehta, G.B.: Representations of Preference Orderings. Lecture Notes in Economics and Mathematical Systems, vol. 422. Springer, Berlin (1995)

Candeal, J., Indurain, E., Zudaire, M.: Numerical representability of semiorders. Math. Soc. Sci. **43**, 61–77 (2002)

Cantor, G.: Contributions to the Founding of the Theory of Transfinite Numbers. Dover, New York (1955)

Debreu, G.: Representation of preference ordering by a numerical function. Decis. Process. **3**, 159–165 (1954)

Estévez, M., Hervés, C.: On the existence of continuous preference orderings without utility representations. J. Math. Econ. **24**, 305–309 (1995)

Fagin, R., Kumar, R., Mahdian, M., Sivakumar, D., Vee, E.: Comparing partial rank. SIAM J. Discret. Math. **20**(3), 628–648 (2006)

Fishburn, P.: Utility Theory of Decision Making. Wiley, New York (1970a)

Fishburn, P.: Intransitive indifference with unequal - indifference intervals. J. Math. Psychol. **7**, 144–149 (1970b)

Fishburn, P.: Interval representation of interval orders and semiorders. J. Math. Psychol. **10**, 91–105 (1973)

Fishburn, P.: Interval-Orders and Interval Graphs. Wiley, New York (1985)

Fishburn, P.: Preference structures and their numerical representations. Theor. Comput. Sci. **217**, 359–383 (1999)

Herrero, C., Subiza, B.: A characterization of acyclic preferences on countable sets. A Discussion WP - AD 91 - 01. Instituto Valenciano de Investigaciones Econmicas (1991)

Jaminson, D.T., Lau, L.J.: Semiorders and the theory of choice. Econometrica **41**, 901–912 (1973); A Correct. Econ. **43**, 975–977

Krantz, D.H., Luce, R.D., Suppes, T., Tversky, A.: Foundations of Measurement Volume 1 (Additive and Polynomial Representations). Academic, New York (1971)

Luce, R.D.: Semiorders and a theory of utility discrimination. Econometrica **24**, 178–191 (1956)

Monteiro, P.K.: Some results on the existence of utility functions on path connected spaces. J. Math. Econ. **16**, 147–156 (1987)

Nakamura, Y.: Real interval representations. J. Math. Psychol. **46**, 140–177 (2002)

Oloriz, E., Candeal, J.C., Indurain, E.: Representability of interval orders. J. Econ. Theory **78**, 219–227 (1998)

Pirlot, M.: Minimal representation of a semiorder. Theory Decis. **28**, 109–141 (1990)

Pirlot, M., Vinche, P.: Semiorders, Properties, Representations an Applications. Kluwer, Dordrecht (1997)

Plata, L.: A numerical representation of acyclic preferences when non comparability and indifference are concepts with different meaning. Econoquantum **1**, 17–23 (2004)

Rodriguez-Palmero, C.: A representation of acyclic preferences. Econ. Lett. **54**, 143–146 (1997)

Sen, A.K.: Collective Choice and Social Welfare. Holden Day, San Francisco (1970)

Scott, D., Suppes, F.: Foundational aspects of theories of measurement. J. Symb. Log. **23**, 113–128 (1958)

Subiza, B.: Numerical representations for acyclic preferences. J. Math. Psychol. **38**(4), 467–476 (1994)

Suppes, P., Krantz, D., Luce, R.D., Tversky, A.: Foundations of Measurement Volume 2 (Geometrical, Threshold and Probabilistic Representations). Academic, San Diego (1989)

Chapter 14
Symmetrical Core and Shapley Value of an Information Transferal Game

Patricia Lucia Galdeano and Luis Guillermo Quintas

Abstract In this paper we study some properties and we characterize the Symmetrical Core. We analyze the relation of the Symmetrical Core with the Shapley value of a game modeling information transferal in a cooperative environment. This type of game was introduced by Galdeano et al. (Int Game Theor Rev 12(1):19–35, 2010) and it was also studied by Hou and Driessen (J Appl Math 2012:1–12, 2012). It consists of an information market game involving identical firms and an innovator having relevant information for the firms (e.g., a new technology).

We analyze how the symmetrical part of the Core varies according with the initial information level of the firms and the value of the information. We also present conditions in order that the Shapley value belongs to the Symmetrical Core. We compare the cooperative outcomes with the noncooperative equilibrium of the game studied by Quintas (Modell Meas Control D 11:11–28, 1995).

Keywords Cooperative game theory • Symmetrical Core • Shapley value • Information transferal

Mathematics Subject Classification: 91A80

P.L. Galdeano
Facultad de Ciencias Físico Matemáticas y Naturales, Departamento de Matematicas,
Universidad Nacional de San Luis, Ejercito de los Andes 950, D5700 San Luis, Argentina
e-mail: patriciagaldeano@gmail.com

L.G. Quintas (✉)
Facultad de Ciencias Físico Matemáticas y Naturales, Departamento de Matematicas,
Universidad Nacional de San Luis, Ejercito de los Andes 950, D5700 San Luis, Argentina

Departamento de Matematicas, IMASL (UNSL-CONICET), Universidad Nacional de San Luis,
Ejercito de los Andes 950, D5700 San Luis, Argentina
e-mail: lu6quintas@gmail.com

© Springer International Publishing Switzerland 2016
A.A. Pinto et al. (eds.), *Trends in Mathematical Economics*,
DOI 10.1007/978-3-319-32543-9_14

14.1 Introduction

Following the pioneer work by Arrow (1962), many studies on information markets appeared in the literature, analyzing economic effects of patent licensing or protection (Taylor and Silberston 1973; Kamien and Tauman 1986; Katz and Shapiro 1986; Gilbert and Shapiro 1990; Muto 1993; Quintas 1995; Nakayama et al. 1991; Wang 1998, 2002). More recently several studies have been done modeling the interaction of an innovator and n firms in an industry (Poddar and Sinha 2004; Sen and Tauman 2007; Tauman and Watanabe 2007; Schmidt 2008; Stamatopoulos and Tauman 2008; Galdeano et al. 2010; Hou and Driessen 2012).

We follow the approach presented by Galdeano et al. (2010). Besides n firms with identical characteristics, there exists an agent called the innovator, having relevant information for the firms. The innovator is not going to use the information for himself, but this information can be sold to the firms. Any firm that decides to acquire the new information (e.g., a new technology) is supposed to make use of the information. The n potential users of the information are the same before and after the innovator offers the new technology. The problem is modeled as a $(n+1)$-players cooperative game.

Galdeano et al. (2010) characterized the Shapley value of this game requiring 0-monotonicity. We now show that the game is also superadditive, we present a new formulation for the Shapley value, and we give conditions in order that the Shapley value belongs to the symmetrical part of the Core.

Hou and Driessen (2012) studied the nucleolus of this game. They also showed the equivalence between the nonemptiness of the Symmetrical Core and one of each conditions: supperadditivity, 0-monotonicity, or monotonicity.

In the present article, we present an explicit characterization of the Symmetrical Core, and we analyze how it varies depending on the initial information level of the firms and the value of the information. We also compare the cooperative outcomes with the noncooperative equilibrium studied by Quintas (1995).

The paper is organized as follows: in Sect. 14.2 we describe the information market and we define the corresponding game. In Sect. 14.3 we present some results on the symmetrical part of the Core, and we give conditions for the Shapley value to be in the symmetrical part of the Core. In Sect. 14.4 we study how the Symmetrical Core varies when firms have more or less prior information. We analyze some limit cases: when firms have full or no prior information and the variation of the value of the information. In Sect. 14.5 we present some conclusions and possible extensions.

14.2 The Information Market

We consider a market with n firms ($n \geq 2$) and an innovator who possess a patent or an information.

The set of agents will be denoted by $N = \{1, 2, \ldots, n + 1\}$, where $I = \{1\}$ (the innovator) is the agent having a new information and $U = \{2, \ldots, n + 1\}$ (the users) are firms who could be willing to obtain the new information.

The n users or firms interact in the same market, producing or performing the same activity with the same technology or the same information. Thus, all the users have the same incentives for the acquisition of the new information or technology. We will make the following assumptions about the problem we want to study:

S.1: The n information users are the same before and after the information holder offers the new technology. This indicates that there are no exits or incoming agents in the market.

S.2: All the players that acquire the new information make use of it.

S.3: In order to compute the utilities of the players, we use a conservative criteria, assuming that the uninformed agents make the right choice.

S.1 and S.2 are natural assumptions. S.3 avoids dealing with externalities (Macho-Stadler et al. 2006; De Clippel and Serrano 2008). As in several other papers (Amer et al. 2008; Belenky 2002; Sandholm et al. 1999), we adopt a principle of prudence: each coalition is assigned a utility corresponding to the worst possible scenario.

We will take into account these conditions in order to define an $(n + 1)$-person cooperative game.

14.2.1 The Cooperative Game

Definition 1. An $(n + 1)$-person game in characteristic function form is given by (N, v), where $N = \{1, 2, \ldots, n + 1\}$ is the set of players, and $v : 2^N \to \mathbb{R}$ is the characteristic function.

Following Galdeano et al. (2010), we consider n firms with identical characteristics (the users) and an agent called the innovator, having relevant information for the firms. The innovator is not going to use the information for himself, but this information can be sold to the firms. Any firm that decides to acquire the new information (e.g., a new technology) is supposed to make use of the information. The n potential users of the information are the same before and after the innovator offers the new technology. The firms acquiring the information will be better than before obtaining it, while their utilities are computed under a conservator point of view, assuming that for any uninformed firm, the probability of making a right decision can be described by a *binomial probability distribution*, being $0 \le c \le 1$ the uniform probability of having success. We will first consider the case when $0 < c < 1$, and we will then consider the cases $c = 0$ and $c = 1$ in Sect. 14.4.

The probability that k among s firms take the right decision is given by $\binom{s}{k} c^k (1 - c)^{s-k}$, and hence, the expected aggregated utility of k firms having success is $k \binom{s}{k} c^k (1 - c)^{s-k} a_k$. Here $a_k \ge 0$ represents the utility if k firms make a right decision. Throughout the paper, the utility function is monotonic decreasing because when the number of firms taking a right decision increases, each firm receives a lower utility level, i.e., $a_{k+1} \leq a_k$ for all $k \geq 1$. We normalized it assuming that $a_1 = 1$.

Throughout the paper, the size ($|S|$, or cardinality) of any coalition $S \subseteq N$ is denoted by s, $0 \leq s \leq n + 1$. In case coalition S contains the innovator, then $v(S) = (s - 1)a_n$ because any member of S, different from the innovator, made a right decision rewarding the expected utility a_n since the $n - s$ uninformed firms outside S are assumed to take a right decisions too.

Definition 2. The $(n + 1)$-person information market game (N, v) in characteristic function form is given by,

$$
\begin{aligned}
v(\emptyset) &= 0 \\
v(S) &= (s - 1)\, a_n & \text{if } 1 \in S. \\
v(S) &= w(s) = \sum_{j=0}^{s} j\binom{s}{j} c^j (1 - c)^{s-j} a_{n-s+j} & \text{if } 1 \notin S
\end{aligned}
\tag{14.1}
$$

for all $S \subseteq N$, $S \neq \emptyset$ and $s = |S|$

14.2.1.1 Properties Fulfilled by the Characteristic Function v

A usual assumption is that the game is superadditive:

Definition 3. A game (N, v) is *superadditive* if for all sets $A \subseteq N$ and $B \subseteq N$ with $A \cap B = \phi$, we have that $v(A \cup B) \geq v(A) + v(B)$.

In superadditive games, the players have incentives to form coalitions.

Definition 4. By a superadditive $(n + 1)$-person game in characteristic function form, we mean a real-valued function v, defined on the subsets of N, satisfying $v(\phi) = 0$ and *superadditivity*.

We will consider a weaker version of the superadditivity property, which will be fulfilled by the games we study here.

Definition 5. A game (N, v) is zero-monotonic If, for all sets $A \subseteq N$ and for all $i \notin A$, we have that $v(A \cup \{i\}) \geq v(A) + v(\{i\})$.

The following statements (Theorem 1, Proposition 1, Theorems 2 and 3) were given in Galdeano et al. (2010) in order to prove that the game was zero-monotonic. We will use them and we will now prove that the game is superadditive.

Theorem 1. *If the innovator is not in the coalition $S (1 \notin S)$ and he belongs to $T (1 \in T)$ such that $S \cap T = \phi$, then $v(S \cup T) \geq v(S) + v(T)$ if and only if*

$$
v(S) \leq v(S \cup \{1\})
\tag{14.2}
$$

Remark 1. The players in an uninformed coalition $S \subseteq U$ have incentives to join an informed coalition $T \subseteq N$, if the utility they obtain is less than they would obtain buying the information. We do not need a restriction on the set T because by assumption S.3, for the computation of the characteristic function $v(T)$, we assumed that the uninformed agents outside the coalition take the right decision. Thus it is always better for them to join the coalition.

It was analyzed the restrictions (14.2) depending on the number of agents in the market and it was obtained the following results:

Proposition 1. *If the innovator is not in the coalition S $(1 \notin S)$ and he belongs to T $(1 \in T)$ such that $S \cap T = \phi$, then $v (S \cup T) \geq v (S) + v (T)$ if and only if*

$$a_n \geq \frac{c(1-c)^{n-2}}{1+c(1-c)^{n-2}} \tag{14.3}$$

Remark 2. Without assuming $a_1 = 1$, condition (14.3) takes the following form:
$$\frac{a_n}{a_1} \geq \frac{c(1-c)^{n-2}}{1+c(1-c)^{n-2}}.$$

The following theorem shows that the function v is *zero − monotonic*

Theorem 2. *For any coalition S and any user $i \notin S$: $v (S \cup \{i\}) \geq v (S) + v (\{i\})$.*

Now we show under what conditions the game is *superadditive*. We will first prove the following lemma:

Lemma 1. *If $S \subseteq U$ and $T = \{i\} \in U \backslash S$ then*

$$\frac{w(s)}{s} \leq \frac{w(s \cup \{i\})}{s+1}$$

Proof. The proof easily follows by using Definition 2 and basic properties of combinatoric numbers. ∎

We will complete the proof of the superadditivity condition:
For disjoint, nonempty coalitions $S, T \subseteq N \backslash \{1\}$, by Definition 2 and by Lemma 1, we have that for $1 \leq s \leq n-1$ it holds $\dfrac{w(s)}{s} \leq \dfrac{w(s+1)}{s+1}$ then

$$v(\{i\}) = w(1) \leq \frac{w(2)}{2} \leq \ldots \leq \frac{w(n)}{n} = \frac{v(U)}{n}$$

We can assume without loss of generality $1 \leq s \leq t \leq s + t \leq n$ then

$$\frac{w(s)}{s} \leq \frac{w(s+1)}{s+1} \leq \frac{w(s+2)}{s+2} \leq \ldots \leq \frac{w(t)}{t} \leq \ldots \leq \frac{w(s+t)}{s+t} \tag{14.4}$$

Therefore

$$\frac{w(s+t)}{s+t} \geq \frac{w(s)}{s} \tag{14.5}$$

Theorem 3. v is superadditive if and only if $a_n \geq \dfrac{c(1-c)^{n-2}}{1+c(1-c)^{n-2}}$

Proof. By Proposition 1, if the innovator is not in the coalition S ($1 \notin S$) and he belongs to T ($1 \in T$) such that $S \cap T = \phi$, then $v(S \cup T) \geq v(S) + v(T)$ if and only if $a_n \geq \dfrac{c(1-c)^{n-2}}{1+c(1-c)^{n-2}}$.

Now we should analyze the supperadditivity for the case of two disjoint nonempty coalitions $S, T \subseteq N \setminus \{1\}$. Using Lemma 1 and operating in (14.5), we get $w(s+t) \geq (s+t)\dfrac{w(t)}{t} = \dfrac{sw(t)}{t} + w(t)$ and by (14.4) $\dfrac{w(t)}{t} \geq \dfrac{w(s)}{s}$ then $w(s+t) \geq w(s) + w(t)$. Therefore $v(S \cup T) \geq v(S) + v(T)$. ∎

14.3   Cooperative Solutions of the Game

In this section we give the definition of the Core (Gillies 1953) and the Symmetrical Core of the game. We analyze how the Symmetrical Core varies for different values of c and a_i. We also study conditions for the Shapley value (Shapley 1953) to be in the Symmetrical Core.

14.3.1   The Symmetrical Core

Definition 6. An imputation or payoff distribution for the game (N, v) is a vector $x = (x_1, \ldots, x_{n+1})$ satisfying $\sum\limits_{i \in N} x_i = v(N)$ and $x_i \geq v(\{i\})$ for each $i \in N$.

The Core allocations are selected through efficiency and group rationality. Besides the appealing motivation for the definition of Core allocations, we might wonder if this set is nonempty. We will prove that for v given by (14.1), the Core is nonempty.

Definition 7. The Core is the set

$$C(v) = \left\{ (x_1, x_2, \ldots, x_{n+1}) : \sum_{i \in N} x_i = v(N) \text{ and } \sum_{i \in S} x_i \geq v(S) \text{ for each } S \subseteq N \right\}$$

The Core, however, is a set-valued solution concept which fails to satisfy the symmetry property in that users of the same type receive identical payoffs according to Core allocations. By symmetry (Condition S.1) we have $x_2 = x_3 = \cdots = x_{n+1}$, and then the Symmetrical Core allocations require equal payoffs to users, that is:

$$SymC(v) = \{(x_1, x_2, \ldots, x_{n+1}) \in C(v) : x_2 = x_3 = \cdots = x_{n+1}\}$$

The following lemma gives a necessary condition in order that an imputation belongs to the Symmetrical Core.

Lemma 2. *Given a game (N, v), with v defined by (14.1) and a_n fulfilling (14.3), if $(x_1, x_2, \ldots, x_2) \in SymC(v)$, then $x_2 = a_n - \frac{1}{n}x_1$ with $ca_n \leq x_2 \leq a_n$.*

Proof. If $(x_1, x_2, \ldots, x_2) \in SymC(v)$ then $x_2 = x_3 = \cdots = x_{n+1}$ and $\sum_{i \in N} x_i = v(N)$.

Now using (14.1) we obtain

$$na_n = x_1 + nx_2 = (x_1 + sx_2) + (n - s)x_2 \tag{14.6}$$

If $1 \in S$ we obtain

$$x_1 + \sum_{i \in S \setminus \{1\}} x_i = x_1 + sx_2 \geq sa_n \text{ with } |S| = s + 1 \tag{14.7}$$

Now using (14.6) in (14.7), we obtain $a_n \geq x_2$.

On the other hand, if $1 \notin S$ then we obtain $ca_n \leq x_2$ therefore $ca_n \leq x_2 \leq a_n$. ∎

Remark 3. This condition is not sufficient because, for example, if $(x_1, x_2, \ldots, x_2) \in SymC(v)$, then $2x_2 \geq v(\{2, 3\})$, that is to say, $2x_2 \geq 2c(1 - c)a_{n-1} + 2c^2 a_n$. Now we assume it fulfills that $x_2 = ca_n$, then ($0 < c < 1$), we obtain $a_{n-1} < a_n$ which contradicts the general conditions of the game (a_j is decreasing).

Theorem 4. *Given a game (N, v), then*
$$SymC(v) = \left\{(x_1, x_2, \ldots, x_2) \in C(v) : x_1 = na(n) - nx_2 \wedge \frac{v(U)}{n} \leq x_2 \leq \frac{v(N)}{n}\right\}.$$

Proof. On one hand, if $(x_1, x_2, \ldots, x_{n+1}) \in SymC(v)$, then $x_2 = x_3 = \cdots = x_{n+1}$ and $(x_1, x_2, \ldots, x_{n+1}) \in C(v)$. Now using (14.1) we obtain a system with $2\binom{n}{s}$ inequalities:

$$\begin{cases} x_2 \geq a_n - \frac{1}{n}x_1 \\ x_2 \geq \sum_{j=0}^{s} \frac{j}{s}\binom{s}{j}c^j(1-c)^{s-j}a_{n-s+j} \end{cases} \text{ with } s = 1, 2, \ldots, n \tag{14.8}$$

We obtain the following equivalent system:

$$a_n \geq \frac{\sum_{j=0}^{s-1} \frac{j}{s}\binom{s}{j}c^j(1-c)^{s-j}a_{n-s+j}}{1 - c^s}, \quad \text{with } s = 1, 2, \ldots, n$$

In Galdeano et al. (2010), it was proven that the solution of this system in $a_n, a_{n-1}, \ldots, a_2$ is $a_n \geq \dfrac{c(1-c)^{n-2}}{1 + c(1-c)^{n-2}}$.

Moreover for each s in (14.8), we have a lower bound for x_2, $\sum\limits_{j=0}^{s} \frac{i}{s} \binom{s}{j} c^j (1 - c)^{s-j} a_{n-s+j} \leq x_2$.

To determine the largest of these lower bounds for x_2, using Lemmas 1 and 2, we have

$ca_n \leq \dfrac{v(S)}{s} \leq \dfrac{v(U)}{n} \leq x_2 \leq a_n = \dfrac{v(N)}{n}$ with $S \subseteq U$, $|S| = s$ and $|U| = n$. Then we obtain

$\dfrac{v(U)}{n} \leq x_2 \leq a_n = \dfrac{v(N)}{n}$ and $x_1 = na_n - nx_2$.

We now prove the reverse implication , i.e.,

on the other hand, if $(x_1, x_2, \ldots, x_{n+1})$ is such that $x_1 = na(n) - nx_2 \wedge \dfrac{v(U)}{n} \leq x_2 \leq a_n$, then $x_2 = x_3 = \cdots = x_{n+1}$, and using (14.1), we obtain

$$\sum_{i \in N} x_i = na_n = v(N)$$

If $S \subseteq U$ then by Lemma 1, $\dfrac{v(S)}{s} \leq \dfrac{v(U)}{n}$, and $x_2 = x_3 = \cdots = x_{n+1}$, we have $v(S) \leq sx_2$.

If $S \cup \{1\} \subseteq N$ then

$$\sum_{i \in S \cup \{1\}} x_i = x_1 + sx_2 = (na_n - nx_2) + sx_2 = na_n + (s-n)x_2 . \tag{14.9}$$

with $s \leq n$ and $x_2 \leq a_n$, then $na_n + (s-n)x_2 \geq na_n + (s-n)a_n = sa_n$. Using (14.1) we obtain

$$\sum_{i \in S \cup \{1\}} x_i \geq v(S \cup \{1\})$$

We have for all $S \subseteq N$: $\sum\limits_{i \in S} x_i \geq v(S)$.

Therefore we obtain that

$\mathrm{SymC}(v) = \left\{ (x_1, x_2, \ldots, x_2) \in C(v) : x_1 = na(n) - nx_2 \wedge \dfrac{v(U)}{n} \leq x_2 \leq \dfrac{v(N)}{n} = a(n) \right\}$

■

Corollary 5. *The Symmetrical Core and by extension the Core is nonempty.*

Proof. It is easy to show that $(x_1, x_2, \ldots, x_2) = (0, a_n, \ldots, a_n) \in \mathrm{SymC}(v)$ and $\left(na_n - v(U), \dfrac{v(U)}{n}, \ldots, \dfrac{v(U)}{n} \right) \in \mathrm{SymC}(v)$.   ■

Remark 4. Thus, the imputation $(0, a_n, \ldots, a_n) = (0, \dfrac{v(N)}{n}, \ldots, \dfrac{v(N)}{n}) \in \mathrm{SymC}(v)$ is the most appealing for the users because they get $\dfrac{v(N)}{n}$ and the less appealing for the innovator that obtains 0.

The imputation $\left(na_n - v(U), \frac{v(U)}{n}, \ldots, \frac{v(U)}{n} \right) \in \text{SymC}(v)$ is the less appealing for the users because they get $\frac{v(U)}{n}$ and the most appealing to the innovator that obtains $na_n - v(U)$.

14.3.2 Shapley Value and Symmetrical Core

Definition 8. Given a game (N, v), the Shapley value (Shapley 1953) is defined by the following vector $\varphi(v) = (\varphi_1(v), \ldots, \varphi_{n+1}(v))$ where

$$\varphi_i(v) = \sum_{S \subseteq N - \{i\}} \frac{s!\,(n - s)!}{(n + 1)!} [v(S \cup \{i\}) - v(S)]$$

with $|S| = s$ and $|N| = n + 1$.

In Galdeano et al. (2010), it was given a formulation of the Shapley value. Now we present a more concise expression:

Theorem 6. Given a game (N, v), with v defined by (14.1) fulfilling (14.3) and $0 < c < 1$, then the Shapley value $\varphi(v) = (\varphi_1(v), \ldots, \varphi_{n+1}(v))$ is an imputation for the game (N, v) with

$$\varphi_i(v) = \varphi_2(v) = \varphi_3(v) = \cdots = \varphi_{n+1}(v) = \frac{a_n}{2} + \frac{1}{n(n+1)} \sum_{s=0}^{n} w(s),$$

$$\varphi_1(v) = \frac{1}{2}v(N) - \frac{1}{n+1} \sum_{s=0}^{n} v(s) = \frac{1}{2}na_n - \frac{1}{n+1} \sum_{s=0}^{n} w(s) \text{ with } s = |S|, 1 \notin S \text{ and}$$

$i \neq 1$.

The proof follows by splitting the sums considering informed and uninformed coalitions, respectively.

The following lemma shows that the Shapley value fulfills the necessary condition of the Symmetrical Core.

Lemma 3. If $(\varphi_1(v), \varphi_2(v), \ldots, \varphi_2(v))$ is as in Theorem 6, then $0 \leq \varphi_1(v) \leq n(1 - c)a_n$ and $c\,a_n \leq \varphi_2(v) \leq a_n$.

The proof is similar to Lemma 2.

The following example shows that the Shapley value could be outside the Symmetrical Core.

Example 1. For $n = 2$ the Symmetrical Core is in a line in $\mathbb{R}^3$ given by

$$\text{SymC}(v) = \begin{cases} x_1 = 2a_2 - 2x_2 \\ x_2 = x_2 \\ x_3 = x_2 \end{cases} \quad \text{with } c^2 a_2 + c(1 - c) \leq x_2 \leq a_2,$$

The Shapley value is $(\varphi_1(v), \varphi_2(v), \varphi_2(v)) \in \mathbb{R}^3$, with

$$\varphi_1(v) = a_2 - \frac{1}{3}ca_2 - \frac{2}{3}c + \frac{2}{3}c^2 - \frac{2}{3}c^2a_2.$$

$$\varphi_2(v) = \frac{1}{2}a_2 + \frac{1}{6}ca_2 + \frac{1}{3}c - \frac{1}{3}c^2 + \frac{1}{3}c^2a_2$$

If $c = \frac{1}{2}$ and $a_2 = 0.667$, then $(0.277, 0.528, 0.528) \in \text{SymC}(v)$.

If $c = \frac{1}{2}$ and $a_2 = 0.369$, then $(0.078, 0.328, 0.328) \notin \text{SymC}(v)$.

The following theorem gives necessary and sufficient conditions for the Shapley value to be in the Symmetrical Core.

Theorem 7. *The Shapley value is in the Symmetrical Core if* $v(U) \leq \frac{1}{2}(n+1)a_n + \frac{1}{n}\sum_{s=1}^{n-1} v(s)$ *with* $s = |S|$ *and* $1 \notin S$.

Proof. Let $(\varphi_1(v), \varphi_2(v), \ldots, \varphi_2(v))$ be a solution for the game (N, v) given by the Shapley value, and then by Theorem 4, the Shapley value will be in the Symmetrical Core if

$$\varphi_1(v) + n\varphi_2(v) = v(N) = na_n \text{ with } \frac{v(U)}{n} \leq \varphi_2(v) \leq a_n$$

The Shapley value verifies: $\sum_{i=1}^{n+1} \varphi_i(v) = \varphi_1(v) + n\varphi_2(v) = v(N)$. As $\varphi_1(v) \geq 0$, then $n\varphi_2(v) \leq v(N)$. Thus $\varphi_2(v) \leq \frac{v(N)}{n} = a_n$.

Now we analyze when the following condition holds: $\frac{v(U)}{n} \leq \varphi_2(v)$.

By Theorem 6, $\varphi_2(v) = \frac{a_n}{2} + \frac{1}{n(n+1)}\sum_{s=0}^{n} v(s)$. Then $\frac{v(U)}{n} \leq \varphi_2(v)$ if and only if

$$\frac{v(U)}{n} \leq \frac{a_n}{2} + \frac{1}{n(n+1)}\sum_{s=0}^{n} v(s) \tag{14.10}$$

Operating in (14.10) we obtain $v(U) \leq \frac{1}{2}(n+1)a_n + \frac{1}{n}\sum_{s=1}^{n-1} v(s)$. ∎

Remark 5. If $n = 2$ then the Shapley value is in the Symmetrical Core if $a_2 \geq \frac{4c}{(4c+3)}$.

As $a_2 \geq \frac{c}{1+c}$ and $\frac{4c}{(4c+3)} \geq \frac{c}{1+c}$, we have a critical zone $\frac{c}{1+c} \leq a_2 < \frac{4c}{(4c+3)}$ where the Shapley value is not in the Symmetrical Core.

It is shown in the following graphic, where $\frac{4c}{(4c+3)}$ (solid) and $\frac{c}{1+c}$ (dots).

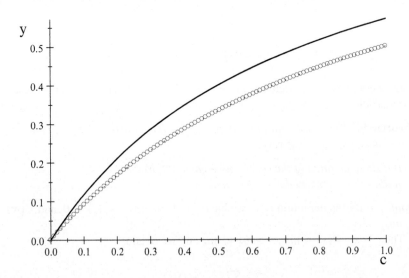

If $n \geq 3$ the Shapley value is in the Symmetrical Core if

$$a_n \geq \frac{n\sum_{j=1}^{n-1}j\binom{n}{j}c^j(1-c)^{n-j-1}a(j)-\sum_{s=2}^{n-1}\sum_{j=1}^{s-1}j\binom{s}{j}c^j(1-c)^{s-j-1}a(n-s+j)}{\sum_{s=0}^{n-1}\left[\left(n^2-sn+\frac{s(s+1)}{2}\right)\right]c^{n-1-s}}$$

Increasing n, the critical zone where the Shapley value is not in the Symmetrical Core shrinks.

14.3.3 Cooperative and Noncooperative Model

The cooperative game studied in this paper was analyzed by Quintas (1995) from a noncooperative point of view. It was observed that the innovator obtained a neat profile by selling the information to the n firms. However the situation was not so appealing for the buyers. The expected utility each one finally obtained after buying the information was that one he would have obtained if he was the only uninformed agent. Nevertheless they couldn't ignore the existence of the information and they should buy it.

The main result of the noncooperative study mentioned above states as follows:

Theorem 8. *The price P that the innovator can ask to the n users such that all of them acquire the information is determined by the unique Nash equilibrium of the noncooperative game.*

This price is $P = (1-c)a_n - \varepsilon$, *with* $\varepsilon \geq 0$ *arbitrarily small, and the payoff n-tupla is*

$$((1-c)na_n + n\varepsilon, ca_n + \varepsilon, \ldots, ca_n + \varepsilon)$$

and $\varepsilon \to 0$, then the payoff n-tupla is

$$((1-c)na_n, ca_n, \ldots, ca_n)$$

Let's compare the expected utility given in Theorem 8 with the results presented in our article.

Theorem 9. *The Nash equilibrium payoff of the noncooperative game,* $P = ((1-c)na_n, ca_n, \ldots, ca_n)$ *verifies:*

1) It is an imputation of the cooperative game (N, v).
2) It does not belong to the Symmetrical Core.

Proof. 1) By the definition of v, we have $v(\{i\}) = c\,a_n$, $v(\{1\}) = 0$, and $v(N) = na_n$.

Then $\sum_{i \in N} x_i = (1-c)na_n + nca_n = na_n$ and we obtain

$$\sum_{i \in N} x_i = v(N) \tag{14.11}$$

If $i = 1$ then $v(\{1\}) = 0 \le (1-c)na_n = x_1$, and if $i \ne 1$ then $v(\{i\}) = ca_n = x_2$, and we have

$$x_i \ge v(\{i\}) \quad \text{for all } i \in N \tag{14.12}$$

From (14.12) and (14.11) we conclude that P is an imputation for the game (N, v).

Now we prove 2).

If $P = ((1-c)na_n, ca_n, \ldots, ca_n) \in \mathrm{SymC}(v)$, then $2x_2 \ge v(\{2, 3\})$, and we obtain

$2x_2 \ge 2c(1-c)a_{n-1} + 2c^2a_n$. As $0 < c < 1$ and $x_2 = ca_n$, then we obtain $a_{n-1} < a_n$. It is impossible, because a_j increases. Then $P \notin \mathrm{SymC}(v)$. ∎

Remark 6. In Galdeano et al. (2010), it was proved that $\varphi_i(v) \ge ca_n$ with i being a user. By Lemma 2, the Symmetrical Core imputations for the users verify $ca_n \le x_i \le a_n$. By Theorem 8, ca_n is the equilibrium outcome for the users. Thus, we conclude that the users are better off in the cooperative environment and an opposite situation results for the innovator.

14.4 Limit Cases

In Theorem 6 it was given a characterization of the Shapley value and in Theorem 4, it was presented a characterization of the Symmetrical Core, being $0 < c < 1$ and

$$\frac{c(1-c)^{n-2}}{1+c(1-c)^{n-2}} \le a_n \le a_{n-1} \le \ldots a_2 \le 1.$$

Now we analyze properties of some limit cases. These limit cases correspond to completely uninformed users ($c = 0$) or completely informed users ($c = 1$). The other extreme cases result when $a_n = a_{n-1} = \cdots = a_2 = \dfrac{c(1-c)^{n-2}}{1+c(1-c)^{n-2}}$ and $a_n = a_{n-1} = \cdots = a_2 = 1$

14.4.1 Users with No Prior Information ($c = 0$): Big Boss Games

We will show that in this case the game (N, v) is a Big Boss Game (Muto et al. 1988).

Definition 9. A monotonic game (N, v) is called a Big Boss Game if there is one player, denoted by i^*, satisfying the following two conditions:

B1) $v(S) = 0$ if $i^* \notin S$ and **B2)** $v(N) - v(S) \geq \sum_{i \in N \setminus S} (v(N) - v(N - \{i\}))$ if $i^* \in S$.

The Big Boss Games are denoted by BBG^N.

B1 implies that one player i^* is very powerful. Coalitions not containing i^* cannot get anything.

B2 implies that for every coalition not containing i^*, its contribution to the grand coalition is not less than the sum of the contributions of its players to the grand coalitions. Hence, weak players may increase their influence by forming coalitions. We also notice that a Big Boss Game v is *superadditive* (Definition 3), because of the monotonicity of v and B1.

When $c = 0$, the characteristic function $v : 2^N \to \mathbb{R}$ results

$$v(S) = \begin{cases} (s-1)\, a_n & \text{if } 1 \in S \\ 0 & \text{if } 1 \notin S \end{cases} \quad \text{for all } S \subseteq N \text{ and } |S| = s \qquad (14.13)$$

Here player i^* is the innovator $i^* = 1$. As a consequence of (14.13), we have:

Theorem 10. *Let (N, v) be a game with v given by (14.13), then $(N, v) \in BBG^N$.*

Proof. We must prove that (N, v) is monotonic, that is, $v(S) \leq v(T)$ for all $S \subseteq T$ and $|S| = s$, $|T| = t$.
   If $1 \in S$ then $1 \in T$, and by (14.13) $v(S) = (s-1)\, a_n \leq (t-1)\, a_n = v(T)$.
   If $1 \notin S$ then we have two cases: $1 \in T$ or $1 \notin T$, then:

i) $1 \notin S$ and $1 \in T$, by (14.13) $v(S) = 0 \leq (t-1)\, a_n = v(T)$.
ii) $1 \notin S$ and $1 \notin T$, $v(S) = 0 = v(T)$. Then (N, v) is monotonic.

Now we must also prove that $B1$ and $B2$ hold.
Condition $B1$ immediately follows from (14.13).

Let be $1 \in S$ and $i \neq 1$; then by (14.13), we have

$$\sum_{i \in N \setminus S} (v(N) - v(N - \{i\})) = \sum_{i \in N \setminus S} (na_n - (n-1)a_n) = \sum_{i \in N \setminus S} a_n = (n+1-s)a_n$$

at the same time $v(N) - v(S) = n\,a_n - (s-1)\,a_n = (n+1-s)\,a_n$ thus

$$v(N) - v(S) = \sum_{i \in N \setminus S} (v(N) - v(N - \{i\})).$$

Therefore B2 holds with equality.   ∎

Now we compute the Shapley value and the Symmetrical Core.

Theorem 11. *1. The Shapley value for the users is given by* $\varphi_i(v) = \frac{1}{2}a_n$ *for*
$i \neq 1$ and for the innovator is given by $\varphi_1(v) = \frac{1}{2}v(N) = \dfrac{na_n}{2}$.
2. The Symmetrical Core is given by $SymC(v) = \{(x_1, x_2, \ldots, x_2) : 0 \leq x_2 \leq a_n$
with $x_1 + nx_2 = na_n\}$.

Proof. 1. Let be $i \neq 1$. By Definition 6, splitting the sum between
informed and uninformed coalitions, and by (14.13) we have $\varphi_i(v) =$
$$\sum_{S \subseteq N \setminus \{i\}} \frac{s!\,(n-s)!}{(n+1)!} [v(S \cup \{i\}) - v(S)] =$$

$$\sum_{\substack{S \subseteq N \setminus \{i\} \\ 1 \in S}} \frac{s!\,(n-s)!}{(n+1)!} [sa_n - (s-1)a_n] + \sum_{\substack{S \subseteq N \setminus \{i\} \\ 1 \notin S}} \frac{s!\,(n-s)!}{(n+1)!} 0 \text{ then}$$

$$\varphi_i(v) = \sum_{\substack{S \subseteq N \setminus \{i\} \\ 1 \in S}} \frac{s!\,(n-s)!}{(n+1)!} a_n \tag{14.14}$$

Now we analyze how many subsets S are in each sum of (14.14).

If $S \subseteq N \setminus \{i\}$, with $1 \in S$, we count how many subsets of the type $S \setminus \{1\} \subseteq$
$N \setminus \{1, i\}$ we have (the innovator is a fixed player in all the coalitions S we could
form), they are $\binom{n-1}{s-1}$.

Then $\binom{n-1}{s-1} \frac{s!\,(n-s)!}{(n+1)!} = \frac{s}{n(n+1)}$, with $|S| = s = 1, \ldots, n$.

As the function $v(S)$ depends only on the cardinality s of the set S, we have
for $i \neq 1$

$$\varphi_i(v) = a_n \sum_{\substack{s=1 \\ 1 \in S}}^{n} \frac{s}{n(n+1)} \tag{14.15}$$

and $\sum_{s=1}^{n} \dfrac{s}{n(n+1)} = 1/2$; thus, we have

$$\varphi_i(v) = \frac{1}{2}a_n \text{ with } i \neq 1 \tag{14.16}$$

Fix $i = 1$. As $\sum_{i=1}^{n} \varphi_i(v) = v(N)$. Then by (14.16) and by (14.13), we have
$\varphi_1(v) = na_n - n\frac{1}{2}a_n = \frac{n}{2}a_n$.

Thus the Shapley value is

$$(\varphi_1(v), \varphi_2(v), \ldots, \varphi_2(v)) = \left(\frac{n}{2}a_n, \frac{1}{2}a_n, \ldots, \frac{1}{2}a_n\right)$$

2. The proof of this part is similar to Theorem 4.   ∎

Corollary 12. *1. The Shapley value is in the Symmetrical Core and it is the midpoint of the segment.*
2. The payoff $(na_n, 0, \ldots, 0)$ found in Quintas (1995) in the noncooperative game is an extreme point in the Symmetrical Core.

14.4.2   Completely Informed Users ($c = 1$)

In the case $c = 1$, the characteristic function $v : 2^N \to \mathbb{R}$ becomes

$$v(S) = \begin{cases} (s-1)a_n & \text{if } 1 \in S \\ sa_n & \text{if } 1 \notin S \end{cases} \quad \text{for all } S \subseteq N \text{ and } |S| = s \qquad (14.17)$$

As an immediate consequence of (14.17), we have:

Theorem 13. *1. The Shapley value for the users is given by $\varphi_2(v) = \frac{1}{n}v(N) = a_n$ and $\varphi_1(v) = 0$*
2. The Symmetrical Core is given by $SymC(v) = \{(x_1, x_2, \ldots, x_2) : x_1 = \varphi_1(v) = 0 \wedge x_2 = \varphi_i(v) = a_n\}$.

The proof is similar to the case $c = 0$.

Remark 7. 1. In this case the Symmetrical Core is a single point set and it coincides with the Shapley value.
2. From (14.17) the innovator is a dummy player. Thus the users have no incentives to form coalitions with the innovator, but they do have incentives to form coalitions among themselves because if $i \neq 1$ then $v(S) = v(S \cup \{1\})$ and $v(S) < v(S \cup \{i\})$.
3. The users in this case have complete information and the payoff outcome $(0, a_n, \ldots, a_n)$ coincides with the result obtained by Quintas (1995) in the noncooperative game.

On the other hand, if we denote by $SymC(v, c)$ the Symmetrical Core corresponds to value of c. It is easy to verify that:

Theorem 14. *If $0 \leq c_1 \leq c_2 \leq \ldots \leq c_n \leq 1$ then*

$$SymC(v, 0) \supseteq SymC(v, c_1) \supseteq SymC(v, c_2) \supseteq \ldots \supseteq SymC(v, c_n) \supseteq SymC(v, 1)$$

Proof. It follows from the characterization of the Symmetrical Core in the general case $0 < c < 1$ and the cases $c = 0$ and $c = 1$.   ∎

14.4.3 Extreme Values of a_j

Now for each c fixed, we analyze the Symmetrical Core and the Shapley value for extreme values of a_j. Namely, we analyze the cases $a_n = 1$ and $a_n = a_{n-1} = \cdots = a_2 = \frac{c(1-c)^{n-2}}{1+c(1-c)^{n-2}}$.

In the case $a_n = 1$, the characteristic function $v : 2^N \to \mathbb{R}$ becomes

$$v(S) = \begin{cases} (s-1) & \text{if } 1 \in S \\ sc & \text{if } 1 \notin S \end{cases} \quad \text{for all } S \subseteq N \text{ and } |S| = s \qquad (14.18)$$

As $a_n = 1$ then $a_j = 1$, for all $j = 1, ..n$.

Lemma 4. *For v given by (14.18):*

1. *The Symmetrical Core is the segment given by:* $SymC(v) = \{(n(1-x_2), x_2, \ldots, x_2) : c \le x_2 \le 1\}$.
2. *The Shapley value* $\left(\frac{1}{2}n(1-c), \frac{c+1}{2}, \ldots \frac{c+1}{2}\right)$ *is in the Symmetrical Core and it is the midpoint of the segment.*
3. *The equilibrium payoff* $((1-c)n, c, \ldots, c)$ *found by Quintas (1995) for the noncooperative game is an extreme point of the Symmetrical Core.*

Proof. Let's prove 1.

By (14.18) we have

$$v(N) = n \text{ and } v(U) = nc \qquad (14.19)$$

Using (14.19), and Theorem 4, we have

$$SymC(v) = \{(n(1-x_2), x_2, \ldots, x_2) : c \le x_2 \le 1\}$$

Now we prove 2.

The Shapley value for the users is given by

$$\varphi_i(v) = \frac{v(n+1)}{2n} + \frac{1}{n(n+1)} \sum_{s=0}^{n} v(s) \qquad (14.20)$$

By (14.18) we have

$$v(S) = sc \text{ with } 1 \notin S \text{ and } |S| = s \qquad (14.21)$$

Using (14.21), (14.19), and Theorem 6, we have $\varphi_i(v) = \frac{v(n+1)}{2n} + \frac{1}{n(n+1)} \sum_{s=0}^{n} sc$.

As $\sum_{s=0}^{n} s = \frac{n(n+1)}{2}$, it results

$$\varphi_i(v) = \frac{n}{2n} + \frac{1}{n(n+1)} \frac{n(n+1)c}{2} = \frac{1+c}{2} \qquad (14.22)$$

As $\varphi_1(v) + n\varphi_i(v) = v(N)$, then $\varphi_1(v) = na_n - n\varphi_i(v)$.

Therefore by (14.19) and (14.22), we obtain $\varphi_1(v) = n - n\left(\frac{1+c}{2}\right) = n\left(\frac{1+c}{2}\right)$.

Then the Shapley value $\left(\frac{1}{2}n(1-c), \frac{c+1}{2}, \ldots, \frac{c+1}{2}\right)$ is in the Symmetrical Core, and it is the midpoint of the segment.

In order to prove 3, let's consider the payoff of the noncooperative game $((1-c)na_n, ca_n, \ldots, ca_n)$ and $a_n = 1$, then it becomes: $((1-c)n, c, \ldots, c)$. It's an extreme point of the Symmetrical Core.   ∎

Now let's analyze the case $a_n = a_{n-1} = \cdots = a_2 = \frac{c(1-c)^{n-2}}{1+c(1-c)^{n-2}}$

Lemma 5. *If* $a_n = a_{n-1} = \cdots = a_2 = \frac{c(1-c)^{n-2}}{1+c(1-c)^{n-2}}$, *then* $v(N) = v(U)$.

Proof. By (14.1) we have

$$v(U) = \sum_{j=1}^{n} j \binom{n}{j} c^j (1-c)^{n-j} a_j \qquad (14.23)$$

Using that $a_n = a_{n-1} = \cdots = a_2 = \frac{c(1-c)^{n-2}}{1+c(1-c)^{n-2}}$, $a_1 = 1$ and by (14.1), we have

$$v(U) = \sum_{j=1}^{n} j \binom{n}{j} c^j (1-c)^{n-j} a_j = nc(1-c)^{n-1} + \left(\sum_{j=2}^{n} j \binom{n}{j} c^j (1-c)^{n-j}\right) \frac{c(1-c)^{n-2}}{1+c(1-c)^{n-2}}$$

As we have $\sum_{j=2}^{n} j \binom{n}{j} c^j (1-c)^{n-j} = nc\left(1 - (1-c)^{n-1}\right)$ then

$$v(U) = nc(1-c)^{n-1} + \left(nc - nc(1-c)^{n-1}\right) \frac{c(1-c)^{n-2}}{1+c(1-c)^{n-2}} \qquad (14.24)$$

Operating in (14.24) we obtain $v(U) = n\frac{c(1-c)^{n-2}}{1+c(1-c)^{n-2}}$, and by (14.1) $v(N) = na_n$, then $v(U) = v(N)$.   ∎

Corollary 15. *The Symmetrical Core is a single point set*

$$SymC(v) = \left\{\left(0, \frac{c(1-c)^{n-2}}{1+c(1-c)^{n-2}}, \ldots, \frac{c(1-c)^{n-2}}{1+c(1-c)^{n-2}}\right)\right\}$$

Remark 8. 1. The Shapley value is not in the Symmetrical Core.

2. The equilibrium payoff $\left((1-c)n\frac{c(1-c)^{n-2}}{1+c(1-c)^{n-2}}, c\frac{c(1-c)^{n-2}}{1+c(1-c)^{n-2}}, \ldots, c\frac{c(1-c)^{n-2}}{1+c(1-c)^{n-2}}\right)$ found by Quintas (1995) for the noncooperative game is not in the Symmetrical Core.

14.5 Conclusions

We studied several properties of the game modeling information transferal. We computed the Symmetrical Core. We showed the Symmetrical Core (and also the Core) is nonempty . We characterized the Symmetrical Core and we analyzed the relation with the Shapley value. We showed examples of cases when the Shapley value is not in the Symmetrical Core. We presented conditions for the Shapley value to be in the Symmetrical Core.

We compared the cooperative outcomes with the noncooperative outcomes. The Nash equilibrium found in Quintas (1995) in the noncooperative game is an imputation for the game but in the general cases is not in the Symmetrical Core (Theorem 9) .The Nash equilibrium gives the users a worse payoff than the Shapley value and the Symmetrical Core allocations (Remark 6).

We also analyzed some limit cases.

In the case of users with no prior information, the game was a Big Boss Game (Muto et al. 1988). The innovator had a huge power, and the payoff corresponding to the noncooperative equilibrium found in Quintas (1995) in the noncooperative game was an extreme point in the Symmetrical Core, giving the best outcome to the innovator and no utilities to the users. The Shapley value was in the Symmetrical Core. It is a segment of a line, being the Shapley value the midpoint of the segment.

In the case of fully informed users, the role of the innovator was irrelevant (it is a "dummy" player), and the Symmetrical Core is a single point.

In these cases both the Shapley value and the noncooperative outcome were in the Symmetrical Core.

Our approach was different from that introduced in the Bi-Form Games (Brandemburger and Stuart 2007), where it is considered a hybrid noncooperative-cooperative model. Instead of that, we made a comparison of the outcomes in noncooperative and cooperative scenarios because both models have a role to play in understanding business strategy, and many times it is not known beforehand if the game is going to be played with or without cooperation among the agents. As it was expected, the users were better off in a cooperative environment, and we explicitly compared the cooperative solution outcomes with the noncooperative equilibrium.

References

Amer, R., Carreras, F., Magaña, A.: Applications of cooperative games to business activities. J. Intangible Cap. **4**(2), 102–142 (2008)

Arrow, K.: Economic welfare and the allocation of resources for invention. In: Nelson, R.R. (ed.) The Rate and Direction of Inventive Activity, pp. 609–625. Princeton University Press, Princeton (1962)

Belenky A.S.: Cooperative games of choosing partners and forming coalitions in the marketplace. Math. Comput. Modell. **36**, 1279–1291 (2002)

Brandemburger, A., Stuart, H.W.: Biform games. Manag. Sci. **53**(4), 537–549 (2007)

De Clippel, G., Serrano, R.: Marginal contributions and externalities in the value. Econometrica **76**, 1413–1436 (2008)

Galdeano, P., Oviedo, J., Quintas, L.: Shapley value in a model of information transferal. Int. Game Theory Rev. **12**(1), 19–35 (2010)

Gilbert, G., Shapiro, C.: Optimal patent length and breadth. RAND J. Econ. **21**, 106–112 (1990)

Gillies, D.B.: Some theorem on n-person games. Ph.D. thesis, Princeton University Press, Princeton, New Jersey (1953)

Hou, D., Driessen, T.: The core and nucleolus in a model of information transferal. J. Appl. Math. **2012**, 1–12 (2012)

Kamien, M.I., Tauman, Y.: Fees versus royalties and the private value of a patent. Q. J. Econ. **101**, 471–491 (1986)

Katz, M.L., Shapiro, C.: How to license intangible property. Q. J. Econ. **101**, 567–589 (1986)

Macho-Stadler, I., Pérez Castrillo, D., Wettstein, D.: Efficient bidding with externalities. Games Econ. Behav. **57**, 304–320 (2006)

Muto, S.: On licensing policies in Bertrand competition. Games Econ. Behav. **5**, 257–267 (1993)

Muto, S., Nakayama, N., Potters, J., Tijs, S.: On big boss games. Econ. Stud. Q. **39**, 303–321 (1988)

Nakayama, M., Quintas, L., Muto, S.: Resale-proof trades of information. Econ. Stud. Q. **42**, 292–302 (1991)

Poddar, S., Sinha, U.B.: On patent licensing in spatial competition. Econ. Rec. **80**, 208–218 (2004)

Quintas, L.G.: How to sell private information. Modell. Meas. Control D **11**, 11–28 (1995)

Sandholm, T., Larson, K., Andersson, M., Shehory, O., Tohmé, F.: Coalition structure generation with worst case guarantees. Artif. Intell. **111**(1–2), 209–238 (1999)

Schmidt, F.: Innovation contests with temporary and endogenous monopoly rents. Rev. Econ. Des. **12**, 189–208 (2008)

Sen, D., Tauman, Y.: General licensing schemes for a cost-reducing innovation. Games Econ. Behav. **59**, 163–186 (2007)

Shapley, L.S.: A value for n-person games. Ann. Math. Study **28**, 307–317 (1953)

Stamatopoulos, G., Tauman, Y.: Licensing of a quality-improving innovation. Math. Soc. Sci. **56**(3), 410–438 (2008)

Tauman, Y., Watanabe, N.: The Shapley value of a patent licensing game: the asymptotic equivalence to non-cooperative results. Econ. Theory **30**, 135–149 (2007)

Taylor, C., Silberston, Z.: The Economic Impact of the Patent System. Cambridge University Press, Cambridge (1973)

Wang, X.H.: Fee versus royalty licensing in a Cournot duopoly model. Econ. Lett. **60**, 55–62 (1998)

Wang, X.H.: Fee versus royalty licensing in a differentiated Cournot duopoly. J. Econ. Bus. **54**, 253–266 (2002)

Chapter 15
Marginal Contributions in Games with Externalities

Joss Sánchez-Pérez

Abstract In this work we explore games with externalities, where our basic approach is rooted in the concept of marginal contributions of players to coalitions. We considered the general case where a player (in a coalition S) may join another coalition after leaving S. We then show that the standard translation of Shapley's four axioms to games with externalities is not sufficient to obtain a unique value. Finally, we provide an axiomatic characterization for the family of solutions for games with externalities satisfying those axioms that traditionally are used to characterize the Shapley value in the absence of externalities. In particular, we show that every such solution is a linear combination of marginal contributions of players and provide an interpretation as a bargaining process.

Keywords Marginal contributions • Cooperative games • Externalities • Shapley value • Partitions

15.1 Introduction

One of the main purposes of cooperative game theory is to study how to divide the joint profits among players when they cooperate together. A value for coalitional games is a solution which provides an allocation for players' payoffs. Shapley (1953) suggests an axiomatic approach to this issue. In his characterization, the axiom linearity, symmetry, efficiency, and the nullity property determine uniquely a value. Young (1985) shows that the axioms of marginality, efficiency, and symmetry also yield an axiomatic characterization of the Shapley value. The Shapley value

*Part of this chapter is based on the paper "A note on a class of solutions for games with externalities generalizing the Shapley value" (2015).

J. Sánchez-Pérez (✉)
Facultad de Economía, UASLP, San Luis Potosí, Mexico

Facultad de Economía, UASLP, Av. Pintores s/n, Col. B. del Estado 78213, San Luis Potosí, Mexico
e-mail: joss.sanchez@uaslp.mx

© Springer International Publishing Switzerland 2016
A.A. Pinto et al. (eds.), *Trends in Mathematical Economics*,
DOI 10.1007/978-3-319-32543-9_15

possesses many desirable properties and becomes one of the most famous solutions for games with no externalities. Nevertheless, the Shapley value cannot give a recommendation for the allocation of payoffs of players in the situation where externalities across coalitions are present. Lucas and Thrall (1963) introduce games with externalities, where the worth of a coalition is described conditional to how outside members of the coalition form coalitions.

Based on the axioms which characterize the Shapley value for games with no externalities, there are apparently many ways to extend it to games with externalities. For instance, the first paper that proposed a value concept for these was Myerson (1977) and then Bolger (1989) derived an efficient value which assigns zero to null players and assigns nonnegative values to players in monotone simple games. More recently, Albizuri et al. (2005), Macho-Stadler et al. (2007), De Clippel and Serrano (2008), Hu and Yang (2010), and Grabisch and Funaki (2012) are contributions to this line of research. All of them are in some way extensions of the Shapley value for games with externalities, where a remarkable difference among each other is the definition of a null player.

We take particular attention to the concept of null player in environments with externalities. A natural requirement for a fair division scheme is that it remunerates the players of a coalitional game taking into account their contribution to the surplus generated via cooperation. Indeed, in Shapley's axiomatization, the nullity axiom requires that no share be allocated to players with zero contribution to any possible coalition that could be created in the coalitional game. The key issue, then, is how such contribution should be measured. Although not explicitly, Shapley's nullity axiom relies on the concept of marginal contribution, one of the fundamental notions in economic theory.

In the cooperative game context, the marginal contribution of a player to a coalition is the difference between the value of this coalition with and without the player. It can be also understood as a loss incurred by the remaining players should the player leave a given coalition. Considering this latter intuition, the Shapley value can be formulated as the weighted average of players' marginal contributions to all coalitions. In games with no externalities, the marginal contribution of a particular player is assigned deterministically as it does not play a role in what a player does after leaving a coalition. This is, however, not the case in games with externalities, where the definition of the marginal contribution becomes much more intricate.

When externalities are present, the worth of the coalition that a player has left may be influenced by which coalition, if any, this player subsequently joins. In other words, the choice of action after leaving a coalition may result in different values of the player's marginal contribution to it. The model we shall employ is that in which the worth of a coalition S may vary with how the players not in S cooperate. In the model, $w(S, Q)$ is the worth of S when the coalition structure is Q, S being an element of Q. To define player i's marginal contribution to coalition S—a trivial task in the absence of externalities—it is now crucial to describe what happens after i leaves S. We consider the general case where a player may join another coalition

T (in Q) after leaving S. The total effect on S of i's move is the difference $w(S, Q) - w(S_{-i}, \{S_{-i}, T_{+i}\} \cup Q_{-S,-T})$.[1]

In games with externalities, not only the definition of the marginal contribution but also the axiomatization of the value becomes more involved. In this work we show that the standard translation of Shapley's four axioms to games with externalities is not sufficient to obtain a unique value. We then provide an axiomatic characterization for the family of solutions for games with externalities satisfying those axioms that traditionally are used to characterize the Shapley value in the absence of externalities. In particular, we show that every such solution is a linear combination of marginal contributions of players and provide an interpretation as a bargaining process.

The chapter is organized as follows. We first recall the main basic features of games with externalities in the next section. In Sect. 15.3 we formally introduce the concept of marginal contribution and axioms. A characterization of all linear, symmetric, efficient, and null solutions is introduced in Sect. 15.4. In Sect. 15.5 we relate previous results with solutions in the literature. Finally, we discuss relationships between different axiomatizations.

15.2 Basic Definitions and Notation

In this section we give some concepts and notations related to n-person games with externalities (including the basic axioms that are considered in this work), as well as a brief subsection of preliminaries related to integer partitions, since it is a key subject in subsequent developments.

15.2.1 Games with Externalities

Let $N = \{1, 2, \ldots, n\}$ be a fixed nonempty finite set, and let the members of N be interpreted as players in some game situation. Given N, let CL be the set of all coalitions (nonempty subsets) of N, $CL = \{S \mid S \subseteq N, S \neq \varnothing\} = 2^N \setminus \{\varnothing\}$. Let PT be the set of partitions of N, so

$$\{S_1, S_2, \ldots, S_m\} \in \text{PT iff } \bigcup_{i=1}^{m} S_i = N, \ S_j \cap S_k = \varnothing \ \forall j \neq k$$

By convention, $\{\varnothing\} \in Q$ for every $Q \in$ PT and $|Q|$ will denote the number of nonempty sets in Q. Also, let $EC = \{(S, Q) \mid S \in Q \in \text{PT}\}$ be the set of *embedded coalitions*, that is, the set of coalitions together with specifications as to how the other players are aligned. The embedded coalition (S, Q) is called *nontrivial* if $S \neq \varnothing$.

---

[1]The precise definitions will be provided in Sect. 15.3.

Definition 1. A game with externalities is a mapping

$$w : \text{EC} \to \mathbb{R}$$

with the property that $w(\varnothing, Q) = 0$ for every $Q \in \text{PT}$. The set of games with externalities with player set N is denoted by G, i.e.,

$$G = G^{(n)} = \{w : \text{EC} \to \mathbb{R} \mid w(\varnothing, Q) = 0 \; \forall Q \in \text{PT}\}$$

The value $w(S, Q)$ represents the payoff of coalition S, given the coalition structure Q forms. In this kind of games, the worth of some coalition depends not only on what the players of such coalition can jointly obtain but also on the way the other players are organized. We assume that, in any game situation, the universal coalition N (embedded in $\{N\}$) will actually form, so that the players will have $w(N, \{N\})$ to divide among themselves. But we also anticipate that the actual allocation of this worth will depend on all the other potential worths $w(S, Q)$, as they influence the relative bargaining strengths of the players.

For $Q \in \text{PT}$, $S \in Q$, and $i, k \in N$, we define $Q_{-S} = Q \backslash \{S\}$, $S_{-k} = S \backslash \{k\}$, and $S_{+k} = S \cup \{k\}$, and Q^i denotes the member of Q where i belongs. Additionally, we will denote the cardinality of a set by its corresponding lower-case letter, for instance, $n = |N|$, $s = |S|$, $q = |Q|$, and so on.

Given $w_1, w_2 \in G$ and $c \in \mathbb{R}$, we define the sum $w_1 + w_2$ and the product cw_1, in G, in the usual form, i.e.,

$$(w_1 + w_2)(S, Q) = w_1(S, Q) + w_2(S, Q) \quad \text{and} \quad (cw_1)(S, Q) = cw_1(S, Q),$$

respectively. It is easy to verify that G is a vector space with these operations. For example, the collection of games $\{u_{(S,Q)} \mid (S, Q) \in \text{EC}, S \neq \varnothing\}$ defined by

$$u_{(S,Q)}(T, P) = \begin{cases} 1 \text{ if } (T, P) = (S, Q) \\ 0 \text{ otherwise} \end{cases}$$

constitutes a basis of the space of games with externalities. Notice that the dimension of G equals the number of nontrivial embedded coalitions.

A *solution* is a function $\varphi : G \to \mathbb{R}^n$. If φ is a solution and $w \in G$, then we can interpret $\varphi_i(w)$ as the utility payoff which player i should expect from the game w.

Now, the group of permutations of N, $S_n = \{\theta : N \to N \mid \theta \text{ is bijective}\}$, acts on CL and on EC in the natural way, i.e., for $\theta \in S_n$:

$$\theta(S) = \{\theta(i) \mid i \in S\}$$

$$\theta(S_1, \{S_1, S_2, \ldots, S_l\}) = (\theta(S_1), \{\theta(S_1), \theta(S_2), \ldots, \theta(S_l)\})$$

And also, S_n acts on the space of payoff vectors, $\mathbb{R}^n$:

$$\theta(x_1, x_2, \ldots, x_n) = \left(x_{\theta^{-1}(1)}, x_{\theta^{-1}(2)}, \ldots, x_{\theta^{-1}(n)} \right)$$

Definition 2. Let $(S, Q) \in EC$ and $i \in S$. Consider a partition Q' obtained by moving player i from S to some other (possible empty) member T of Q. The mapping $\alpha_{iT} : Q \to Q'$ defined by

$$\alpha_{iT}(S) = S_{-i}$$
$$\alpha_{iT}(T) = T_{+i}$$
$$\alpha_{iT}(S') = S' \quad \text{for } S' \in Q_{-S,-T}$$

is called a move for player i. Notice that $\alpha_{iT}(Q) = \{S_{-i}, T_{+i}\} \cup Q_{-S,-T}$.

15.2.2 Integer Partitions

A partition of a nonnegative integer is a way of expressing it as the unordered sum of other positive integers, and it is often written in tuple notation. Formally,

Definition 3. $\lambda = [\lambda_1, \lambda_2, \dots, \lambda_l]$ is a partition of n (denoted as $\lambda \vdash n$) if $\lambda_1, \lambda_2, \dots, \lambda_l$ are positive integers and $\lambda_1 + \lambda_2 + \cdots + \lambda_l = n$. Two partitions which only differ in the order of their elements are considered to be the same partition.

The set of all partitions of n will be denoted by $\Pi(n)$, and, if $\lambda \vdash n$, $|\lambda|$ is the number of elements of λ.

For example, the partitions of $n = 4$ are $[1, 1, 1, 1]$, $[2, 1, 1]$, $[2, 2]$, $[3, 1]$, and $[4]$. We will abbreviate this notation by dropping the commas, so $[2, 1, 1]$ becomes $[211]$.

If $Q \in PT$, there is a unique partition $\lambda_Q \vdash n$, associated with Q, where the elements of λ_Q are exactly the cardinalities of the elements of Q. In other words, if $Q = \{S_1, S_2, \dots, S_m\} \in PT$, then $\lambda_Q = [s_1, s_2, \dots, s_m]$.

For a given $\lambda \vdash n$, we represent by λ° the set of numbers determined by the λ_i's and by $m_{\lambda_j}^{\lambda}$ the multiplicity of λ_j in λ. So, if $\lambda = [4, 4, 2, 1, 1, 1]$, then $\lambda^{\circ} = \{1, 2, 4\}$, $m_1^{\lambda} = 3$, $m_2^{\lambda} = 1$, and $m_4^{\lambda} = 2$. By convention, $m_0^{\lambda} = 1$ for every $\lambda \in \Pi(n)$.

Additionally, if $[\lambda_1, \lambda_2, \dots, \lambda_l] \vdash n$, for $l > k \geq 1$, we define $[\lambda_1, \lambda_2, \dots, \lambda_l] - [\lambda_1, \lambda_2, \dots, \lambda_k] = [\lambda_{k+1}, \lambda_{k+2}, \dots, \lambda_l]$. For example, $[4, 3, 2, 1, 1, 1] - [3, 1, 1] = [4, 2, 1]$.

If $\lambda \in \Pi(n)$ and $\lambda' \in \Pi(m)$, then we can form a partition $\lambda + \lambda'$ in $\Pi(n + m)$ by combining all elements of such partitions. For example, $[4, 3, 2, 1, 1, 1] + [3, 1, 1] = [4, 3, 3, 2, 1, 1, 1, 1, 1]$.

For $\lambda \in \Pi(n)$ and $z, r \in \lambda^{\circ}$ ($r \neq 1$), we define $\lambda_z^r = \lambda - [r, z] + [r - 1, z + 1]$.

Finally, we need to define certain sets which are used in the sequel.

Definition 4. Let E_n be a set of pairs:

$$E_n = \{(\lambda, s) \mid \lambda \in \Pi(n), s \in \lambda^{\circ} \setminus \{1, n\}\}$$

and similarly, define the set of triples:

$$B_n = \{(\lambda, s, t) \mid \lambda \in \Pi(n) \backslash \{[n]\}, s \in \lambda^{\circ}, t \in (\lambda - [s])^{\circ}\}$$

Example 1. If $n = 4$, then

$$E_4 = \{([211], 2), ([22], 2), ([31], 3)\}$$

and

$$B_4 = \{([1111], 1, 1), ([211], 1, 1), ([211], 1, 2), ([211], 2, 1), ([22], 2, 2), ([31], 1, 3), ([31], 3, 1)\}$$

15.3 Marginal Contributions and Axioms

In this section we shall present the axioms that are asked solutions to satisfy in the cooperative game theory framework. Reasonable requirements to impose on a value are those underlying the construction of the Shapley value in the absence of externalities, namely, the axioms of linearity, symmetry, efficiency, and nullity axioms.

Axiom 1 (Linearity). *The solution φ is linear if $\varphi(w_1 + w_2) = \varphi(w_1) + \varphi(w_2)$ and $\varphi(cw_1) = c\varphi(w_1)$, for all $w_1, w_2 \in G$, and $c \in \mathbb{R}$.*

The axiom of linearity means that when a group of players shares the benefits (or costs) stemming from two different issues, how much each player obtains does not depend on whether they consider the two issues together or one by one. Hence, the agenda does not affect the final outcome. Also, the sharing does not depend on the unit used to measure the benefits.

Axiom 2 (Symmetry). *The solution φ is said to be symmetric if and only if $\varphi(\theta \cdot w) = \theta \cdot \varphi(w)$ for every $\theta \in S_n$ and $w \in G$, where the game $\theta \cdot w$ is defined as*

$$(\theta \cdot w)(S, Q) = w[\theta^{-1}(S, Q)]$$

Symmetry means that player's payoffs do not depend on their names. The payoff of a player is only derived from his influence on the worth of the coalitions.

Axiom 3 (Efficiency). *The solution φ is efficient if $\sum_{i \in N} \varphi_i(w) = w(N, \{N\})$ for all $w \in G$.*

We assume that the grand coalition forms and we leave issues of coalition formation out of this paper. Efficiency then simply means that the value must be feasible and exhaust all the benefits from cooperation, given that everyone cooperates.

Next, we turn to our discussion of marginal contributions in environments with externalities, central in our work. For characteristic functions, the marginal contribution of a player i within a coalition S is defined as the loss incurred by the other members of S if i leaves the group. This number could depend on the organization of the players not in S when there are externalities. It is natural, therefore, to define the marginal contribution of a player within each embedded coalition, and we consider the general case where a player may join another coalition T after leaving S.

Formally, let i be a player, let $(S, Q) \in EC$ such that $S \ni i$, and let $T \in Q_{-S}$. Then the marginal contribution of i to (S, Q) when i joins T is given by

$$\mathrm{MC}_{i,(S,Q),T}(w) = w(S, Q) - w(S_{-i}, \alpha_{iT}(Q))$$

Definition 5. $i \in N$ is called a null player in the game w if

$$\mathrm{MC}_{i,(S,Q),T}(w) = 0$$

for each embedded coalition (S, Q) such that $i \in S$ and each $T \in Q_{-S}$.[2]

Axiom 4 (Nullity). *Let $i \in N$ and let $w \in G$. If i is a null player in w, the $\varphi_i(w) = 0$.*

Notice that for a player to be a null player, it must be the case that he alone receives zero for any organization of the other players and has no effect on the worth of any coalition S. The nullity axiom only makes sure that a player with absolutely no influence on the gains that any coalition can obtain should not receive nor pay anything.

Shapley (1953) proved that these four axioms characterize a unique value in the class of games with no externalities.[3] If v denotes a game with no externalities (where $v : 2^N \to \mathbb{R}$ is a function that gives the worth of each coalition, independently of the partition structure), then the Shapley value Sh is defined as

$$\mathrm{Sh}_i(v) = \sum_{\{S \subseteq N : i \notin S\}} \frac{s!(n - s - 1)!}{n!} [v(S \cup \{i\}) - v(S)]$$

for each player $i \in N$ and each characteristic function v.

---

[2]De Clippel and Serrano (2008) have called it null player in the strong sense. This definition of a null player agrees with the definition presented in Bolger (1989) and Macho-Stadler et al. (2007), and it is different than the one considered in Myerson (1977) and Albizuri et al. (2005).

[3]A game is with no externalities if and only if the payoff that the players in a coalition S can jointly obtain if this coalition is formed is independent of the way the other players are organized. This means that in a game with no externalities, the characteristic function satisfies $w(S, Q) = w(S, Q')$ for any two partitions $Q, Q' \in PT$ and any coalition S which belongs both to Q and Q' Hence, the worth of a coalition S can be written without reference to the organization of the remaining players, $w(S) := w(S, Q)$ for all $Q \ni S, Q \in PT$.

15.4 Axiomatic Characterization

The purpose of this section is to present a characterization for the whole class of values satisfying the axioms discussed above. As mentioned before, the key ingredient in the definition of the nullity axiom is the way we define the marginal contribution of a player within each embedded coalition.

Example 2. For $N = \{i, j, k\}$, all possible marginal contributions of player i are

$$M_1(w) = w(N, \{N\}) - w(\{j, k\}, \{\{i\}, \{j, k\}\})$$

$$M_2(w) = w(\{i, j\}, \{\{k\}, \{i, j\}\}) - w(\{j\}, \{\{j\}, \{i, k\}\})$$

$$M_3(w) = w(\{i, j\}, \{\{k\}, \{i, j\}\}) - w(\{j\}, \{\{i\}, \{j\}, \{k\}\})$$

$$M_4(w) = w(\{i, k\}, \{\{j\}, \{i, k\}\}) - w(\{k\}, \{\{k\}, \{i, j\}\})$$

$$M_5(w) = w(\{i, k\}, \{\{j\}, \{i, k\}\}) - w(\{k\}, \{\{i\}, \{j\}, \{k\}\})$$

$$M_6(w) = w(\{i\}, \{\{i\}, \{j, k\}\})$$

$$M_7(w) = w(\{i\}, \{\{i\}, \{j\}, \{k\}\})$$

For general games with externalities, there is no hope to get a uniqueness result of a value with linearity, symmetry, efficiency, and nullity. To see this, consider the following example.

Example 3. For $N = \{i, j, k\}$, three different values satisfy the axioms of linearity, symmetry, efficiency, and nullity:

$$\psi_i(w) = \frac{1}{3}M_1(w) + \frac{1}{6}M_2(w) + \frac{1}{6}M_4 + \frac{1}{3}M_6(w)$$

$$\psi_i(w) = \frac{1}{3}M_1(w) + \frac{1}{6}M_3(w) + \frac{1}{6}M_5 + \frac{1}{3}M_7(w)$$

$$\psi_i(w) = \frac{1}{3}M_1(w) + \frac{1}{9}M_2(w) + \frac{1}{18}M_3 + \frac{1}{9}M_4(w) + \frac{1}{18}M_5(w) + \frac{2}{9}M_6(w) + \frac{1}{9}M_7(w)$$

As the case of the Shapley value for games with no externalities, what we want to do here is to characterize a family of values that are linear combinations of marginal contributions.[4] It turns out that such values are precisely those that satisfies the axioms discussed in the previous section.

Now, we present the characterization for all linear, symmetric, efficient, and null solutions, which establishes the main result of this work.

Theorem 1. *The solution* $\varphi : G \to \mathbb{R}^n$ *satisfies linearity, symmetry, efficiency, and nullity axioms if and only if it is of the form*

---

[4]More precisely, the Shapley value for games with no externalities happens to be calculated as the weighted average of marginal contributions of players to coalitions.

$$\varphi_i(w) = \frac{w(N,\{N\})}{n} + \sum_{(\lambda,s,t)\in B_n} \beta_{(\lambda,s,t)} \left[\sum_{\substack{(S,Q)\in EC \\ S\ni i,|S|=s \\ \lambda_Q=\lambda}} \sum_{\substack{T\in Q_{-S} \\ |T|=t}} tw(S,Q) - \sum_{\substack{(S,Q)\in EC \\ S\not\ni i,|S|=s \\ \lambda_Q=\lambda,|Q^i|=t}} sw(S,Q) \right]$$

(15.1)

for real numbers $\{\beta_{(\lambda,s,t)} \mid (\lambda,s,t) \in B_n\}$ *such that*

i)

$$\beta_{([n-1,1],n-1,1)} = \frac{1}{n(n-1)}$$

(15.2)

   and

ii) *for every* $(\lambda,r) \in E_n$

$$\left(m_r^\lambda - 1\right)\left[r\beta_{(\lambda,r,r)} - (r-1)\beta_{(\lambda_r^r,r-1,r+1)} \right]$$

$$+ \sum_{\substack{z\in\lambda^\circ\cup\{0\} \\ z\neq r}} \left[zm_z^\lambda \beta_{(\lambda,r,z)} - (r-1)m_z^\lambda \beta_{(\lambda_z^r,r-1,z+1)} \right] = 0 \qquad (15.3)$$

Moreover, such representation is unique.

Proof. From Hernández-Lamoneda et al. (2009, Theorem 4),

$$\varphi_i(w) = \frac{w(N,\{N\})}{n} + \sum_{(\lambda,s,t)\in B_n} \beta_{(\lambda,s,t)} \left[\sum_{\substack{(S,Q)\in EC \\ S\ni i,|S|=s \\ \lambda_Q=\lambda}} \sum_{\substack{T\in Q_{-S} \\ |T|=t}} tw(S,Q) - \sum_{\substack{(S,Q)\in EC \\ S\not\ni i,|S|=s \\ \lambda_Q=\lambda,|Q^i|=t}} sw(S,Q) \right]$$

is a linear, symmetric, and efficient solution for arbitrary constants $\{\beta_{(\lambda,s,t)} \mid (\lambda,s,t) \in B_n\}$. Suppose $i \in N$ is a null player in $u_{(S,Q)}$ for every $(S,Q) \in EC$, that is, $u_{(S,Q)}(R,P) = u_{(S,Q)}(R_{-i},\alpha_{iT}(P))$ for each (R,P) such that $i \in R$ and each $T \in P_{-R}$.

   Nullity implies

1.

$$0 = \varphi_i(u_{(N,\{N\})}) = \frac{1}{n} - (n-1)\beta_{([n-1,1],n-1,1)}$$

Hence,

$$\beta_{([n-1,1],n-1,1)} = \frac{1}{n(n-1)}$$

2.

$$0 = \varphi_i(u_{(S,Q)}) = \sum_{T \in Q_{-s}} \left[t\beta_{(\lambda_Q,s,t)} - (s-1)\beta_{(\lambda_{\alpha_{iT}(Q)},s-1,t+1)} \right] \tag{15.4}$$

for every pair (S, Q) such that $s \notin \{1, n\}$. Notice that the above relation yields many repeated equations. In particular, relation (15.4) provides the same equation for (S, Q) and (S', Q'), if $s = s'$ and $\lambda_Q = \lambda_{Q'}$. Thus, the number of distinct equations derived from (15.4) coincides with the number of elements in E_n. Now, for a fixed (S, Q) such that $s \notin \{1, n\}$, it holds

$$\sum_{T \in Q_{-s}} t\beta_{(\lambda_Q,s,t)} = \begin{cases} \left(m_s^{\lambda_Q} - 1 \right) s\beta_{(\lambda_Q,s,s)} & \text{if } t = s \\ \sum_{\substack{z \in \lambda_Q^\circ \cup \{0\} \\ z \neq s}} z m_z^{\lambda_Q} \beta_{(\lambda_Q,s,z)} & \text{if } t \neq s \end{cases} \tag{15.5}$$

and

$$\sum_{T \in Q_{-s}} \beta_{(\lambda_{\alpha_{iT}(Q)},s-1,t+1)} = \begin{cases} \left(m_s^{\lambda_Q} - 1 \right) \beta_{((\lambda_Q)_s^s,s-1,s+1)} & \text{if } t = s \\ \sum_{\substack{z \in \lambda_Q^\circ \cup \{0\} \\ z \neq s}} m_z^{\lambda_Q} \beta_{((\lambda_Q)_z^s,s-1,z+1)} & \text{if } t \neq s \end{cases} \tag{15.6}$$

The system (15.3) follows from the substitution of equalities (15.5) and (15.6) in relation (15.4).

The converse is a straightforward computation in view of the equalities in the first part of the proof.

Finally, to check uniqueness it is enough to prove that if

$$0 = \frac{w(N, \{N\})}{n} + \sum_{(\lambda,s,t) \in B_n} \beta_{(\lambda,s,t)} \left[\sum_{\substack{(S,Q) \in EC \\ S \ni i, |S| = s \\ \lambda_Q = \lambda}} \sum_{T \in Q_{-s}} tw(S, Q) - \sum_{\substack{(S,Q) \in EC \\ S \not\ni i, |S| = s \\ \lambda_Q = \lambda, |Q^i| = t}} sw(S, Q) \right]$$

($\beta_{(\lambda,s,t)}$'s satisfying conditions (15.2) and (15.3) for $(\lambda, r) \in E_n$) for every game w and for every player i, then every $\beta_{(\lambda,s,t)}$ vanish.

Thus, for given $(\lambda, s, t) \in B_n$, let $S = \{1, \dots, s\}$ and Q be any partition such that $S \in Q$ and $\lambda_Q = \lambda$. Also let $T \in Q$ such that $|T| = t$. Let $w = u_{(S,Q)}$ and pick any $i \in T$. Then the above sum reduces to

$$0 = \beta_{(\lambda,s,t)}$$

∎

Remark 1. The intuition behind the expression derived in Theorem 1 has an interpretation as a bargaining process:

1. We allocate $\frac{w(N,\{N\})}{n}$ to each player.
2. For each $(S,Q) \in EC$ and each $T \in Q\backslash\{S\}$, we keep going with one transfer from T to S:

   i) Every player in S receives (from every player in T) the fraction $\beta_{(\lambda_Q,s,t)}$ of the worth $w(S,Q)$:

   $$t\beta_{(\lambda_Q,s,t)}w(S,Q)$$

   ii) Every player in T pays (to every player in S) a fraction $\beta_{(\lambda_Q,s,t)}$ of the worth $w(S,Q)$:

   $$s\beta_{(\lambda_Q,s,t)}w(S,Q)$$

3. Finally, these transfers must satisfy $\beta_{([n-1,1],n-1,1)} = \frac{1}{n(n-1)}$ and

$$\sum_{T \in Q-s} t\beta_{(\lambda_Q,s,t)} = \sum_{T \in Q-s} (s-1)\beta_{(\lambda_{\alpha_{iT}(Q)},s-1,t+1)}$$

for every $(S,Q) \in EC$ such that $S \ni i$, $s \neq 1$ and $s \neq n$. Here, $\beta_{(\lambda_{\alpha_{iT}(Q)},s-1,t+1)}$ represents the fraction of the worth $w(S_{-i},\alpha_{iT}(Q))$ that receives each player in S_{-i} from player i.

Remark 2. It is no difficult to show that any linear, symmetric, efficient, and null solution can be written as a linear combination of marginal contributions:

$$\varphi_i(w) = \sum_{\substack{(S,Q)\in EC \\ S\ni i,s\neq 1}}\sum_{T\in Q-s}(s-1)\beta_{(\lambda_{\alpha_{iT}(Q)},s-1,t+1)}MC_{i,(S,Q),T}(w) + \sum_{\substack{(S,Q)\in EC \\ S\ni i,s=1}}\sum_{T\in Q-s}t\beta_{(\lambda_Q,s,t)}MC_{i,(S,Q),T}(w)$$

Example 4. As an illustration for $N = \{i,j,k\}$, any linear, symmetric, efficient, and null solution is of the form (for player i)

$$\varphi_i(w) = \frac{1}{3}M_1(w) + \beta_{([21],1,2)}[M_2(w) + M_4(w) + 2M_6(w)] + \beta_{([111],1,1)}[M_3(w) + M_5(w) + 2M_7(w)]$$

For any choice of real numbers such that

$$\beta_{([111],1,1)} + \beta_{([21],1,2)} = \frac{1}{6}$$

Corollary 1. *The space of all linear, symmetric, efficient, and null solutions in n players has dimension $|B_n| - |E_n| - 1$.*

Remark 3. The solutions we have characterized in Theorem 1 are extensions of the Shapley value for games with no externalities, since our four axioms coincide with the set of axioms that traditionally are used to characterize the Shapley value in the absence of externalities. First, notice that the subset of G formed by the games w such that $w(S, Q) = w(S) \; \forall (S, Q) \in EC$ can be identified with the set of games with no externalities. So, every linear, symmetric, efficient, and null solution coincides with the Shapley value for these games.

Example 5. In Example 4 we obtained an expression of any linear, symmetric, efficient, and null solution (for $n = 3$). Applying such general solution to a game w such that $w(S, Q) = w(S) \; \forall (S, Q) \in EC$:

$$\varphi_i(w) = \frac{1}{3} [w(N) - w(\{j, k\})]$$
$$+ \beta_{([21],1,2)} [w(\{i, j\}) - w(\{j\})]$$
$$+ \beta_{([111],1,1)} [w(\{i, j\}) - w(\{j\})]$$
$$+ \beta_{([21],1,2)} [w(\{i, k\}) - w(\{k\})]$$
$$+ \beta_{([111],1,1)} [w(\{i, k\}) - w(\{k\})]$$
$$+ 2\beta_{([111],1,1)} w(\{i\}) + 2\beta_{([21],1,2)} w(\{i\})$$

Finally, according to the condition $\beta_{([111],1,1)} + \beta_{([21],1,2)} = \frac{1}{6}$, then $\varphi_i(w) = Sh_i(w)$.

15.5 Some Examples

In this section we briefly provide three examples of linear, symmetric, efficient, and null solutions, as well as two values that fail to satisfy our nullity axiom.

15.5.1 Bolger Value

Bolger (1989) obtains a unique value characterized by our properties of linearity, symmetry, efficiency, and nullity and an additional requirement based on the behavior of the value in simple games (the worth of any coalition is either one or zero). Now, consider an embedded coalition $(S_{-i}, \alpha_{iT}(Q))$ obtained from (S, Q) by a move for player i (from S to $T \in Q_{-S}$). Such a move is called a pivot move if S wins with respect to (S, Q) and S_{-i} loses with respect to $(S_{-i}, \alpha_{iT}(Q))$. The additional property that Bolger introduced states that for simple games, a player i obtains the same payoff in two games w_1 and w_2 if he has the same number of pivot moves in both games.

There is no closed-form expression for this value, but it must be of the form (15.1) satisfying (15.2) and (15.3). As an example for $N = \{i, j, k\}$, the Bolger value for player i is

$$\psi_i(w) = \frac{1}{3}M_1(w) + \frac{1}{12}\left[M_2(w) + M_3(w) + M_4(w) + M_5(w)\right] + \frac{1}{6}\left[M_6(w) + M_7(w)\right]$$

15.5.2   The Value of Macho-Stadler et al.

Macho-Stadler et al. (2007, Theorem 1) showed that any solution that satisfies our version of nullity axiom, as well as the axioms of efficiency, linearity, and a strong version of symmetry[5] is a Shapley value of a characteristic function that is obtained by performing averages of the partition function. They also characterized a unique solution by adding an axiom of similar influence. Such a value is given by

$$\psi_i(w) = \sum_{\substack{(S,Q) \in EC \\ i \in S}} \frac{(s-1)! \prod_{T \in Q \setminus \{S\}} (t-1)!}{n!} w(S, Q) - \sum_{\substack{(S,Q) \in EC \\ i \notin S}} \frac{s! \prod_{T \in Q \setminus \{S\}} (t-1)!}{(n-s)n!} w(S, Q)$$

Which we get when we choose $\beta_{(\lambda, s, t)} = \dfrac{(s-1)! \prod_{t \in (\lambda - [s])^{\circ}} (t-1)!}{(n-s)n!}$.

15.5.3   Externality-Free Value

De Clippel and Serrano (2008) define an extension of the Shapley value to the class of games with externalities as

$$\psi_i(w) = Sh_i(v)$$

for each $i \in N$ and each $w \in G$, where Sh is the Shapley value operator for games with no externalities and the characteristic function v is defined as follows:

$$v(S) = w\left(S, \{S, \{j\}_{j \in N \setminus S}\}\right)$$

for each $S \subseteq N$.

---

[5] However, their strong version of symmetry implies our symmetry axiom. It strengthens the symmetry axiom by requiring that the payoff of a player should not change after permutations in the set of players in $N \setminus S$, for any embedded coalition structure (S, Q).

This solution satisfies the axioms of linearity, symmetry, efficiency, and nullity, so it is of the form (15.1) where the $\beta_{(\lambda,s,t)}$'s satisfies (15.2) and (15.3):

$$\beta_{(\lambda,s,t)} = \begin{cases} \frac{(s-1)!(n-s-1)!}{n!} & \text{if } \lambda \in \{[m, \overbrace{1, \ldots, 1}^{n-m}]\}_{m=1}^{n-1}, s = m \text{ and } t = 1 \\ 0 & \text{otherwise} \end{cases}$$

15.5.4 Myerson Value

Myerson (1977) proceeds axiomatically and proposes a value that extends the well-known Shapley value (Shapley 1953), which is defined for games with no externalities. The three axioms that uniquely characterize the Myerson's extension are linearity, symmetry, and a carrier axiom requiring that the surplus is shared only among the members of the carrier. The carrier axiom implies both efficiency and a nullity concept different from the one assumed in our analysis. The Myerson value of a player is given by

$$\psi_i(w) = \sum_{(S,Q)\in EC} (-1)^{q-1}(q-1)! \left(\frac{1}{n} - \sum_{T\in Q\setminus\{S\},i\notin T} \frac{1}{(q-1)(n-t)} \right) w(S,Q)$$

which we get with the parameters

$$\beta_{(\lambda,s,t)} = \frac{(-1)^{|\lambda|}(|\lambda|-1)!}{s} \left(\frac{1}{n} - \sum_{t\in(\lambda-[r,s])^\circ} \frac{1}{(|\lambda|-1)(n-t)} \right)$$

However, those parameters do not satisfy conditions (15.2) and (15.3).

15.5.5 The Value of Albizuri et al.

As a final example, Albizuri et al. (2005) obtain a unique value characterized by the properties of linearity, symmetry, efficiency, oligarchy, and an additional requirement of symmetry with respect to the embedded coalitions. They define the value for a player as

$$\psi_i(w) = \sum_{\substack{(S,Q)\in EC \\ i\in S}} \frac{(s-1)!(n-s)!}{n!P(S,N)} w(S,Q) - \sum_{\substack{(S,Q)\in EC \\ i\in S}} \frac{s!(n-s-1)!}{n!P(S,N)} w(S,Q)$$

where $P(S, N) = |\{(T, Q) \in EC \mid T = S\}|$. In fact, they notice that $P(S, N) = p(n-s)$ where $p(k)$ represents the number of partitions of any set K with cardinality k.

This solution is also of the form (15.1) and the corresponding parameters are $\beta_{(\lambda, s, t)} = \frac{(n-s-1)!(s-1)!}{n! \cdot p(n-s)}$. Such parameters do not satisfy (15.2) and (15.3), since the value fail to satisfy the nullity axiom.

15.6 Related Literature

First, two works proposed definitions of marginality[6] (in terms of marginal contributions) for games with externalities and proved uniqueness based on Shapley's standard axiomatization (Hu and Yang 2010; PhamDo and Norde 2007). Some other authors used Young's axiomatization: Bolger (1989) modified it by adding an additional nullity axiom to derive his value and De Clippel and Serrano (2008) in their analysis of externality-free value. These results for Young's axiomatization were generalized by Fujinaka (2004). He was the first to propose a general formula for marginal contribution as the affine combination of elementary marginal contributions. Fujinaka proved that Young's axiomatization parameterized by any weights implies a unique value.

Macho-Stadler et al. (2007) proposed the average approach, where the authors provided a value using Shapley's axioms together with strong symmetry and similar influence. This latter axiom says that, if we exchange the values of two embedded coalitions in which players i and j appear in the first one together and, in the second one, as singletons, then their payoffs should not change. Although axiomatization departed from marginality, the authors introduce a definition of marginal contribution and note that value can be transformed as the weighted average of player's marginal contributions.

Myerson (1977) was the first to propose a new extension of the Shapley value to games with externalities. He based his value on the concept of carrier. We say that a set C is a carrier if the value of any embedded coalition is determined by a partition of players from C. Now, Carrier implies that if C is a carrier, then the payoff of the grand coalition is divided between players from C. Against this, Myerson showed that there exists a unique value that satisfies symmetry, additivity, and carrier. As the set of all players, N is clearly a carrier, and if i is a null player, then $N \setminus \{i\}$ is also a carrier; we have that carrier implies both efficiency and the nullity axiom. This means that Myerson's value satisfies all four of Shapley's axioms.

Other authors proposed values that are rather far from Shapley's understanding of fairness. Albizuri et al. (2005) argued that, in a game with externalities, a

---

[6]Originally provided by Young (1985) for games with no externalities. He formulated the marginality principle as an axiom, that is, that the solution should pay the same to a player in two games if his or her marginal contributions to coalitions are the same in both games. Marginality is an idea with a strong tradition in economic theory.

coalition should be evaluated by the set of values it has, regardless of which partitions these values correspond to. The authors combined this principle, called embedded coalition anonymity, with the oligarchy axiom (which can be understood as the weakened Myerson axiom) and three of Shapley's original axioms: efficiency, additivity, and symmetry. The resulting value can be derived as the Shapley value for a game without externalities calculated by assigning to every coalition an arithmetic average of all its values in games with externalities.

Finally, in a stochastic process, players leave the grand coalition one by one. Grabisch and Funaki (2012) formulate a different process. They take as a starting point the partition containing singletons of all players and consider all possible sequences of mergers which result in the grand coalition. That said, the contribution of a player is evaluated as the effect that the player merging with other coalitions makes on their values. If a player enters some coalition alone, he is rewarded with the whole change of its value, i.e., with the marginal contribution; but if he is already a part of a coalition that merges with another one, the authors argue that the change of the value of the coalition they merge with should be divided equally between him and other members of the coalition. This contradicts the nullity axiom, as a null player is rewarded with a payoff even though the coalition without him would cause the same impact on the merged coalition.

15.7 Conclusion

This paper has explored games with externalities. Our basic approach is rooted in the concept of marginal contributions of players to coalitions. In problems involving externalities, we considered the general case where a player (in a coalition S) may join another coalition after leaving S.

The paper follows an axiomatic methodology and presumes that the grand coalition has exogenously formed. Then the implications of linearity, symmetry, efficiency, and nullity are explored, leading to two main results: the first one establishes that the standard translation of Shapley's four axioms to games with externalities is not sufficient to obtain a unique value, and therefore, we provided an axiomatic characterization for the family of solutions for games with externalities satisfying those axioms that traditionally are used to characterize the Shapley value in the absence of externalities. Additionally, we showed that every such solution can be written as a linear combination of marginal contributions of players and provided an interpretation as a bargaining process.

Based in the analysis presented in this work, one can consider additional restrictions (or axioms) in order to obtain a uniqueness result. Finally, other approaches to the concept of marginal contribution can be studied.

Acknowledgements I thank the participants of the XV Latin-American Workshop on Economic Theory (JOLATE) for comments, interesting discussions, and encouragement. J. Sánchez-Pérez acknowledges financial support from CONACYT research grant 130515.

References

Albizuri, M.J., Arin, J., Rubio, J.: An axiom system for a value for games in partition function form. Int. Game Theory Rev. **7**(1), 63–72 (2005)

Bolger, E.M.: A set of axioms for a value for partition function games. Int. J. Game Theory **18**(1), 37–44 (1989)

De Clippel, G., Serrano, R.: Marginal contributions and externalities in the value. Econometrica **6**, 1413–1436 (2008)

Fujinaka, Y.: On the marginality principle in partition function form games, Unpublished Manuscript, Graduate School of Economics, Kobe University (2004)

Grabisch, M., Funaki, Y.: A coalition formation value for games in partition function form. Eur. J. Oper. Res. **221**(1), 175–185 (2012)

Hernández-Lamoneda, L., Sánchez-Pérez, J., Sánchez-Sánchez, F.: The class of efficient linear symmetric values for games in partition function form. Int. Game Theory Rev. **11**(3), 369–382 (2009)

Hu, C.C., Yang, Y.Y.: An axiomatic characterization of a value for games in partition function form. SERIEs **1**(4), 475–487 (2010)

Lucas, W.F., Thrall, R.M.: n-Person games in partition function form. Nav. Res. Logist. Q. **10**, 281–298 (1963)

Macho-Stadler, I., Pérez-Castrillo, D., Wettstein, D.: Sharing the surplus: an extension of the Shapley value for environments with externalities. J. Econ. Theory **135**, 339–356 (2007)

Myerson, R.B.: Values of games in partition function form. Int. J. Game Theory **6**(1), 23–31 (1977)

PhamDo, K., Norde, H.: The Shapley value for partition function form games. Int. Game Theory Rev. **9**(2), 353–360 (2007)

Sánchez-Pérez, J.: A note on a class of solutions for games with externalities generalizing the Shapley value. Int. Game Theory Rev. **17**(3), 1–12 (2015)

Shapley, L.: A value for n-person games. Contrib. Theory Games **2**, 307–317 (1953)

Young, H.P.: Monotonic solutions of cooperative games. Int. J. Game Theory **14**, 65–72 (1985)

Chapter 16
Approximation of Optimal Stopping Problems and Variational Inequalities Involving Multiple Scales in Economics and Finance

Andrianos E. Tsekrekos and Athanasios N. Yannacopoulos

Abstract Many interesting decision-making problems in economics and finance can be expressed in terms of variational inequalities, whose well-developed theory provides valuable answers and insights concerning optimal policies. In this chapter, we first provide a brief introduction to the theory of variational inequalities as applied to economic decision-making, before focusing on a particular class (optimal stopping problems) where the underlying Markov process that introduces the uncertainty in the setting presents evolution of multiple time scales. Such problems lead to variational inequalities with fast-varying coefficients which require techniques related to homogenisation theory. Our results establish how, for such problems, approximate solutions to any order and (importantly) in almost closed form can be obtained by a singular perturbation approach. Our example from the *waiting-to-invest* literature in the last section demonstrates the applicability of the results.

Keywords Economic decision-making • Variational inequalities • Optimal stopping problems • Multi-scale volatility

16.1 Introduction

Many interesting problems in economic decision-making can be expressed in terms of variational inequalities. The well-developed theory in this field provides a variety

A.E. Tsekrekos (✉)
Department of Accounting and Finance, School of Business, Athens University of Economics and Business, Athens, Greece
e-mail: tsekrekos@aueb.gr

A.N. Yannacopoulos
Department of Statistics, School of Information Sciences and Technology, Athens University of Economics and Business, Athens, Greece
e-mail: ayannaco@aueb.gr

© Springer International Publishing Switzerland 2016
A.A. Pinto et al. (eds.), *Trends in Mathematical Economics*,
DOI 10.1007/978-3-319-32543-9_16

of analytical and numerical tools that allow one to treat these problems and obtain valuable answers and insight concerning optimal economic policies.

However, there are certain problems for which, even though the analytical framework is well understood, the actual treatment of the resulting variational inequalities is not practical and require the use of specialised tools, motivated by the nature of problems at hand. One such case is when the underlying process which introduces the uncertainty in the model presents evolution of multiple time scales, leading to variational inequalities with fast-varying coefficients.

Such problems are not easy to treat from the numerical point of view, on account of various problems resulting from the fact that we need to resolve the solution at a fine scale, and require techniques related to the theory of homogenisation (see, e.g. Cioranescu and Donato 1999). It is interesting that the application of such methods allows one to obtain approximate solutions to any order and, most importantly, in almost closed form that lend themselves easily to comparative statics that are important for the economic decision maker for policy purposes. It is the aim of the present short presentation to provide a brief introduction to the theory of variational inequalities as applied to economic decision-making (and in particular to optimal stopping problems that are common in economics and finance) and introduce a singular perturbation approach for the closed-form approximation of the solution of such variational inequalities, for the particular class of models that exhibit multi-scale (fast/slow) dynamics. This class of problems is motivated by ample empirical evidence and has already been addressed in the literature within the context of optimal control theory (e.g. Bardi and Cesaroni 2011), financial option pricing (e.g. Fouque et al. 2011, and the references therein), among other areas.

In the section that follows, we start with a brief introduction to the theory of variational inequalities, with optimal stopping problems from economics and finance in mind, before addressing a wide variety of applied cases where the limiting behaviour of the family of Markov processes that underlie the problem is of interest. In such cases, the solutions to the corresponding family of stopping problems are derived from variational inequalities in a weak sense, and the main result of the section establishes the well posedness of such weak formulations.

However, although the existence of a solution, and hence of a stopping rule, is guaranteed for every member of such family of problems, one may not claim the same for the problem of the limiting behaviour of the value function and the stopping rule. More stringent conditions are needed in order to answer this positively, and in most cases asymptotic arguments related to homogenisation theory have to be employed. Our main result in Sect. 16.3 addresses this issue for a particular class of inequalities that are associated with optimal stopping problems under Markov processes that evolve in different time scales, namely, a "slow" and a "fast" time scale. The interest in such forms of Markov processes is motivated by ample empirical evidence in favour of "fast", mean-reverting stochastic volatility dynamics for many asset classes (see, e.g. the empirical evidence in Alizadeh et al. 2002; Hillebrand 2006, among others). In the concluding section, we demonstrate the applicability of our main results, by extending the classic *waiting-to-invest* model of McDonald and Siegel (1986), to a setting where the investment value follows a geometric Ornstein–Uhlenbeck process, with latent stochastic volatility that mean reverts on a faster time scale than the underlying project value.

16.2   Optimal Stopping Problems in Economics and Finance

The general form of optimal stopping problems, as they frequently appear in economics and finance, is the following: Consider a Markov process $\{X(t)\}_t$ with infinitesimal generator $\mathcal{L}$, taking values in an open subset $D \subset \mathbb{R}^d$. In most economic applications, $X(t)$ is considered to be the value of a particular asset or project at time t. If the process is stopped (or exercised) at time t and $X(t) = x$, then it is usually assumed that the economic agent derives instantaneous utility or pay-off $e^{-rt}U(x)$, where $U : \mathbb{R}^d \to \mathbb{R}$ is a utility or pay-off function. Since the stopping/exercise time is subject to the agent's discretion, it will be chosen so as to maximise the expected present value of the pay-off. It is also usual to assume that the agent has an infinite time horizon available to her in order to decide when to exercise her discretion.

This setting is equivalent to the following optimisation problem

$$\sup_{\tau > 0} \mathbb{E}\left[e^{-r\tau} U[X(\tau)]\right] =: V(x), \tag{16.1}$$

where the optimisation is performed over all stopping times τ for the process $X(t)$, which is conditioned such that $X(0) = x$. The first stopping time τ^* for which $\mathbb{E}[e^{-r\tau^*} U[X(\tau^*)]] = V(x)$ is called the *optimal stopping time*, and the function V, defined as above, is called the *value function*. In most practical applications, we need to specify the stopping rule in terms of a feedback strategy, i.e. define the stopping time τ^* as the first time that the process $X(t)$ reaches a certain level, which is called the *optimal exercise boundary*. In Eq. (16.1), r plays the role of the agent's discount rate, which is often taken equal to the risk-free interest rate whenever the agent is risk neutral or whenever a replicating (spanning) argument related to the uncertainty in $X(t)$ can be invoked.

Problems of the form in (16.1) have been studied extensively in the literature, using either probabilistic methods based on the concept of the Snell envelope or by variational inequality methods, which most of the time are better suited for calculations. Typical examples include the well-studied problem of financial option pricing (Black and Scholes 1973; Merton 1973, among others), optimal switching problems (see, e.g. Duckworth and Zervos 2001; Brekke and Øksendal 1994) and applications in optimal capacity choice (e.g. Pindyck 1988) and natural resource investments (Brennan and Schwartz 1985) among others.

The variational inequality approach for problem (16.1) relies on the use of Itô's formula and the Markovian properties of the process.[1] It can be shown that the value

---

[1]The application of the Itô's formula in its standard sense requires smoothness of the value function. It should be noted that the variational inequality approach can hold, even if the value

function solves the variational inequality

$$\mathcal{L}V(x) - rV(x) = 0, \quad V(x) < U(x)$$
$$\mathcal{L}V(x) - rV(x) < 0, \quad V(x) = U(x), \tag{16.2}$$

which can be conveniently reformulated as

$$\max\{\mathcal{L}V(x) - rV(x), U(x) - V(x)\} = 0. \tag{16.3}$$

The exercise boundary for this problem is defined as the set $C = \{x \in \mathbb{R}^d : V(x) = U(x)\}$, meaning that $\tau^* = \inf_{t>0}\{X(t) \in C\}$. The above formulation shows that, at least formally, the optimal stopping time is completely solved in terms of the value function, which is characterised in terms of the variational inequality (16.2).

The characterisation of the treatment above as formal refers to the fact that in most cases of interest, equation (16.2) does not have solutions of the required degree of regularity for the formulation to hold in a classical sense. This leads to the need of restating the variational inequalities in a weak sense or, more generally, in a viscosity solution sense. Such weak formulations are well posed and lead to well-defined and practical solutions for the corresponding economic problem.

In a variety of interesting economic problems, the Markov process which models the value process depends upon a parameter, and the pricing often has to be performed for the whole parametric family of value processes. Let us denote this parameter by ϵ and use the notation $\{X_\epsilon(t)\}$ for the whole family of processes for any value of the parameter ϵ and $\mathcal{L}_\epsilon$ for the corresponding generator. This leads to a whole family of variational inequalities

$$\max\{\mathcal{L}_\epsilon V_\epsilon(x) - rV_\epsilon(x), U_\epsilon(x) - V_\epsilon(x)\} = 0, \quad \epsilon > 0, \tag{16.4}$$

where the assumption is made that the utility function U may also depend on the same parameter ϵ (meaning that U is a function of the instantaneous position of the process at time t, $X_\epsilon(t) = x$, as well as of the parameter ϵ). In such cases, an interesting problem that is often encountered is the behaviour of the whole family of solutions at a relevant limit, e.g. as $\epsilon \to 0$.

A few concrete examples of the above general formulation could be (a) the discretisation of the process $X(t)$, either in state space or in time, where ϵ corresponds to the level of discretisation, (b) parametric models of additional stochastic factors (e.g. stochastic volatility, stochastic short rate, etc.) where ϵ plays the role of a particular parameter whose value $\epsilon = 0$ corresponds to some benchmark model [e.g. the constant volatility and constant short rate Black–Scholes (Black and Scholes

---

function is not C^2, in which case generalisations of Itô's formula based on convexity properties of value functions can be used (Itô–Meyer formula, see, e.g. Föllmer et al. 1995) or the concept of viscosity solutions (see Fleming and Soner 2006). The treatment of such problems within the present framework is in progress, but clearly beyond the scope of this work.

1973) model] or (c) multi-scale stochastic volatility models (Fouque et al. 2003a) that will also be treated in our example below.

A convenient formulation of the family of problems in (16.4) is as follows: Define the set $K_\epsilon = \{V \in \mathbb{X} : V(x) \geq U_\epsilon(x), x \text{ a.e.} \in \mathbb{R}^d\}$, where $\mathbb{X}$ is a conveniently chosen function space (most usually a Sobolev space). For U_ϵ continuous, K_ϵ is a closed, convex subset of $\mathbb{X}$. Consider any function $\phi \in \mathbb{Y}$, where $\mathbb{Y}$ is a set of smooth functions, dense in $\mathbb{X}$. We multiply the inequality by any such ϕ, we integrate over $\mathbb{R}^d$, and, using the density, we obtain the so-called weak form which reads

$$\langle \mathcal{A}_\epsilon V_\epsilon, \phi - V_\epsilon \rangle \geq 0, \quad \forall \phi \in K_\epsilon, \tag{16.5}$$

where $\mathcal{A}_\epsilon = -\mathcal{L}_\epsilon + rI$, and by $\langle \cdot \rangle$ we denote the duality pairing $\langle \phi, \psi \rangle = \int_D \phi(x)\psi(x)dx$ between $\mathbb{X}$ and its dual $\mathbb{X}^*$. We may simplify notation by using the family of bilinear forms $\alpha_\epsilon : \mathbb{X} \times \mathbb{X} \to \mathbb{R}$, defined by $\alpha_\epsilon(\phi, \psi) := \langle \mathcal{A}_\epsilon \phi, \psi \rangle$, to express inequality (16.5) as

$$\alpha_\epsilon(V_\epsilon, \phi - V_\epsilon) \geq 0, \quad \forall \phi \in K_\epsilon \tag{16.6}$$

We say that the bilinear form α_ϵ is continuous on $\mathbb{X}$ if there exists a constant c such that $\alpha_\epsilon(\phi, \psi) \leq c\|\phi\|_\mathbb{X}\|\psi\|_\mathbb{X}$ and coercive on $\mathbb{X}$ if there exists a constant $C > 0$ such that $\alpha_\epsilon(\phi, \phi) \geq C\|\phi\|_\mathbb{X}^2$.

The following guarantees the well posedness of the weak form in (16.5):

Proposition 16.2.1. *Assume that for any $\epsilon > 0$, the family of bilinear forms α_ϵ : $\mathbb{X} \times \mathbb{X} \to \mathbb{R}$ is continuous and coercive. Then for any $\epsilon > 0$, there exists a unique solution of the variational inequality (16.6).*

Proof. The proof is based on an application of the Lax–Milgram–Lions–Stampacchia theorem, which uses a fixed-point argument based on Banach's fixed-point theorem, along with the properties of the projection operator for closed, convex subsets of Hilbert spaces (see Chipot 2012). In particular, let $P_{K_\epsilon} : \mathbb{X} \to K_\epsilon$ be the projection operator from $\mathbb{X}$ to the closed convex K_ϵ, which is well known to be a pseudo-contraction characterised by the variation inequality $\langle \phi - P_{K_\epsilon}\phi, \psi - P_{K_\epsilon}\psi \rangle$ for any $\psi \in K_\epsilon$. Using the projection operator and fixing a $\rho > 0$, we may express (16.6) in the equivalent form

$$V_\epsilon = P_{K_\epsilon}(-\rho\mathcal{A}_\epsilon V_\epsilon + V_\epsilon). \tag{16.7}$$

Consider the family of maps $R_\rho : K_\epsilon \to K_\epsilon$, defined by $R_\rho(\psi) = P_{K_\epsilon}(-\rho\mathcal{A}_\epsilon\psi + \psi)$ for every $\psi \in K_\epsilon$, and note that (16.7) is equivalent to the existence of a fixed point for the map R_ρ. By continuity and coercivity, it is possible to show that for a proper choice of ρ, the mapping R_ρ is a contraction. Hence, by an application of the Banach fixed-point theorem, we obtain the existence of a unique solution.

Furthermore, if α_ϵ is symmetric, the variational inequality (16.6) is equivalent to the family of minimisation problems $\min_{\phi \in K_\epsilon} J_\epsilon(\phi)$, where $J_\epsilon(\phi) = \frac{1}{2}\alpha_\epsilon(\phi, \phi)$. ∎

The assumptions of Proposition 16.2.1 are fairly natural and are valid for a wide class of Markov processes, which are frequently encountered in applications. For example, they hold if $\mathcal{L}_\epsilon$ is the generator of an Itô process driven by an $\mathbb{R}^m$ Brownian motion for a covariance matrix σ, such that $\sigma\sigma'$ is invertible, and for mild enough drift or for a variety of Lévy processes (see, e.g. Øksendal and Sulem 2005; Peskir and Shiryaev 2006, and the references therein).

While the above proposition provides the existence of a solution, and hence of a stopping rule for every ϵ in a quite general and easy way, we may not say the same for the problem of the limiting behaviour of the value function and the stopping rule as ϵ goes to zero. This is understandable when one realises the connection of the variational inequality (16.6) if α_ϵ is symmetric, with a family of minimisation problems, and recalls that in general, when given a family of functionals, the family of minimisers does not converge to the minimiser of the limit functional. More detailed analysis is needed to guarantee such behaviour, and this leads us to problems related to homogenisation theory (see Cioranescu and Donato 1999, for an introduction). More stringent conditions are needed in order to answer this problem positively, and in most cases asymptotic arguments have to be employed in order to provide closed or semi-closed-form solutions for the solution of the variational inequality as $\epsilon \to 0$. We will deal with this problem, for a particular class of inequalities, in the section that follows.

16.3 A Class of Variational Inequalities Related to Multi-scale Volatility Models

In many applications, the process $X_\epsilon(t)$ can be split in two components, which evolve in different time scales, a "slow" and a "fast" one. Let us denote by S and Y the slow and fast component of X_ϵ, respectively, and assume that $X_\epsilon = (S, Y)$ is modelled by the Itô process

$$dS_t = b(S_t, Y_t)dt + f(S_t, Y_t)dW_t^S, \quad S_0 = s \tag{16.8}$$

$$dY_t = \delta^{-2}\zeta(Y_t)dt + \delta^{-1}\xi(Y_t)dW_t^Y, \quad Y_0 = y, \tag{16.9}$$

where the correlation structure between dW_t^S and dW_t^Y is described by

$$\begin{bmatrix} W_t^S \\ W_t^Y \end{bmatrix} = \begin{bmatrix} 1 & 0 \\ \rho & \sqrt{1-\rho^2} \end{bmatrix} \mathbf{W}_t, \tag{16.10}$$

with $\mathbf{W}_t$ a standard two-dimensional Wiener process on a complete filtered probability space satisfying the usual conditions (e.g. see Karatzas and Shreve 1991) and $|\rho| < 1$ a constant correlation coefficient.

In the above, Y_t is a latent stochastic factor that affects the dynamics of the observed component S_t through appropriately chosen functions b and f. The small parameter δ, which is related to ϵ by $\epsilon = \delta^2$, models the fact that the latent factor Y follows a process, whose dynamics are on a "faster" time scale than the dynamics of S. We will also use the notation $x = (s, y)$.

We will assume that the "fast" component Y reaches a statistical equilibrium at a time scale which is shorter than the scale of evolution of S. Therefore, after the passage of an initial transition period, the Y part of the process X_ϵ will have reached an equilibrium density, thus allowing a decoupling of the system of Eqs. (16.8) and (16.9), by replacing Y in (16.8) by a constant random variable distributed by the equilibrium density, hence leading to a single equation for S. This simplistic argument can be turned into a rigorous mathematical approach, leading to a perturbative, semi-closed-form solution for the class of variational inequalities associated with Markov processes $X_\epsilon = (S, Y)$ of the above form.

This form of Markov processes is very commonly used in the modelling of stochastic volatility, and it is motivated by empirical research providing ample evidence in favour of "fast", mean-reverting stochastic volatility dynamics for many asset classes (see the empirical evidence in Alizadeh et al. 2002; Fouque et al. 2003b; Hillebrand 2006; Hikspoor and Jaimungal 2008, among others). For Markov processes X_ϵ of this particular form, the generator operator is of the form

$$\mathcal{L}^\delta = \delta^{-2}\mathcal{L}_0 + \delta^{-1}\mathcal{L}_1 + \mathcal{L}_2$$

with

$$\mathcal{L}_0 = \zeta(y)\frac{\partial}{\partial y} + \frac{1}{2}\xi^2(y)\frac{\partial^2}{\partial y^2}$$

$$\mathcal{L}_1 = \rho\xi(y)f(s, y)\frac{\partial^2}{\partial s\partial y}$$

$$\mathcal{L}_2 = \frac{1}{2}f^2(s, y)\frac{\partial^2}{\partial s^2} + b(s, y)\frac{\partial}{\partial s},$$

leading to an expansion for the operator $\mathcal{A}_\delta$ as $\mathcal{A}_\delta = \delta^{-2}\mathcal{A}_0 + \delta^{-1}\mathcal{A}_1 + \mathcal{A}_2$, where $\mathcal{A}_0 = \mathcal{L}_0$, $\mathcal{A}_1 = \mathcal{L}_1$ and $\mathcal{A}_2 = \mathcal{L}_2 - rI$, with I the identity operator. For simplicity, we assume no explicit dependence of the function $U : \mathbb{R}^d \to \mathbb{R}$ on δ. We further assume that ζ and ξ are chosen so that the only acceptable solution of $\mathcal{L}_0\psi = 0$ is independent of y (Liouville property).

Under the above assumptions, it can be shown that the variational inequality

$$\max\{-\mathcal{A}_\delta V_\delta(x), U(x) - V_\delta(x)\} = 0, \quad \delta > 0, \quad x = (s, y) \tag{16.11}$$

admits a solution of the form

$$V_\delta(x) = \sum_{n=0}^{\infty} \delta^n V_n(x), \tag{16.12}$$

and the stopping rule is defined in terms of the solution of the equation $U(x) = V_\delta(x)$.

The proposition below provides a constructive scheme for the solution, for which the following definitions will be needed: Let φ be the solution of the differential equation $\mathcal{L}_0^* \varphi = 0$, where $*$ denotes the adjoint operator (φ corresponds to the invariant density for the fast component Y) and let $\overline{\mathcal{A}}_2$ be the "averaged" operator $\overline{\mathcal{A}}_2 = -\frac{1}{2} \left(\int f^2(s,y) \varphi(y) dy \right) \frac{\partial^2}{\partial s^2} - \left(\int b(s,y) \varphi(y) dy \right) \frac{\partial}{\partial s} + rI$. Define $F(s,y)$ as the solution of

$$\mathcal{A}_0 F(s,y) = (\overline{\mathcal{A}}_2 - \mathcal{A}_2) G(s), \tag{16.13}$$

with $G(s)$ a function of the s coordinate only; then $F(s,y)$ can be written as

$$F(s,y) = \frac{1}{2} [\phi(y) + c(s)] s^2 \frac{\partial^2 G(s)}{\partial s^2} + [\gamma(y) + g(s)] s \frac{\partial G(s)}{\partial s}, \tag{16.14}$$

where the exact expressions of $\phi(y)$, $\gamma(y)$, $c(s)$ and $g(s)$ depend on the form of $f(s,y)$ and $b(s,y)$. Finally, define

$$\Phi(s) = \rho \int \xi(y) f(s,y) \phi'(y) \varphi(y) dy \text{ and } \Gamma(s) = \rho \int \xi(y) f(s,y) \gamma'(y) \varphi(y) dy, \tag{16.15}$$

with the prime denoting the derivative with respect to y. Then the following proposition can be proven.

Proposition 16.3.1. *For $n \in \{0,1\}$, $V_n(s,y) = V_n(s)$ and V_0 solves the "averaged" variational inequality*

$$\max[-\overline{\mathcal{A}}_2 V_0, U - V_0] = 0, \tag{16.16}$$

while V_1 solves the "averaged" equation

$$\overline{\mathcal{A}}_2 V_1 = -\frac{1}{2} \Phi(s) s^2 \frac{\partial^3 V_0}{\partial s^3} - [\Phi(s) + \Gamma(s)] s \frac{\partial^2 V_0}{\partial s^2} - \Gamma(s) \frac{\partial V_0}{\partial s} \tag{16.17}$$

which depend only on the coordinate s. All the higher-order terms depend on both s and y, and solve differential equations (rather than variational inequalities) that are explicitly derived (but not reproduced here for the sake of brevity).

Proof. Introducing the expansion (16.12) into the inequality and separating orders, we see that the leading order (δ^{-2}) yields $\mathcal{A}_0 V_0 = 0$, which by assumption leads to the result that V_0 depends only on s, while at the next order (δ^{-1}), we get $\mathcal{A}_1 V_0 +$

$\mathcal{A}_0 V_1 = \mathcal{A}_0 V_1 = 0$, which again by the assumption leads to the result that V_1 depends only on s. At the next order (δ^0), we get the differential inequality

$$\max[-(\mathcal{A}_2 V_0 + \mathcal{A}_1 V_1 + \mathcal{A}_0 V_2), U - V_0] = 0,$$

and at the continuation region, by straightforward application of the Fredholm alternative, we are led to equation (16.16), which specifies completely V_0 as a function of s. Moreover, from the application of the Fredholm alternative, we get

$$\mathcal{A}_0 V_2(s, y) = (\overline{\mathcal{A}}_0 - \mathcal{A}_0) V_0(s),$$

which is of the form in (16.13), and thus $V_2(s, y)$ can be expressed in the form of (16.14). At the next order, we get

$$\max[-(\mathcal{A}_2 V_1 + \mathcal{A}_1 V_2 + \mathcal{A}_0 V_3), -V_1] = 0,$$

which by the properties of $\mathcal{A}_2$ leads to $\mathcal{A}_2 V_1 + \mathcal{A}_1 V_2 + \mathcal{A}_0 V_3 = 0$, where by an application of the Fredholm alternative and by what has been established regarding V_2, we get (16.17). The next-order corrections follow in a similar fashion. ∎

16.4 An Optimal Stopping Example Under "Fast" Mean-Reverting Stochastic Volatility

The example we illustrate in this section is a modified version of the classic *waiting-to-invest* model in McDonald and Siegel (1986). From a mathematical perspective, the problem corresponds to a free boundary differential equation associated with the *option to delay* an irreversible investment, so as to undertake it at some future point in time where its expected net present value is maximised. Such *real options* to defer investment in the face of uncertainty and cost irreversibility are typically encountered in many contexts (see, inter alia, Brennan and Schwartz 1985; Paddock et al. 1988; Grenadier 1995; Schwartz and Moon 2000, for applications in natural resources, real estate and R&D investments) where managerial flexibility over the timing of the investment can be exerted. They are usually formulated and treated as the optimal exercise of American-style option contracts (see Dixit and Pindyck 1994; Trigeorgis 1996, for reviews of the real options approach).

In order to fix ideas, consider a risk-neutral decision maker contemplating a new project that can be launched instantaneously, at any point in time once a constant fixed cost $I > 0$ is incurred, and suppose that the estimated value of the project at time t, denoted S_t, evolves according to the Itô process

$$dS_t = \eta(\kappa - S_t)S_t dt + f(Y_t) S_t dW_t^S, \quad S_0 = s \qquad (16.18)$$

$$dY_t = \delta^{-2} (m - Y_t) \, dt + \frac{v\sqrt{2}}{\delta} dW_t^Y, \quad Y_0 = y, \tag{16.19}$$

which is a particular case of the general class of processes introduced in the previous section.[2] In (16.18) and (16.19) the latent "fast" component Y_t follows an Ornstein–Uhlenbeck (OU) process that depends on the small parameter δ, while the observable "slow" component S_t evolves according to a geometric OU process, whose variance is affected by the latent factor via the $f(\cdot)$ function.[3]

For any initial state (s, y), the maximised expected present value of the project is

$$V(s, y) = \sup_{\tau > 0} \mathbb{E}\left[e^{-r\tau} [S(\tau) - I] \right]. \tag{16.20}$$

For the particular problem, $\varphi(y) = \frac{1}{\sqrt{2\pi v}} e^{-\frac{(m-y)^2}{2v^2}}$, and the "averaged" operator is

$$\overline{\mathcal{A}}_2 = -\frac{1}{2}\overline{f}^2 s^2 \frac{\partial^2}{\partial s^2} - \eta(\kappa - s)s\frac{\partial}{\partial s} + rI,$$

with

$$\overline{f}^2 = \int_{-\infty}^{+\infty} f^2(y)\varphi(y)dy.$$

For any particular choice of $f(y)$, the above can be explicitly calculated, analytically or numerically.

An application of Proposition 16.3.1 leads to the conclusion that the leading-order term V_0 is the solution of the variational inequality

$$\max\left[\frac{1}{2}\overline{f}^2 s^2 \frac{\partial^2 V_0}{\partial s^2} + \eta(\kappa - s)s\frac{\partial V_0}{\partial s} - rV_0, I - V_0 \right] = 0$$

which yields (given that S equals zero is an absorbing state and $\lim_{s \to 0} V_0(s) = 0$)

$$V_0(s) = As^\beta H\left[\beta, Z(\beta), \frac{2\eta}{\overline{f}^2} s \right], \tag{16.21}$$

with

---

[2]The assumption of a risk-neutral decision maker is not crucial, as the extension to risk aversion is fairly straightforward, only at the cost of additional notation.

[3]See Pindyck (1991) and Metcalf and Hassett (1995) for investment models in economics that employ the geometric Ornstein–Uhlenbeck process.

$$\beta = \frac{\bar{f}^2 - 2\eta\kappa + \sqrt{8\bar{f}^2 r + (2\eta\kappa - \bar{f}^2)}}{2\bar{f}^2} > 0, \text{ and } Z(\beta) = 2\beta + \frac{2\eta\kappa}{\bar{f}^2}, \qquad (16.22)$$

and $H[a, b, x]$ is the confluent hypergeometric function (see Abramowitz and Stegun 1972). The constant A in (16.21) is determined jointly with the leading-order term of the stopping rule threshold S_0^* by the following system of equations:

$$V_0(S_0^*) = S_0^* - I$$
$$V_0'(S_0^*) = 1 \qquad (16.23)$$

Finally, for this particular problem

$$\Phi(s) = \rho v s \sqrt{2} \int f(y)\phi'(y)\varphi(y)dy =: \Omega s$$

$$\Gamma(s) = \rho v s \sqrt{2} \int f(s)\gamma'(y)\varphi(y)dy =: \Theta s, \qquad (16.24)$$

and from Proposition 16.3.1, the first-order correction term V_1 is given by solving

$$\overline{\mathcal{A}}_2 V_1 = -\frac{1}{2}\Omega s^3 \frac{\partial^3 V_0}{\partial s^3} - [\Omega + \Theta]s^2 \frac{\partial^2 V_0}{\partial s^2} - \Theta s \frac{\partial V_0}{\partial s}, \qquad (16.25)$$

where the right-hand side is a known function of the s coordinate only.

As a simple numerical demonstration of our main result in Proposition 16.3.1, assume that the latent "fast" component Y_t affects the dynamics of the project value in (16.18) via $f(Y_t) = \exp(Y_t)$. For this particular choice of $f(y)$, one gets $\bar{f} = \exp(m + v^2)$, $\Theta = 0$ and $\Omega = \frac{\rho}{v\sqrt{2}}\left(e^{\frac{5}{2}v^2 + 3m} - e^{3m + \frac{9}{2}v^2}\right)$. We fix the parameters at $I = \kappa = 1, r = \eta = 0.10, \delta = 1/\sqrt{200}, v = 1/\sqrt{2}, m = \ln 0.1$ and $\rho = -0.3$. All parameters are as in Fouque et al. (2003c) and McDonald and Siegel (1986).

For these parameter values, the blue solid line in Fig. 16.1 plots the leading-order value term $V_0(s)$ in (16.21) for $s \in [0, S_0^* = 1.34374]$. The dashed blue lines plot the "corrected" value function $V(s) + \delta V_1(s)$ over the $s \in [0, S_0^* + \delta S_1^* = 1.38226]$ region. "Fast" stochastic volatility that is negatively correlated with project value (as is usually the case) decreases the value of the option to invest and postpones optimal investment timing, and the effect is more pronounced the more negative the correlation between the two. The red and green lines demonstrate the effect of increases in m, which increase the "averaged" volatility $\bar{f}$, making the option to invest more valuable, ceteris paribus.

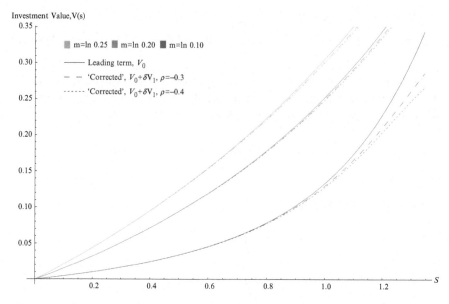

Investment Value, V(s)

m=ln 0.25 ■ m=ln 0.20 ■ m=ln 0.10

——— Leading term, V_0

– – 'Corrected', $V_0 + \delta V_1$, $\rho = -0.3$

······ 'Corrected', $V_0 + \delta V_1$, $\rho = -0.4$

Fig. 16.1 Plots (*solid lines*) the leading-order value term $V_0(s)$ in Eq. (16.21) and the "corrected" value $V(s) + \delta V_1(s)$ (*dashed lines*) for different parameter values. Lines in *blue* ($m = \ln 0.1$), *red* ($m = \ln 0.2$) and *green* ($m = \ln 0.25$) correspond to different long-run levels for the "fast" stochastic volatility latent factor. The rest of the parameters are $I = \kappa = 1$, $r = \eta = 0.10$, $\delta = 1/\sqrt{200}$, $v = 1/\sqrt{2}$ and $f(y)$ is equal to $\exp(y)$

References

Abramowitz, M., Stegun, I.A.: Handbook of Mathematical Functions. Dover Publications, New York (1972)

Alizadeh, S., Brandt, M., Diebold, F.: Range–based estimation of stochastic volatility models. J. Finance **57**(3), 1047–1091 (2002)

Bardi, M., Cesaroni, A.: Optimal control with random parameters: a multiscale approach. Eur. J. Control **17**(1), 30–45 (2011)

Black, F., Scholes, M.: The pricing of options and corporate liabilities. J. Polit. Econ. **81**(3), 637–654 (1973)

Brekke, K.A., Øksendal, B.: Optimal switching in an economic activity under uncertainty. SIAM J. Control Optim. **32**(4), 1021–1036 (1994)

Brennan, M.J., Schwartz, E.S. Evaluating natural resource investments. J. Bus. **58**(2), 135–157 (1985)

Chipot, M.: Elements of Nonlinear Analysis. Birkhäuser, Basel (2012)

Cioranescu, D., Donato, P.: An Introduction to Homogenization. Oxford Lecture Series in Mathematics and Its Applications. The Clarendon Press/Oxford University Press, New York (1999)

Dixit, A.K., Pindyck, R.S.: Investment Under Uncertainty. Princeton University Press, Princeton, NJ (1994)

Duckworth, J.K., Zervos, M.: A model for investment decisions with switching costs. Ann. Appl. Probab. **11**(1), 239–260 (2001)

Fleming, W.H., Soner, H.M.: Controlled Markov Processes and Viscosity Solutions, vol. 25. Springer Science & Business Media, New York (2006)

Föllmer, H., Protter, P., Shiryayev, A.N.: Quadratic covariation and an extension of Itô's formula. Bernoulli, 1(1/2), 149–169 (1995)

Fouque, J.-P., Papanicolaou, G., Sircar, R., Sølna, K.: Multi–scale stochastic volatility asymptotics. Multi-Scale Model. Simul. 2(1), 22–42 (2003a)

Fouque, J.-P., Papanicolaou, G., Sircar, R., Sølna, K.: Short time-scale in S&P 500 volatility. J. Comput. Finance 6(4), 1–23 (2003b)

Fouque, J.-P., Papanicolaou, G., Sircar, R., Sølna, K.: Singular perturbation in option pricing. SIAM J. Appl. Math. 63(5), 1648–1665 (2003c)

Fouque, J.-P., Papanicolaou, G., Sircar, R., Sølna, K.: Multiscale Stochastic Volatility for Equity, Interest Rate, and Credit Derivatives. Cambridge University Press, Cambridge (2011)

Grenadier, S.R.: Valuing lease contracts: a real options approach. J. Financ. Econ. 38, 297–331 (1995)

Hikspoor, S., Jaimungal, S.: Asymptotic pricing of commodity derivative using stochastic volatility spot models. Appl. Math. Finance 15(5 and 6), 449–477 (2008)

Hillebrand, E.: Overlaying time scales in financial volatility data. In: Fomby, T.B., Terrell, D. (eds.) Econometric Analysis of Financial and Economic Time Series, pp. 153–178. Elsevier, Amsterdam (2006)

Karatzas, I., Shreve, S.E.: Brownian Motion and Stochastic Calculus. Springer, Berlin/Heidelberg/New York (1991)

McDonald, R.L., Siegel, D.R.: The value of waiting to invest. Q. J. Econ. 101(4), 707–727 (1986)

Merton, R.C.: The theory of rational option pricing. Bell J. Econ. 4(1), 141–183 (1973)

Metcalf, G.E., Hassett, K.A.: Investment under alternative return assumptions: comparing random walks and mean reversion. J. Econ. Dyn. Control 19, 1471–1488 (1995)

Øksendal, B.K., Sulem, A.: Applied Stochastic Control of Jump Diffusions, vol. 498. Springer, Berlin (2005)

Paddock, J.L., Siegel, D.R., Smith, J.L.: Option valuation of claims on physical assets: the case of offshore petroleum leases. Q. J. Econ. 103(3), 479–508 (1988)

Peskir, G., Shiryaev, A.: Optimal Stopping and Free-Boundary Problems. Springer, New York (2006)

Pindyck, R.S.: Irreversible investment, capacity choice and the value of the firm. Am. Econ. Rev. 78(5), 969–985 (1988)

Pindyck, R.S.: Irreversibility, uncertainty and investment. J. Econ. Lit. 29(3), 1110–1148 (1991)

Schwartz, E.S., Moon, M.: Evaluating research and development investments. In: Brennan, M.J., Trigeorgis, L. (eds.) Project Flexibility, Agency, and Competition : New Developments in the Theory and Application of Real Options, pp. 85–106. Oxford University Press, New York (2000)

Trigeorgis, L.: Real Options: Managerial Flexibility and Strategy in Resource Allocation. MIT Press, Cambridge (1996)

Chapter 17
Modelling the Uruguayan Debt Through Gaussians Models

Ernesto Mordecki and Andrés Sosa

Abstract We model bond price curves corresponding to the sovereign Uruguayan debt nominated in USD, as an alternative to the official bond price publication released by the Central Bank of Uruguay (CBU). Four different Gaussian models are fitted, based on historical data issued by the CBU, corresponding to some of the more frequently traded bonds. The main difficulty we approach is the absence of liquidity in the bond market. Nevertheless the adjustment is relatively good, giving the possibility of non-arbitrage pricing of the whole family of nontraded instruments and also the possibility of pricing derivative securities.

Keywords Uruguayan sovereign bonds • Stochastic processes • Yield curve • Term structure surface • Interest rate models • Forward-rate models • Arbitrage possibilities

17.1 Introduction

Bond price curves (or equivalently yield curves) constitute a major tool in debt analysis and perspective of the sovereign debt of a country. Term structure models have therefore been used in different ways by different classes of market participants. In the monetary policy context, the term structure is an indicator of the market's expectations regarding interest rate and inflation rates. From a financial point of view, the existence of a bond price curve helps the development of the domestic capital market, both for the primary and secondary market.

There are essentially two approaches to model the term structure. The general equilibrium approach starts from a description of the economy and derives the term structure of interest rate endogenously (see for example Cox et al. (1985)). In contrast, the arbitrage approach starts from assumptions about the stochastic evolution of one or more interest rates and derives prices of contingent claims by imposing the no-arbitrage condition; this is, for example, the pioneering approach proposed by Vasicek (1977).

E. Mordecki (✉) • A. Sosa
Mathematics Center, School of Sciences, Universidad de la República, Montevideo, Uruguay
e-mail: mordecki@cmat.edu.uy; asosa@cmat.edu.uy

© Springer International Publishing Switzerland 2016
A.A. Pinto et al. (eds.), *Trends in Mathematical Economics*,
DOI 10.1007/978-3-319-32543-9_17

331

In this first approximation to the problem of the Uruguayan USD nominated debt, we use the second approach, and further we restrict our analysis to four Gaussian models due to their analytical and numerical tractability. Factor models assume that the term structure of interest rate is driven by a set of state variables named as "factors". As one factor generally explains a large proportion of the yield curve movement, it may tempting to reduce the analysis to one factor models. Nevertheless, the consideration of one factor involves, in some cases, perfect correlation between log prices of bonds. Although empirical data shows high correlation, bond prices do not show perfect correlation. Therefore, we use multiple factor model and in particular two factor models.

The first two models we choose are short-rate models, the classical Vasicek model (Vasicek 1977), and the more flexible $G2 + +$ model (Brigo and Mercurio 2006). The parameter estimation is carried out based on maximum likelihood, following (Chen and Scott 1993). The second couple of models belong to the forward-rate model family, based on the methodology proposed by Heath et al. (1992). Here we again choose first the simplest possible one, the Ho–Lee model with constant volatility (Ho and Lee 1986), and second the more flexible Hull–White model with tempered volatility (Hull and White 1993). In these two last cases, we calibrate the models with the help of minimization of squared differences between theoretical and quoted prices of bonds.

Our contribution is to provide an arbitrage-free set of bond prices for the Uruguayan USD debt that can be used for portfolio valuation and derivative pricing.

The rest of the paper is organized as follows. Sections 17.2 and 17.3 describe the short-rate model and the forward-rate models, respectively. Section 17.4 provides information about the Uruguayan debt and the methodology of computation of bond prices used by the CBU. Section 17.5 describes the estimation methodology through maximum likelihood and calibration and presents the results obtained. Section 17.6 concludes.

17.2   Short-Rate Models

When modelling yield curves through stochastic processes, the classical approach consists in modelling the short rate through a certain amount of sources of uncertainty, under the denomination of factors. When these factors are used to construct the short interest rate, we obtain a large variety of models, described in a vast part of the literature in fixed income mathematical finance (Bjork 2009; Brigo and Mercurio 2006; Filipovic 2009; Musiela and Rutkowski 2005).

In this part of the work, we use two Gaussian models, the one proposed by Vasicek in 1977 (Vasicek 1977) that is seminal in the literature. The second model we apply is the G2++ (Brigo and Mercurio 2006), a modification of the previous one, having two factors and including the initial price curve, avoiding in this way arbitrage at the initial time. This model has in fact a more general version, under the denomination of Gn++ model (Di Francesco 2012), including an arbitrary amount of factors. However, the choice of the number of factors involves a compromise between numerically efficient implementation and capability of the model to represent realistic correlation patterns and to fit satisfactorily enough market data.

17.2.1 The Vasicek Model

In his classical paper (Vasicek 1977), Vasicek proposes a model for the short rate through a stochastic differential equation driven by a Wiener process,

$$dr(t) = a(b - r(t))dt + \sigma dW(t) \qquad\qquad r(0) = r_0; \qquad (17.1)$$

where a, b and σ are positive constants and $\{W(t)\}$ is a standard Wiener process defined in a stochastic basis $(\Omega, \mathscr{F}, \{\mathscr{F}_t\}, \mathbf{P})$. The solution to (17.1) is known as the Ornstein–Uhlenbeck process. It defines an elastic random walk around a trend, with a mean-reverting characteristic. Given the set of information at time s, the short rate $r(t)$ is normally distributed with

$$\mathbf{E}\left(r_t | \mathscr{F}_s\right) = r_s e^{-a(t-s)} + b\left(1 - e^{-a(t-s)}\right)$$

$$\mathbf{var}(r_t | \mathscr{F}_s) = \frac{\sigma^2}{2a}\left(1 - e^{-2a(t-s)}\right).$$

The bond price can be obtained by computing the discounted expected terminal value of the bond with respect to a risk-neutral probability measure Q. This quantity can be explicitly computed concluding that the Vasicek model is an affine model whose solution is

$$P(t, T) = A(t, T)e^{-B(t,T)r_t}; \qquad (17.2)$$

where

$$A(t, T) = \exp\left(\left(b - \frac{\sigma^2}{2a^2}\right)(B(t, T) - T + t) - \frac{\sigma^2}{4a}B(t, T)^2\right);$$

$$B(t, T) = \frac{1}{a}\left(1 - e^{-a(T-t)}\right).$$

17.2.2 The G2++ Model

As mentioned above, the price correlation given by the Vasicek model to log prices of different bonds is one. For this reason, as observed prices do not show such high correlations; models with more factors (and more parameters) are considered, expecting to better fit to the observed data. One of these proposals is the Gn++ model, proposed in Di Francesco (2012) that we use with two factors. Another relevant characteristic of this proposal is that it takes into account the whole initial price curve.

The dynamics of the short-rate process in this model is given by

$$r(t) = x(t) + y(t) + \varphi(t), \qquad r(0) = r_0;$$

where the process $x(t)$ and $y(t)$ is driven by a correlated two-dimensional Wiener process $\{W_1(t), W_2(t)\}$, by the equations

$$dx(t) = -ax(t)dt + \sigma dW_1(t), \qquad dy(t) = -by(t)dt + \eta dW_2(t). \qquad (17.3)$$

We have $dW_1(t)dW_2(t) = \rho dt$ where $-1 \le \rho \le 1$ and the constants r_0, a, b, σ, η are positive. The function $\varphi(t)$ is deterministic and well defined in the time interval $[0, T]$, which is added in order to fit exactly the initial zero-coupon curve. Given the set of information at time s, the short rate $r(t)$ is normally distributed with

$$\mathbf{E}\left(r(t)|\mathscr{F}_s\right) = x(s)e^{-a(t-s)} + y(s)e^{-b(t-s)} + \varphi(t);$$

$$\mathbf{var}(r(t)|\mathscr{F}_s) = \frac{\sigma^2}{2a}(1 - e^{-2a(t-s)})$$

$$+ \frac{\eta^2}{2b}(1 - e^{-2b(t-s)}) + \frac{2\rho\sigma\eta}{a+b}(1 - \exp^{-(a+b)(t-s)}).$$

The price at time t of a zero-coupon bond with maturity at time T is

$$P(t, T) = \exp\left(-\int_t^T \varphi(u)du - \frac{1 - e^{-a(T-t)}}{a}x(t)\right.$$

$$\left. - \frac{1 - e^{-b(T-t)}}{b}y(t) + \frac{1}{2}V(t, T)\right); \qquad (17.4)$$

where

$$V(t, T) = \frac{\sigma^2}{a^2}\left(T - t + \frac{2}{a}e^{-a(T-t)} - \frac{1}{2a}e^{-2a(T-t)} - \frac{3}{2a}\right)$$

$$+ \frac{\eta^2}{b^2}\left(T - t + \frac{2}{b}e^{-b(T-t)} - \frac{1}{2b}e^{-2b(T-t)} - \frac{3}{2b}\right)$$

$$+ \frac{2\rho\sigma\eta}{ab}\left(T - t + \frac{e^{-a(T-t)} - 1}{a} + \frac{e^{-b(T-t)} - 1}{b} - \frac{e^{-(a+b)(T-t)} - 1}{a+b}\right). \qquad (17.5)$$

Even though the previous formula (17.4) gives bond prices in the model, it is necessary to estimate the function φ. In order to do this, it is necessary to assume that the initial price curve $T \mapsto P^M(0, T)$ is known. The model fits the currently observed term structure if $P^{Model}(0, T) = P^M(0, T)$; therefore, the price at time t of a zero-coupon bond maturity at time T is

$$P(t, T) = \frac{P^M(0, T)}{P^M(0, t)} \exp(A(t, T)); \qquad (17.6)$$

where

$$A(t, T) = \frac{1}{2}\left(V(t, T) - V(0, T) + V(0, t)\right) - \frac{1 - e^{-a(T-t)}}{a} x(t) - \frac{1 - e^{-b(T-t)}}{b} y(t).$$

17.3 Forward-Rate Models

In the market we do not have a real instantaneous interest rate. In certain cases, the one (or three)-month interest rate series is used as a proxy to estimate this interest rate. It is not convenient to use the overnight rate, as it has very high volatility due to economical factors, as, for instance, the daily liquidity in the market that can give distortions on the structure of the yield curve.

By this reason, and also to avoid arbitrage opportunities in a systematic way, Heath, Jarrow and Morton introduce a new methodology (Heath et al. 1992) (referred to as HJM models) that models the *forward instantaneous rate* at time t by a stochastic differential equation driven by a Wiener process

$$f(s, t) = f(0, t) + \int_0^t \alpha(u, t)du + \int_0^t \sigma(u, t)dW_u;$$

where W_t is a Wiener process.

We depart from the free arbitrage model that assumes that the coefficients $\alpha(u, t)$, $0 \le u \le t$ and $\sigma(u, t), 0 \le u \le t$ are adapted processes defined in an underlying stochastic basis $(\Omega, \mathscr{F}, \{\mathscr{F}_u\}, Q)$, and Q is a risk-neutral probability measure. The key aspect of HJM techniques lies in the recognition that the drift of the no-arbitrage evolution of certain variables can be expressed as functions of their volatilities and the correlations among themselves, giving

$$\alpha(s, t) = \sigma(s, t) \int_s^t \sigma(s, u)du. \qquad (17.7)$$

Therefore, in order to specify an HJM model, the initial forward rate curve $f(0, t)$ and the volatility structure $\sigma(s, t)$ should be given, because no drift estimation is needed. It should be observed that as long as the function σ is deterministic, by condition (17.7), the drift is also deterministic; in consequence, the forward rates are Gaussian.

17.3.1 The Ho–Lee Model and Hull–White Model

In the applications that follow, we first choose the simplest possible alternative for the volatility. We assume $\sigma(s, u) = \sigma > 0$ is a positive constant, and this gives

$$f(s, t) = f(0, t) + \sigma(st - s^2/2) + \sigma W_t;$$

the so-called Ho–Lee model (Ho and Lee 1986). The price of a zero-coupon bond in this model is given by

$$P(t, T) = \frac{P^M(0, T)}{P^M(0, t)} \exp\left((T - t)f^M(0, t) - \frac{\sigma^2}{2}t(T - t)^2 - (T - t)r(t)\right). \quad (17.8)$$

In the second model, we assume that the volatility is also deterministic but in this case is time dependent, given by $\sigma(s, u) = \sigma e^{-a(u-s)}$, where a and σ are positive constants. This gives

$$r(t) = f(0, t) + \frac{\sigma^2}{2a^2}(e^{-at} - 1)^2 + \sigma \int_0^t e^{-a(t-u)} dW_u.$$

The price of a zero-coupon bond in this case is

$$P(t, T) = \frac{P^M(0, T)}{P^M(0, t)} \exp\left(B(t, T)f^M(0, t) - \frac{\sigma^2}{4a}(1 - e^{-2at})B(t, T)^2 - B(t, T)r(t)\right).$$
$$(17.9)$$

where the function $B(t, T)$ is

$$B(t, T) = \frac{1}{a}\left(1 - e^{-a(T-t)}\right).$$

This corresponds to the Hull–White model (Hull and White 1993) with time-dependent parameters.

17.4 About the Uruguayan Debt

The Uruguayan government issues debt through different financial instruments, on different currencies and expirations. In present times, the most relevant circulating instruments according to terms and currencies are:

- the *Treasury Notes* in local currency and linked to CPI (consumer price index),
- the *Local Bonds* (issued in the country) linked to CPI,
- the *Local Bonds* (issued in the country) in USD,
- the *Global Bonds* (issued mainly in United States of America) in USD.

Fig. 17.1 Central
government debt profile

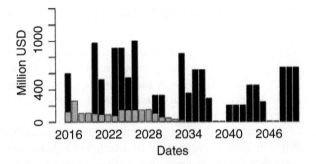

To complete the analysis of the debt, it should be mentioned that the government also holds obligations with multilateral financial institutions, estimated in less than the 10 % of the global debt amount. In Fig. 17.1 we observe the debt's profile of the Uruguayan debt across maturities. The last expiration dates correspond to the 2050 bond, recently issued, paying its face value on thirds, in the years 2048, 2049 and 2050.

Regarding the currency composition, more than one half of the Uruguayan debt is nominated in local currency (part of it linked to CPI), 45 % is issued in USD, and the rest corresponds to bonds issued in Euros and Yens. Almost all of the debt (94 %) pays fixed coupons, and the rest is mainly adjusted to Libor rate.

17.4.1 Curve Price Estimation

In most of the countries, the corresponding regulating agencies and some financial corporations release yield curves corresponding to the sovereign debt. The most popular methods to produce these curves include parametric methods [as the ones proposed by Nelson and Siegel (1987) and Svensson (1994)]. Non-parametric methods are also used.

Regarding Uruguay, the CBU does not release yield curves but daily issues a report with prices for all the financial instruments issued by the government. The two authorized stock exchange institutions release their respective yield curves. The *Electronic Stock Exchange of Uruguay (Bolsa Electrónica de Valores del Uruguay (BEVSA))* uses B-splines to produce the yield curves corresponding to local currency, linked to CPI, and to USD. The *Montevideo Stock Exchange (Bolsa de Valores de Montevideo (BVM))* emits a single curve (linked to CPI) based on the methodology proposed by Svensson (1994).

The purpose of the present paper is to propose an alternative methodology for bond valuation, for the debt issued in USD, to the one used by the CBU and also by the BVM and BEVSA. In the next subsection, we briefly describe the procedure used by the monetary authority (CBU) to compute the prices of the different financial instruments.

17.4.2   CBU Pricing Methodology

The CBU issues daily reports containing prices of all the financial instruments issued by the Uruguayan government. This set of prices is known as the "price vector". The pension funds and insurance companies that are relevant participants in the domestic bond market are obliged by law to use these prices in order to report their respective portfolio values.

This vector of prices is computed through a methodology that includes four different criteria, according to whether the bond has been traded or not and according also to its expiration, and applies with a hierarchical scheme. At the end of each business day, the CBU releases information about bond prices, according to the following procedure:

1. For bonds that have been negotiated in the day (according to certain minimal amounts), the prices are computed as a weighted mean of the respective negotiation prices in the stock exchanges.
2. For bonds with less than a year to expire, an interpolation procedure between the previous price and the face value is applied to compute the prices.
3. An *index* $I(t)$ where t represents the current date is computed in order to compute the prices of the other bonds. This amount represents the value of an "ideal" mean bond.
4. For all other bonds, the new price is obtained from the previous one multiplying by the ratio $I(t)/I(t-1)$.

The most relevant characteristic to be taken into account when analysing this procedure is that the stock exchange of long expiration debt instruments issued by the government (bonds) is really not liquid. This implies that most of the prices published by the CBU result from the application of 4, i.e. are computed instead of negotiated, and this can happen for some instruments consecutively, during a relatively long period of time. This can lead to (theoretical) arbitrage opportunities, for instance, giving larger yields for bonds with smaller maturities, which is equivalent to larger prices of zero-coupon bond with smaller maturity than other zero-coupon bonds. Of course, these arbitrage opportunities do not exist in real negotiations, as they do not follow the released bond prices, but are present, for instance, when evaluating portfolios. Although the intention of the methodology is to follow the movements of the market, one should always take into account this fact.

17.5   Empirical Results

17.5.1   Model Fitting

In relatively developed markets, prices of financial derivatives on bonds are available in order to estimate the parameters of the model; see Brigo and Mercurio (2006),

Filipovic (2009) and the references therein. Our main difficulties are the absence of derivatives market and the lack of liquid market prices of bonds due to the absence of frequent transactions. This is the reason why we develop two different methodologies of model adjustment, first, the classical maximum likelihood estimation, and, second, we propose a calibration procedure for models with fewer parameters that seems to give relatively good results.

Our first approach is closer to Chen and Scott (1993), where model adjustment is carried out departing from bond prices. In our case we do not have the need to introduce measurement errors in order to fit the model, as we develop a first approximation to the data. In the second approach, we perform a conventional calibration model using daily prices of bonds.

17.5.1.1 Maximum Likelihood Estimation

Observe that in formulas (17.2) and (17.6), the log prices are expressed as a linear function of the state variables. In order to determine these state variables in both models, it is necessary to use one bond price time series in Vasicek model and two time series in G2++ model. But to carry out this procedure, the values of the parameters (to be estimated) are necessary. This is why we use the maximum likelihood (ML) method with the help of a change of variables that gives the corresponding Jacobian term in the ML expression. In both models, based on the Markov property of the processes, the joint density is written as a product of conditional densities, each of which has normal distribution, with three parameters in the first case (a, b, σ) and five parameters in the second $(a, b, \sigma, \eta, \rho)$.

Given the panel data set $P_t^M = \left(P^M(t, T^{(i)})\right)$, $i = 1, \ldots, I$ and $t = 1, \ldots, T$, where $P^M(t, T^{(i)})$ is the price at time t of the zero-coupon bond with maturity $T^{(i)}$, denote by $X_t = \left(x_t^{(i)}\right)$, $t = 1, \ldots, T$ the state vector.

The joint density of $P_2^M, \ldots, P_T^M$ satisfies

$$f(P_2^M, \ldots, P_T^M | P_1^M, \Theta) = \prod_{k=2}^{T} f(P_k^M | P_{k-1}^M, \Theta)$$

by the Markov property. Changing variables in each conditional density, we obtain

$$f(P_k^M | P_{k-1}^M, \Theta) = f(X_k | X_{k-1}, \Theta) \left| \frac{\partial X_k}{\partial P_k} \right|$$

$$= f(X_k | X_{k-1}, \Theta) \left| \frac{\partial P_k}{\partial X_k} \right|^{-1} = f(X_k | X_{k-1}, \Theta) \frac{1}{|J_k|},$$

where J_k is the Jacobian of the change of variables.

We find adequate in our situation to use as many bond price time series as number of factors. But this is not strictly necessary. In some cases, when more time series than state variables are used, it is possible to introduce additional random variables in order to estimate the model (Chen and Scott 1993).

17.5.1.2 Calibration

In order to use our models in practice, to price contingent claims or with other aims, we have to calibrate its parameters departing from market data. To do this we adopt a cross-sectional approach. Assume that we observe a cross section of market prices of contingent claims, that is to say, the prices of a set of N zero-coupon bonds corresponding to the same day. For time t, let us denote the vector prices $P_t^M = (P^M(t, T^{(i)}))$, $i = 1, \ldots, I$, where $P^M(t, T^{(i)})$ is the price at time t the zero-coupon bond with maturity at time $T^{(i)}$ and assume that we are able to compute the price vector corresponding to the model, denoted by $P_t^{\text{Model}} = (P^{\text{Model}}(t, T^{(i)}))$, $i = 1, \ldots, I$. Model prices are functions of the parameter vector Θ of the respective models; see (17.8) for the Ho–Lee model and (17.9) for the Hull–White model.

The idea is to find the values for Θ minimizing the difference between market prices and model prices. To do this we have to solve the least-square problem,

$$\min_{\Theta} \frac{1}{I} \sum_{i=1}^{i=I} \left(P^M(t, T^{(i)}) - P^{\text{Model}}(t, T^{(i)}) \right)^2.$$

17.5.2 Results in Short-Rate Models

17.5.2.1 Data

The information used in the construction of the yield curves is the one provided by the CBU, for USD nominated bond prices, traded both in the domestic market and in foreign exchanges. It should be noticed that besides the liquidity problem of the bond market, there are no derivatives on these instruments, so all the available information is provided by the bond prices.

The data to be used in the estimation procedure corresponds to the time period from January 4, 2010 to October 30, 2013. In the Vasicek model, we use the bond *BE330115P*, expiring in January 2033, and, in the G2++ model, we use the same one and add the bond *BE250928F* with expiry in September 2025. We choose these bonds as they are the most frequently traded. Each time series is processed according to the coupon payment scheme (amount and frequency) to obtain the corresponding yields that give the zero-coupon bond prices used in the estimation.

17.5.2.2 Estimation in Vasicek Model

In order to evaluate the influence of the nonliquid computed prices (as explained above), we proceed in this case in estimating the parameters corresponding to the Vasicek model in two situations. We first estimate parameters using all the available information, i.e. we use weekly bond prices taken on Wednesdays. The obtained

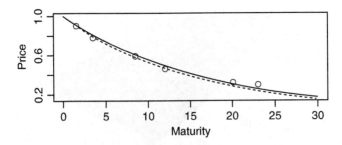

Fig. 17.2 The *solid line* shows the first estimation with Vasicek model and the *dotted* one the second (that uses only negotiated prices). The *dots* correspond to market prices of zero-coupon bonds

parameters in (17.1) are:

$$a = 1.7051; \qquad b = 0.0937; \qquad \sigma = 0.3721.$$

In the second case, we use only negotiated prices (504 observations). This implies that the time intervals between prices are not regular, leading to a slightly more complicated estimation scheme. The obtained parameters in (17.1) are:

$$a = 1.7145; \qquad b = 0.0896; \qquad \sigma = 0.4971.$$

This comparison helps us to verify that the variation in the values of the parameters is not significant (perhaps the most important variation is registered in σ), and for this reason in what follows and in this first approach to the problem, we will use all the available time series. In Fig. 17.2 we observe the price curve corresponding to August 13, 2013, given by the model in both estimations carried out and also the bond prices issued by the CBU.

17.5.2.3 Estimation in G2++ Model

As we mentioned above, we use two bond price time series, with expiration in 2025 and 2033, respectively. We take weekly prices corresponding to Wednesdays. The obtained values for the parameters in (17.3), with the ML estimation method described above, are:

$$a = 0.1300; \quad b = 0.3526; \quad \sigma = 0.2062; \quad \eta = 0.4892; \quad \rho = -0.99.$$

Using these values with the corresponding bond prices, we can obtain the daily time series corresponding to the short rate. This allows us to compute the bond prices for arbitrary maturities in each one of the days. We show the bond price curve for the following maturities: 1, 2, 3, 6 and 9 months and 1, 2, 3, 5, 7, 10, 15, 20 and 25 years.

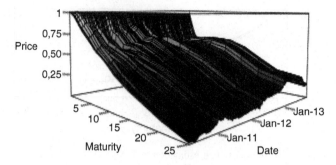

Fig. 17.3 Surface of zero-coupon bond prices using the G2++ model

In Fig. 17.3 we see the daily zero-coupon bond price curve corresponding to the analysed time period issued by the Uruguayan government, as a result of the application of the G2++ model.

If we analyse the data in Fig. 17.3 for the second semester of 2012, the model gives a curve with non-negative slope for some maturities. This fact introduces arbitrage possibilities, as it gives cheaper bonds with smaller maturities than others with larger maturities. More precisely, our model adjustment gives parameters that result in increasing bond prices in some intervals of the maturity T. Our parameter estimations include a correlation very close to -1 and instances of negative values of the factors $x(t)$ and $y(t)$. We will return to this fact in more detail in the next subsection.

17.5.2.4   On Arbitrage Possibilities in G2++ Model

We analyse how formula (17.6) for a zero-coupon bond price in G2++ model can give arbitrage opportunities. Equation (17.5) can be written as

$$V(t, T) = \frac{\sigma^2}{a^2} \int_t^T (1 - e^{-a(T-u)})^2 du + \frac{\eta^2}{b^2} \int_t^T (1 - e^{-b(T-u)})^2 du$$

$$+ \frac{2\rho\sigma\eta}{ab} \int_t^T (1 - e^{-a(T-u)})(1 - e^{-b(T-u)}) du.$$

We then take logarithms and differentiate formula (17.6) w.r.t T, to obtain

$$\frac{\partial}{\partial T} \log P(t, T) = \frac{\partial}{\partial T} \log P^M(0, T) + \frac{\sigma^2}{2a^2} \left((1 - e^{-a(T-t)})^2 - (1 - e^{-aT})^2 \right)$$

$$+ \frac{\eta^2}{2b^2} \left((1 - e^{-b(T-t)})^2 - (1 - e^{-bT})^2 \right)$$

$$+ \frac{\rho\sigma\eta}{ab}\left((1 - e^{-a(T-t)})(1 - e^{-b(T-t)}) - (1 - e^{-aT})(1 - e^{-bT})\right)$$

$$- e^{-a(T-t)}x(t) - e^{-b(T-t)}y(t).$$

In this formula the first three addends are negative. The fourth has the opposite sign of ρ, and the last two depend on $x(t)$ and $y(t)$. In our application ρ is close to -1, which is associated with the fact that $x(t)$ and $y(t)$ take opposite signs; therefore, in some time intervals, the derivative is positive, as shown in Fig. 17.3. Nevertheless the model gives general valuable information about the structure of the debt.

17.5.3   Results for Forward-Rate Models

17.5.3.1   Data

In this analysis we use weekly prices corresponding to the whole year 2014, of 10 USD nominated bonds issued by the CBU. The maximum expiration date corresponds to the *BE451120F* bond, in November 2045. To obtain an approximation of the initial interest rate, we used the reference curve *CUD-BEVSA* daily issued by BEVSA for three months, and, to adjust the initial forward-rate curve, we used the yield curve corresponding to January 5, 2013.

17.5.3.2   Results in Ho–Lee Model and Hull–White Model

Calibrating the Ho–Lee model, we obtain the parameter σ for each day. In Fig. 17.4 we observe then the daily variation of σ. The mean value of the estimation is 0.0232, with a standard deviation of 0.0094.

For the Hull–White model, the same procedure is carried out. In Fig. 17.5 (above), we observe the daily values of the parameter a and in Fig. 17.5 (below) the values of the parameter σ. The mean value for a is 0.0693 with a standard deviation of 0.0257. The corresponding mean and standard deviation for σ are 0.0177 and 0.0079, respectively.

Fig. 17.4 Calibrated Ho–Lee σ parameter over time

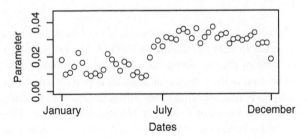

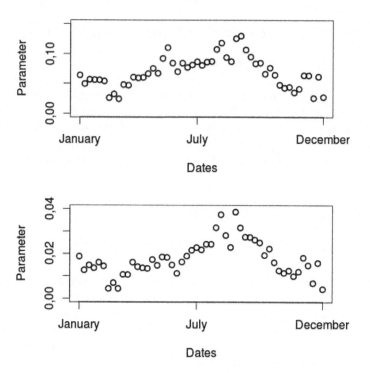

Fig. 17.5 Calibrated Hull–White parameters over time. *Above a, below σ*

The calibration of the Ho–Lee model corresponding to Wednesday, June 25, 2014, gives $\sigma = 0.3071$ and is shown in Fig. 17.6 (above). In Fig. 17.6 (below), we see the calibration for the same day corresponding to the Hull–White model. The parameters in this case are $a = 0.0813$ and $\sigma = 0.0215$.

Both in the Ho–Lee and Hull–White models, the daily calibration allows us to construct term structure of interest rates for the corresponding day. The result is shown in Fig. 17.7.

17.6 Conclusions

The objective of the present work is to present an arbitrage-free model to price the Uruguayan debt nominated in USD. A second purpose is to provide an instrument capable of pricing derivatives on bonds, as interest rate swaps, that are beginning to be used in the Uruguayan market. Based on data from the Uruguayan market for coupon bonds, we adjust four different Gaussian models.

This information is of valuable interest to financial practitioners and policymakers alike. Policymakers monitor expectations of future monetary policy to gauge the effectiveness of their strategy. For practitioners, the availability of accurate interest

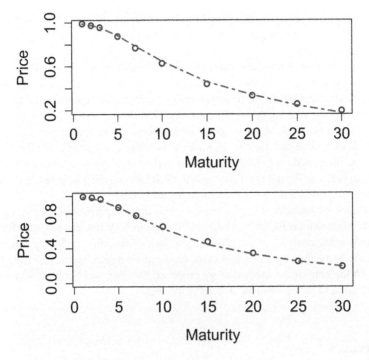

Fig. 17.6 Bond price curve adjusted to market data. *Above* Ho–Lee, *below* Hull–White. In both cases the *dots* represent the market prices

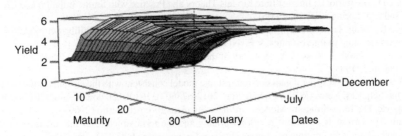

Fig. 17.7 Term structure in Uruguayan market via Hull–White model

rate forecasts can be the key to a successful trading strategy. We hope that these modelling exercises would enrich our understanding of market expectations and improve the understanding of the characteristic behaviour of the term structure of interest rate.

First, we adjust the Vasicek and the G2++ models for interest rates, departing from data from one bond and two bonds, respectively. We follow the methodology proposed by Chen and Scott (1993) that departs from historical data to estimate the parameters of the model through maximum likelihood. The second model is more flexible than the first one, as it involves more parameters, and also adjusts the initial interest rate curve avoiding the possibility of arbitrage. Nevertheless, as we show, when high values of correlation are obtained (and this is our case), the model can give arbitrage situations.

Second, we model the debt through the HJM models. We choose two different situations, assuming first that the volatility is constant, giving rise to the Ho–Lee model, the simplest HJM model, and second we assume an exponentially decreasing volatility, giving rise to the two-parameter Hull–White model. In this case the adjustment seems to be better, as no-arbitrage possibilities appear (the HJM model uses the initial curve and also works under a risk-neutral measure), and the different values for different days in both cases show certain stability. We conclude that HJM models fit better to the data and are capable of describing the whole market structure with information of a series of then more traded bonds.

Further work includes several issues, such as the comparison of the two methods employed (maximum likelihood and calibration) in both interest rate and forward-rate models; the analysis of the effect of the lack of liquidity, including adjustment with respect to transaction volumes in the least-square minimization; the use of more data with the help of the methodology proposed by Chen and Scott (1993); and the consideration of more sophisticated HJM models.

References

Bjork, T.: Arbitrage Theory in Continuous Time, 3rd edn. Oxford Finance, Levin (2009)

Brigo, D., Mercurio, F.: Interest Rate Models Theory and Practice with Smile, Inflation and Credit, 2nd edn. Springer, Berlin (2006)

Chen, R., Scott, L.: Maximum likelihood estimation for a multi factor equilibrium model of the term structure of interest rates. J. Fixed Income 3(3), 14–31 (1993)

Cox, J., Ingersoll, J., Ross, S.A.: A theory of the term structure of interest rates. Econometrica 53, 385–407 (1985)

Di Francesco, M.: A general gaussian interest rate model consistent with the current term structure. Int. Sch. Res. Netw. (ISRN) Probab. Stat. 2012(673607), 16 (2012). doi: 10.5402/2012/673607

Filipovic, D.: Term-Structure Models. Springer Finance. Springer, Berlin (2009)

Heath, D., Jarrow, R., Morton, A.: Bond pricing and the term structure of interest rates: a new methodology for contingent claims valuation. Econometrica 60(1), 77–105 (1992)

Ho, T., Lee, S.: Term structure movements and pricing interest rate contingent claims. J. Finance 41, 1011–1029 (1986)

Hull, J., White, A.: One factor interest rate models and the valuation of interest rate derivative securities. J. Financial Quant. Anal. 28(2), 235–254 (1993)

Musiela, M., Rutkowski, M.: Martingale Methods in Financial Modeling. 2nd edn. Springer, Berlin (2005)

Nelson, C., Siegel, A.: Parsimonious modeling of yield curves. J. Bus. 60(4), 473–489 (1987)

Svensson, L.: Estimating and Interpreting Forward Interest Rates: Sweden 1992–1994. Papers 579. Institute for International Economic Studies, Stockholm (1994)

Vasicek, O.: An equilibrium characterization of the term structure. J. Financ. Econ. 5, 177–188 (1977)

Chapter 18
A Q-Learning Approach for Investment Decisions

Martín Varela, Omar Viera, and Franco Robledo

Abstract This work deals with the application of the Q-learning technique in order to make investment decisions. This implies to give investment recommendations about the convenience of investment on a particular asset. The reinforcement learning system, and particularly Q-learning, allows continuous learning based on decisions proposed by the system itself. This technique has several advantages, like the capability of decision-making independently of the learning stage, capacity of adaptation to the application domain, and a goal-oriented logic. These characteristics are very useful on financial problems.

Results of experiments made to evaluate the learning capacity of the method in the mentioned application domain are presented. Decision-making capacity on this domain is also evaluated.

As a result, a system based on Q-learning that learns from its own decisions in an investment context is obtained. The system presents some limitations when the space of states is big due to the lack of generalization of the Q-learning variant used.

Keywords Reinforcement learning • Q-learning • Portfolio selection • Artificial intelligence script • Machine learning • Finance • Investment decisions • Metaheuristics • Technical analysis

18.1 Introduction

18.1.1 Context

The portfolio selection problem is a highly combinatorial problem with two opposite goals: maximize profit and minimize risk. There are many works that cover this problem with different approaches, being the modern portfolio theory proposed by Harry Markowitz in 1952 (Markowitz 1959), a milestone for further work until today.

M. Varela (✉) • O. Viera • F. Robledo
Facultad de Ingeniería, Universidad de la República, Julio Herrera y Reissig 565, 11300
Montevideo, Uruguay
e-mail: martinv@fing.edu.uy; viera@fing.edu.uy; frobledo@fing.edu.uy

© Springer International Publishing Switzerland 2016   347
A.A. Pinto et al. (eds.), *Trends in Mathematical Economics*,
DOI 10.1007/978-3-319-32543-9_18

From the very existence of the stock exchange, there exists a discussion about the predictability of the market, or more precisely, if the future price of an asset can be predicted. Efficient market hypothesis (Fama 1965, 1970) states that in an efficient market, all existing information is reflected in the price, so the current price is the best measure of its intrinsic value, and as consequence of this, it is impossible to predict its future behavior. Moreover, some argue that the market has enough inefficiencies to get a better profit based on speculation. The technical analysis is a widely used tool by who defend this position, since it provides mechanisms to deal with the historical series of price and volume of transactions, giving more digested information that is used to make market predictions. A strong argument of those who believe that the market behavior can be predicted is the existence of buy and sell behavioral patterns associated with psychological factors.

18.1.2 Motivation

Unlike many decision problems in which obtaining historical information is a difficult obstacle to overcome, in this problem the available information is abundant. It is not only available historical information, but it is possible to obtain near real-time information and for free. Moreover, this availability of information has been rising over time, which has meant that there are many more market participants, not always with a high level of knowledge about financial theories. Furthermore, this widespread growth tends to increase. This fact enforces the theory contrarian to the efficient market hypothesis, in the sense that a major number of participants with limited knowledge about financial theories generate possibilities of more inefficiencies, guided by a behavior more related with human psychology factors that with rationality.

While the application of several techniques and models to solve the portfolio selection problem has been studied, they have not been found in the context of this work, i.e., studies on the application of decision-making techniques that could adapt automatically to market changes. In this sense, reinforcement learning (Sutton and Barto 2000) results in a very interesting technique and its application to this domain seems to be novel.

18.1.3 Scope

This work pretends to advance in the treatment of the complex problem of investment decisions, providing a learning and decision-making tool based on historical and real-time information. Specifically, there is presented a system of recommendation on whether or not invest in a particular asset, being the result of that potential investment the feedback to the system, which allows it to learn and provide better recommendations in the future.

18.1.4   Organization

The work is divided into five sections. In the first one, the application domain is described, covering some important concepts like portfolio management and technical analysis. While portfolio management is mentioned for context reasons, technical analysis plays an important role in this work. In Sect. 18.2, reinforcement learning and particularly Q-learning is explained. In the third section, the aim of this work, the application of Q-learning to investment decisions, is justified and explained with detail. In the fourth, experimental results are discussed, and finally, conclusions and some lines of future work are described.

18.2   About the Application Domain

18.2.1   Portfolio Management Problem

In a market with m stocks, let $v_t = (v_t(1), \ldots, v_t(m))$ be the vector that contains the closing price of the m stocks at the day t. One way to become independent of individual prices of each stock is working with **relative prices** $x_t(j) = v_t(j)/v_{t-1}(j)$, that is, the relationship between the closing prices of two consecutive days. Thus, an investment of \$$d$ in the stock j made between day $t-1$ and day t will result in \$$dx_t(j)$ and is then denoted as $x_t = (x_t(1), \ldots, x_t(m))$ to the vector of relative prices for the m stocks at the day t. A **portfolio** b represents an allocation of weights to stocks, specified proportionally to the amount of money invested. Therefore b is expressed as $b = (b(1), \ldots, b(m))$, where $b(m) \geq 0$ and $\sum_j b(j) = 1$. The **daily return** of a portfolio b subject to the relative prices vector x is $bx = \sum_j b(j)x(j)$, and the **total return** $ret_X(b_1, \ldots, b_n)$ of a sequence of relative prices vectors $X = x_1, \ldots, x_n$ is $\prod_{t=1}^n b_t x_t$. It is called **portfolio selection algorithm** to any strategy to specify a sequence of portfolios (Borodin et al. 2004).

18.2.2   Modern Portfolio Theory

The **modern portfolio theory**, proposed by Harry Markowitz in 1952 (Markowitz 1959), proposes two conflicting objectives: maximize return and minimize risk. In this sense, an investment in a risky asset is only justified by a bigger expected return.

The model assumes that only asset expected return and volatility are relevant to investors. Volatility represents the risk, while the expected return is calculated as the average of historical returns.

The portfolio return is calculated as the weighted sum, according to the relative weight, of the returns of the assets comprising the portfolio. The portfolio volatility is a function of the correlation between assets comprising the portfolio. Expressed in mathematical equations,

$$\text{Expected return: } EV(x_{t+1}(P)) = \sum_i b_t(i)EV(x_{t+1}(i)) \tag{18.1}$$

$$\text{Portfolio variance: } \sigma_t^2(P) = \sum_i \sum_j b_t(i)b_t(j)\sigma_t(i)\sigma_t(j)\rho(i,j) \tag{18.2}$$

$$\text{Portfolio volatility: } \sigma_t(P) = \sqrt{\sigma_t^2(P)} \tag{18.3}$$

The mathematical model of variance minimization proposed by Markowitz is as follows:

$$\begin{cases} \min & \sigma_t^2(P) = \sum_i \sum_j b_t(i)b_t(j)\sigma_t(i)\sigma_t(j)\rho(i,j) \\ \text{s.a.} & \\ & \sum_i b_t(i)EV(x_{t+1}(i)) \geq 0 \\ & \sum_i b_t(i) \leq 0 \\ & b_t(i) \geq 0 , \forall i \end{cases} \tag{18.4}$$

In this model, a quadratic problem must be solved. The objective is to minimize the portfolio variance while the expected return acts as a constraint.

18.2.3 Technical Analysis

Technical analysis is the study of market action, primarily through the use of charts, for the purpose of forecasting future price trends, Murphy (1999).

The market action includes the three main sources of information available for technicians: price, volume, and open interest (this is only used for futures and options).

Technical analysis is based on three premises:

1. Market action discounts everything.
2. Price moves in trends.
3. History tends to repeat itself.

The first premise indicates that any factor (financial, psychological, political, etc.) that could affect an asset price is already reflected on price, so the study of price action is enough to make right forecasts.

Based on the second premise, the objective is to identify trends early, so as to trading in the trend direction.

The study of market behavior is concerned with the study of human psychology. In a century of information about market behavior, behavioral patterns of buying and selling have been identified. Since these patterns worked well in the past, it is assumed that they will work well in the future. These patterns are based on the study of human psychology, which tends not to change. The third premise is based on this foundation.

There are, of course, some criticisms of the technical approach:

1. *Self-fulfilling prophecy*: This criticism suggests that given the growth in the use of these techniques in recent years, there are many traders using them, taking similar decisions massively, affecting the market and thus generating the expected movement. This statement seems somewhat simplistic in that it assumes that based in the same graph, the vast majority of traders will act the same way. Typically, the information resulting from technical analysis is not so clear, and every analyst makes a subjective interpretation. There are too many indicators and techniques as to assume that technical analysts behave in the same way.
2. *Can the past be used to predict the future?* Any known prediction method, whatever its application domain, is based on past data. There not exists another source of information to predict the future, that is, the knowledge about the past. This criticism is equally applicable to any prediction mechanism, including fundamental analysis. Regardless, what determines the price is the relation between supply and demand. Supply and demand respond to belief of investors about the future behavior of the asset. Investor beliefs rely on their knowledge, which is directly related to the past. Therefore, it seems reasonable to think that past experiences are a very good base for predicting future behavior.

18.3   Reinforcement Learning

18.3.1   Classification of Application Domain

Russell and Norvig (1995), Russell and Norvig (2003), and Russell and Norvig (2010) categorize application domains for decision-making systems based in the following properties:

- **Fully observable** or **partially observable**: If it is possible to know the complete state of the system at each point in time, then the domain is fully observable. If, instead, just some relevant information can be accessed, the domain is partially observable. If it is not possible to obtain any information about the domain at all, the domain is **unobservable**.
- **Deterministic** or **stochastic**: If the next state of the system is completely determined by the current state and the taken action, the domain is deterministic. Otherwise, it is stochastic.
- **Episodic** or **sequential**: An application domain is episodic if the experience can be divided into "episodes". Actions executed on an episode do not influence on the states of the following episodes. The last state of an episode is called **terminal state**. In sequential domains, the current decision could affect all future decisions.

- **Static** or **dynamic**: If the system state can change during the decision-making process, the domain is dynamic. Otherwise, it is a static domain.
- **Discrete** or **continuous**: If the union of the sets of state variables and possible actions result in a discrete set, then the domain is also discrete. If it is not the case, the domain is continuous.

The most difficult domains to work are those partially observable (or unobservable), stochastic, sequential, dynamic, and continuous.

18.3.2 Definition

The problem of reinforcement learning (Russell and Norvig 1995, 2003, 2010; Sutton and Barto 2000) is a simplistic abstraction of the interactive learning to achieve a goal. In this model, the apprentice and decision-maker is called agent, while all that which interact with the agent is called **environment**.

Reinforcement learning (Russell and Norvig 1995, 2003, 2010; Sutton and Barto 2000) tries to relate situations with actions in order to maximize a reward. Unlike supervised learning, where learning is based on samples provided by an external supervisor, here the agent is not told about which action must be taken in a determined situation, but is the agent itself who must discover which actions generate better rewards in each state, following a trial-and-error approach. This property and the fact that a reinforcement, positive or negative, is not associated with one only action but with a sequence of actions, are the most important factors to distinguish reinforcement learning from another learning techniques.

A reinforcement learning agent can start to work without previous knowledge, taking random decisions at first and learning from the reinforcements received as result of its actions. To make this possible, it is essential to maintain a balance between exploitation and exploration. In other words, as the system must learn from its own actions, sometimes it has to sacrifice an immediate positive reinforcement in order to explore the space of states with the hope of finding even better reinforcements.

A reinforcement learning system is composed by four main elements:

- **Policy**: Defines the agent behavior at a given time. It is a mapping between states and actions to be taken in those states. The policy may be a simple lookup table or it may involve a complex computational process.
- **Reward function**: The reward function is where the objective of a reinforcement learning problem is defined. This function maps the states or state-action pairs with a numerical value called reward. The intrinsic goal of a reinforcement learning agent is to maximize the total reward it receives in the long run. The reward function must be fixed, but the rewards received by it can be used to modify the policy.
- **Value function**: The value function estimates the goodness of a state in the long run, that is, the expected future accumulated reward starting from a given state.

State values estimation has a central role in reinforcement learning since the agent must try to reach those states with bigger value.

- **Model of the environment** (optional): The model of the environment tries to explain its behavior. Using this model the agent could predict the next state and the reward it will receive. The model of the environment is useful to incorporate planning into the agent decision-making process.

18.3.3   Agent-Environment Interaction

The agent and the environment interact continuously, the agent selecting actions and the environment responding to that actions and presenting new situations to the agent. The environment answer to the actions taken by the agent in form of rewards, numerical values that the agent will seek to maximize in the long run.

Interaction between the agent and the environment can be divided in a sequence of discrete steps. In every time step t, the agent receives a representation of the state of the environment, $s_t \in S$ (S is the set of possible states), and has to choose an action $a_t \in A(s_t)$ ($A(s_t)$ is the set of possible actions that can be taken in the state s_t). In the next time step $(t + 1)$, partially as a consequence of the action a_t, the agent receives a numerical reward, $r_{t+1} \in R$, and a representation of the new state of the environment s_{t+1}. A diagram of this interaction is showed in Fig. 18.1.

The rewards give an indication about how good or bad are the actions taken earlier. The agent implicit objective is to maximize accumulated rewards in the long run. The mechanism of rewards indicates to the agent the goals it must pursue, but not how. This is a distinguishing factor of reinforcement learning.

Formally, the goal of a reinforcement learning agent is to maximize the **expected return**, where the return R_t is defined as a function of the sequence of returns. The simplest example is calculating the return as a sum of rewards: $R_t = r_{t+1} + r_{t+2} + \ldots + r_T$, where T is a final time step. This calculation has sense when the problem can be divided into independent subsequences called episodes. When the problem cannot be divided into episodes and the interaction between the agent and

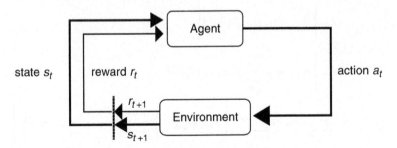

Fig. 18.1 Interaction between the agent and the environment in a reinforcement learning problem (Sutton and Barto 2000)

the environment continues to infinity, this equation is not valid because there is not a final time step T. A solution to this issue is to add the concept of **discount**. Thus, the agent will try to maximize the expected **discounted return**. The discounted return is expressed by the following equation: $R_t = r_{t+1} + \gamma r_{t+2} + \gamma^2 r_{t+3} + \ldots = \sum_{k=0}^{\infty} \gamma^k r_{t+k+1}$, where $\gamma \in [0, 1]$ is the **discount factor**, a parameter that adjusts the value of a future reward depending on the number of time steps to get the reward.

18.3.4 Value Functions

Reinforcement learning algorithms are, in the most, based on estimating **value functions**. Value functions depend on state (or state-action pair) and give an estimation about the goodness of reaching a particular state (or the goodness of taking a particular action in a particular state). The goodness is based on the expected future rewards, which depend on the actions taken by the agent. Therefore, the value functions are defined regarding particular policies.

A policy π maps for every pair of state $s \in S$ and action $a \in A(s)$, the likelihood of taking the action a in the state s. The value of a state s under a policy π, $V^{\pi}(s)$, is the expected return starting in the state s and following from there the policy π.

Formally, $V^{\pi}(s) = E_{\pi}\{R_t \mid s_t = s\} = E_{\pi}\{\sum_{k=0}^{\infty} \gamma^k r_{t+k+1} \mid s_t = s\}$, where $E_{\pi}\{\}$ denotes the expected value following the policy π. V^{π} is called **state-value function for policy** π.

Similarly, the value of selecting a particular action a in a state s under a policy π, $Q^{\pi}(s, a)$, is defined as the expected return starting in the state s, selecting the action a, and then following the policy π: $Q^{\pi}(s, a) = E_{\pi}\{R_t \mid s_t = s, a_t = a\} = E_{\pi}\{\sum_{k=0}^{\infty} \gamma^k r_{t+k+1} \mid s_t = s, a_t = a\}$. Q^{π} is called **action-value function for policy** π.

The value functions V^{π} and Q^{π} can be estimated from experience.

A fundamental property of value functions is that satisfy the **Bellman equation for** V^{π}:

$$V^{\pi}(s) = E_{\pi}\{R_t \mid s_t = s\}$$

$$= E_{\pi}\left\{\sum_{k=0}^{\infty} \gamma^k r_{t+k+1} \mid s_t = s\right\}$$

$$= E_{\pi}\left\{r_{t+1} + \gamma \sum_{k=0}^{\infty} \gamma^k r_{t+k+2} \mid s_t = s\right\}$$

$$= \sum_{a \in A(s)} \pi(s, a) \sum_{s' \in S} P_{ss'}^a \left[R_{ss'}^a + \gamma E_{\pi}\left\{\sum_{k=0}^{\infty} \gamma^k r_{t+k+2} \mid s_{t+1} = s'\right\}\right]$$

$$= \sum_{a \in A(s)} \pi(s, a) \sum_{s' \in S} P_{ss'}^a [R_{ss'}^a + \gamma V^{\pi}(s')] \tag{18.5}$$

where $P_{ss'}^{a} = P\{s_{t+1} = s' \mid s_t = s, a_t = a\}$ is the likelihood of reaching the state s' from the application of the action a in the state s. Similarly, $R_{ss'}^{a} = E\{r_{t+1} \mid s_t = s, a_t = a, s_{t+1} = s'\}$ is the expected value of the reward to obtain for reaching the state s' after applying the action a in the state s.

Bellman equation expresses a relation between the value of a state and the value of the successor states, weighing every possible future reward with the likelihood of obtaining it, considering the discount factor.

18.3.5 Temporal Difference Learning

Temporal difference (TD) (Fama 1965) is a combination of Monte Carlo and dynamic programming ideas.

TD and Monte Carlo have in common the use of experience to adjust its estimation of state values V. Monte Carlo methods wait to know the reward following the visit to a state, to update the value $V(s_t)$ using this return as an objective value of $V(s_t)$. An example of a Monte Carlo method is $V(s_t) \leftarrow V(s_t) + \alpha[R_t - V(s_t)]$, where R_t is the return obtained after time t and α is a learning factor that weighs the influence of a particular learning instance over the estimated state value $V(s_t)$. This method is called *constant-α MC*.

Unlike Monte Carlo methods, which must wait until the end of an episode to update $V(s_t)$ because the value of R_t is unknown till then, TD methods just have to wait until the next time step. The TD method known as **TD(0)** updates the value $V(s_t)$ according to the following equation:

$$V(s_t) \leftarrow V(s_t) + \alpha[r_{t+1} + \gamma V(s_{t+1}) - V(s_t)] \qquad (18.6)$$

As can be observed, TD uses $r_{t+1} + \gamma V(s_{t+1})$ as an alternative of the Monte Carlo method objective R_t, based on the following deductive logic:

$$V^{\pi}(s) = E_{\pi}\{R_t \mid s_t = s\}$$

$$= E_{\pi}\left\{\sum_{k=0}^{\infty} \gamma^k r_{t+k+1} \mid s_t = s\right\}$$

$$= E_{\pi}\left\{r_{t+1} + \gamma \sum_{k=0}^{\infty} \gamma^k r_{t+k+2} \mid s_t = s\right\}$$

$$= E_{\pi}\{r_{t+1} + \gamma V^{\pi}(s_{t+1}) \mid s_t = s\} \qquad (18.7)$$

As $V^{\pi}(s_{t+1})$ is not known at time t, TD uses its current estimation $V_t(s_{t+1})$ instead of it.

So, TD methods combine the sample-based learning of Monte Carlo methods with the iterative logic of dynamic programming, which has the advantage versus Monte Carlo methods of not requiring waiting until a terminal state to learn, and versus dynamic programming of not requiring a model of the environment.

18.3.6 Q-Learning

The Q-learning algorithm (Watkins 1989) is a control algorithm based on TD learning. Instead of estimating every state value, this algorithm estimates every state-action pair value. State-action pair values are updated according to the following equation:

$$Q(s_t, a_t) \leftarrow Q(s_t, a_t) + \alpha \left[r_{t+1} + \gamma \max_{a \in A(s_{t+1})} Q(s_{t+1}, a) - Q(s_t, a_t) \right] \qquad (18.8)$$

The Q-learning algorithm pseudocode is shown below:

Algorithm 1 Q-learning algorithm pseudocode

Initialize $Q(s, a)$ arbitrarily
for each episode **do**
    Initialize s
    repeat
        Take action $a \in A(s)$ based on $Q(a, s)$ values (e.g., ϵ-greedy)
        Observe r and s'
        $Q(s, a) \leftarrow Q(s, a) + \alpha \left[r + \gamma \max_{a' \in A(s')} Q(s', a') - Q(s, a) \right]$
        $s \leftarrow s'$
    until s is terminal
end for

This algorithm converges to optimal policy and optimal state-action pair values insofar as every state-action pair is evaluated infinite times and that the policy converges to the greedy policy (eliminating exploration).

18.4 Q-Learning for Investment Decisions

18.4.1 Introduction

The portfolio selection problem can be divided into several subproblems. Some are:

1. Determining the expected return of a particular asset
2. Determining the risk associated with a particular asset
3. Determining the best portfolio composition in order to maximize the expected return and minimize the risk

Each subproblem mentioned above can imply a lot of work given the complexity of them. The scope of this work is a mix of subproblems 1 and 2, looking to develop a system capable of deciding if it is convenient or not to invest in a particular asset. While this approach differs from subproblems 1 and 2 in that the system does not return an expected return neither a risk associated with the investment, these elements are implicitly considered by the system in order to determine the investment convenience.

In addition to building a system capable to make good recommendations, the learning capability of the system has a big emphasis in this work. It is pretended to build an agent capable of learning from the results of its own recommendations, improving in this way its future decisions.

In this sense, reinforcement learning, and particularly the Q-learning variant, offers an appropriated theoretical framework for the treatment of this problem. Here are some properties of this learning and decision-making mechanism that make it appropriate to be applied in this context:

- It is suitable for stochastic problems
- It is suitable for non-episodic problems
- It is suitable for dynamic problems
- It is suitable for nonstationary problems
- It is possible to take decisions without prior knowledge
- Adaptation capability (continuous learning without between training and execution)

The application of Q-learning for investment decisions implies decisions at three levels: the generation of a model of the environment, determination of a learning mechanism, and determination of a decision-making mechanism.

18.4.2   Model of the Environment

Defining the model of the environment implies, basically, selecting the variables that compose the state of the environment and the set of actions that the agent can take in each state. The system of reward determination can also be considered part of the model of the environment.

In this model it is considered that the agent acts daily, so the time is discrete and every time step represents 1 day.

As the agent will make investment recommendations on a single asset, the state variables are the result of different technical analyses applied to this asset. The available data to define the state variables are those that summarize the activity of a day: Open, Close, Low, High, and Volume.

A lot of technical analysis and even combination of these were analyzed, but finally, after studying the individual behavior of each analysis and the correlation between them, the following set of 18 variables were selected to compose the state of the environment:

- Closing price and moving averages of closing price

  - Closing price is higher or lower than previous closing price.

- The 20-day moving average goes up or down from the previous day.
- The 50-day moving average goes up or down from the previous day.
- The 200-day moving average goes up or down from the previous day.

- Relative strength index (RSI)

  - RSI value is in the interval $[0, 30)$, $[30, 70]$, or $(70, 100]$.
  - RSI value goes up or down from the previous day.

- Moving average convergence/divergence (MACD)

  - A range is determined from the minimum and maximum MACD values in the last year. MACD value is positive and greater than 80 % of the defined range, is positive and less than 80 % of the defined range, is negative and greater than 20 % of the defined range, or is negative and less than 20 % of the defined range. That is, from the range, MACD value is in the interval $[-\infty, \text{range} \times 0.2)$, $[\text{range} \times 0.2, 0)$, $[0, \text{range} \times 0.8]$, or $[\text{range} \times 0.8, \infty)$.
  - MACD value goes up or down from the previous day.
  - MACD histogram value goes up or down from the previous day.
  - MACD signal line value goes up or down from the previous day.

- Stochastic oscillator (K%D)

  - Slow %K value is in the interval $[0, 20)$, $[20, 50)$, $[50, 80]$, or $(80, 100]$.
  - Slow %K value goes up or down from the previous day.
  - Slow %D value goes up or down from the previous day.

- Slopes of closing price moving averages (slopes are calculated as the difference between current value and 10 days before value)

  - The 10-day moving average of the slope of the 20-day moving average of closing price goes up or down from the previous day.
  - The 25-day moving average of the slope of the 50-day moving average of closing price goes up or down from the previous day.
  - The 100-day moving average of the slope of the 200-day moving average of closing price goes up or down from the previous day.

- Open, Close, Low, High

  - Close is higher or lower than Open.

- Volume

  - Volume of transactions goes up or down from the previous day.

An additional variable is added to indicate if the agent has money invested on the asset or it has not.

The actions that the agent can take are three:

- Buying: Implies investing all available money on the asset
- Selling: Implies selling the totality of the asset shares
- Do nothing: The agent maintains the number of shares of the previous step

At each time step, the agent can choose an action depending on the investment state. If the agent has the money invested on the asset, it can sell the shares it owns or do nothing. If the agent has not invested the money, it can buy shares or do nothing. So, at each time step, the agent can choose one of two possible actions: *sell, donothing* or *buy, donothing*.

The reward system selected is based on the investment return. It is not considered the money that the agent win or lose, but the return of the investment measured as $ret_t \leftarrow (Close_t/Close_{t-1}) - 1$. The agent is rewarded based on this equation only when it owns shares of the asset.

18.4.3   Learning Mechanism

The learning mechanism emerges from the subject of this work: Q-learning. However, there are alternatives for Q-learning application. The classical variant allows to work only with a discrete space of states and actions, while the problem being solved is continuous in nature. While there is research on the adaptation of Q-learning for the treatment of continuous problems (i.e., the use of neural networks for estimating Q-values), it was understood in this work that, for a first approach to this problem using Q-learning, it was more convenient to apply the classical variant of this learning technique, which is based on the storage of the Q-value of each state-action pair, considering that it has a stronger theoretical framework, taking into account its limitations for continuity and state generalization treatment. So, the learning of each state-action pair is based on the following equation:

$$Q(s_t, a_t) \leftarrow Q(s_t, a_t) + \alpha \left[r_{t+1} + \gamma \max_{a \in A(s_{t+1})} Q(s_{t+1}, a) - Q(s_t, a_t) \right] \quad (18.9)$$

where s_t and a_t are the state and the chosen action at time t, respectively, r_{t+1} is the reward obtained at time $t + 1$, and s_{t+1} and $A(s_{t+1})$ are the state and the set of possible actions to be taken at time $t + 1$, respectively, α the learning factor and γ the discount factor.

18.4.4   Decision-Making Mechanism

There are several alternatives for decision-making, being the equilibrium between exploitation and exploration a fundamental factor. It was selected in this work as an adaptation of an algorithm based on Ant Colony Systems (Dorigo and Gambardella 1997), which are widely used to solve transportation problems and their learning mechanism can be classified as reinforcement learning.

The adaptation made comes to solve the issue of negative reinforcements, which are not considered on its original formulation. The adaptation consists of dividing the set of possible actions in two sets, one containing the actions with positive Q-values and the other containing the actions with negative Q-values. Then, with a probability p (a system parameter), a positive Q-value action is selected, and with probability $1 - p$, a negative Q-value action is selected. On each case, positive or negative, the action is taken with likelihood proportional to its Q-value. Dorigo and Gambardella (1997) also propose on their work a heuristic factor designed for transportation problem. This factor is not considered in this work.

The pseudocode of the decision-making algorithm proposed on this work is shown below:

Algorithm 2 Decision-making algorithm pseudocode proposed in this work

---

$r = $ A random number between 0 and 1

if $r \leq 1 - \epsilon$ **then**

  Select $a \in A(s)$ such that $Q(s, a) = \max_{a' \in A(s)} Q(s, a')$

else

  Select $a \in A(s)$ as follows

  $A^+(s) = \{a \in A(s) \mid Q(s, a) \geq 0\}$ (set of actions with positive Q-value)

  $A^-(s) = \{a \in A(s) \mid Q(s, a) < 0\}$ (set of actions with negative Q-value)

  if $A^+(s)$ and $A^-(s)$ are not empty **then**

    With probability p select $a \in A^+(s)$ with probability $Q(s, a) / \sum_{a' \in A^+(s)} Q(s, a')$

    With probability $1 - p$ select $a \in A^-(s)$ with probability $\frac{1}{-Q(s,a)} / \sum_{a' \in A^-(s)} \frac{1}{-Q(s,a')}$

  else

    if $A^-(s)$ is empty **then**

      Select $a \in A^+(s)$ with probability $Q(s, a) / \sum_{a' \in A^+(s)} Q(s, a')$

    else

      Select $a \in A^-(s)$ with probability $\frac{1}{-Q(s,a)} / \sum_{a' \in A^-(s)} \frac{1}{-Q(s,a')}$

    end if

  end if

end if

As can be seen, parameters ϵ and p determine the exploration level of the agent.

18.5 Experimental Results

The tests performed on the developed system had two main objectives:

1. Evaluating the learning capacity
2. Evaluating the system ability to take good decisions

For the first objective, the analysis is centered on the agent behavior while it is acquiring more experience. To make this analysis, a graph of the average of rewards obtained as a function of experience is used.

For the second objective, it is necessary to make a comparison against some measure that belongs to the application domain. In this sense, a widely used measure is the result of applying the B&H strategy. The graph used to make this analysis shows the economic profit made by the Q-learning agent (Q-agent) versus the economic profit following the B&H strategy.

Some assumptions were made in order to simplify evaluation:

- Actions taken by the agent are always executed.
- Transaction costs were not considered.
- The asset price does not change while the action of buying and selling is being executed.

Preliminary tests were performed in order to determine the values of the Q-agent parameters. The results showed here are based on the following parameter values: $\alpha = 0.1, \gamma = 0.9, \epsilon = 0.5$, and $p = 0.9$.

The historical data used correspond to the Dow Jones Industrial Index between October 2, 1930 and December 12, 2010. The data was obtained from the Yahoo Finance Database (2011).

The tests were performed on a Toshiba Satellite L305-SP6924R Laptop, with an Intel Pentium Dual CPU T3400 2.16 GHz processor and 4 GB RAM, using Windows Vista 64 bits as operative system.

18.5.1   First Evaluation: Simultaneous Execution and Learning

In the first evaluation of the system, the agent began to take decisions from the first data of the time series (October 2, 1930), learning dynamically from the results of its own decisions. Therefore, the complete set of historical data was used in this evaluation.

Figure 18.2 shows the average of rewards obtained as a function of the experience.

As can be observed, the agent begins taking actions with negative reward, and as it gains experience, the rewards increase. A very important observation is that the average of rewards is positive once it stabilizes, so the agent achieves profit.

Figure 18.3 compares the accumulated return obtained by the Q-agent versus the return obtained by the B&H strategy.

A first hypothesis about the difference between both mechanisms could be that the Q-agent begins without knowledge, taking bad initial decisions and affecting in this way the accumulated return. But nevertheless, if the graph is cautiously observed, it can be seen that the accumulated returns do not show big differences until some point after the middle of the considered period. At this point the average of rewards received by the agent was already stable.

Analyzing more deeply the agent Q behavior, it was detected that the agent did not repeat each state many times, affecting clearly its learning capacity.

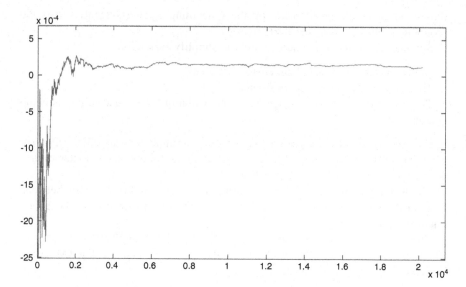

Fig. 18.2 Average of rewards obtained by the agent as a function of the experience

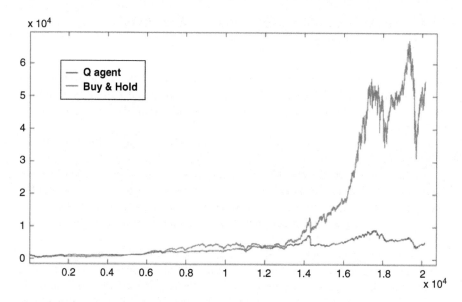

Fig. 18.3 Return obtained by the Q-agent against B&H

18.5.2   Second Evaluation: Training and Execution

As a way to solve the issue of the very little repeat level of states achieved by the agent, the historical data series was divided in two subsets: training data (10/02/1930–12/31/2004) and validation data (01/03/2005–12/31/2010).

This evaluation implies using the training set several times, with the goal of achieving a greater level of state repeat, and thus being able to learn more. This procedure introduces a risk of overtraining. That is why an independent set of data (validation data) is used to evaluate the agent behavior.

The training was therefore divided in episodes. In each episode the agent goes through the training set taking decisions and learning. The agent behavior was analyzed as a function of the number of training episodes.

Figure 18.4 shows the average of rewards received by the agents as a function of experience, for different numbers of training episodes.

As can be observed, as the number of training episodes increases, also the average of rewards increases. This fact shows that the agent is learning.

Figure 18.5 shows the accumulated returns obtained by the agent at different numbers of training episodes. These returns are compared against the B&H strategy.

As can be observed, the accumulated return obtained by the Q-agent fails to overcome the accumulated return obtained by B&H, but it does as the number of episodes increases. The behavior of the accumulated return as a function of the number of episodes is another indicator that the agent achieves learning.

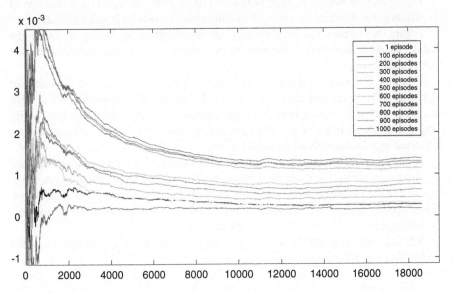

Fig. 18.4 Average of rewards as a function of experience, according to the number of training episodes

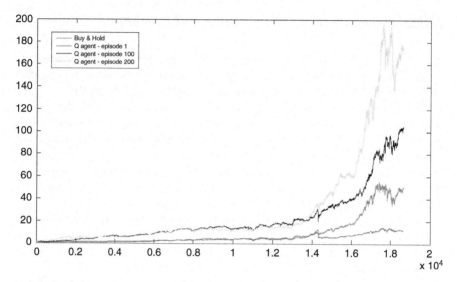

Fig. 18.5 Accumulated returns obtained by the Q-agent (at episodes 1, 100, and 200) compared against B&H

After verifying that the agent learns to make better decisions as it trains many times using the same set of data, the next step is to evaluate the system using the validation data.

Figure 18.6 shows the average of rewards received by the agents as a function of experience, for different numbers of training episodes using the validation data.

Figure 18.7 shows the accumulated returns obtained by the agent using the validation data, according to the number of training episodes and their comparison with B&H strategy.

From the observation of the last two graphs, it can be deduced that the behavior achieved with the training data is not achieved over the validation data. Unlike what happens with the training data, here the average of returns does not necessarily improve as the number of training episodes increases. The same behavior can be observed in relation with accumulated return.

A deeper analysis about the agent behavior showed that the most states reached by the agent during the evaluation phase were unknown to the agent (were never reached in the training phase), so the agent is making decisions without knowledge. This explains the excellent behavior achieved over the training data and the average performance obtained over the validation data.

The explanation about the unknown states is on the lack of generalization of the Q-learning algorithm used.

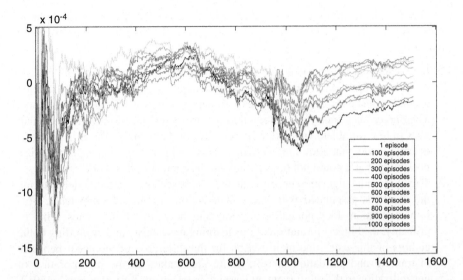

Fig. 18.6 Average of rewards as a function of experience, according to the number of training episodes, using validation data

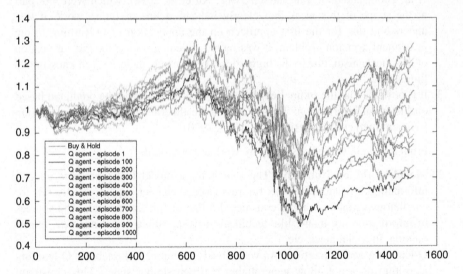

Fig. 18.7 Accumulated returns obtained by the Q-agent using validation data compared against B&H

18.6   Conclusions and Further Research

It was built as a learning and decision-making system based on Q-learning. This system was applied in an investment recommendation context using a discrete model. It was empirically shown that the agent achieves learning when it is applied with models that contain enough information of the environment.

However, it was not possible to generalize the learning achieved by the agent to future situations. According to the authors of this work, this may be due to several reasons:

1. It was not possible to generate a discrete model of the environment generic enough (composed for a little number of variables) to represent the information needed to make a forecast. In the experiments made, many models were tested. The bigger ones (many variables) did not allow generalization, while the small ones (a few variables) were not representative of the domain. The existence of a discrete model that allows the system to learn and generalize between states is a possibility, but it could not be generated in the context of this work.
2. The Q-learning algorithm used in this work is based on discrete models, storing the Q-values associated with every state-action pair in a lookup table. This implementation does not allow generalizing between similar states. This is, probably, the biggest limitation of this learning mechanism and, according to the authors of this work, the main reason for the difference of the agent behavior in training and validation phases. There are studies on the incorporation of generalization to the reinforcement learning mechanism (Bertsekas and Yu 2012; Maei et al. 2010; Precup et al. 2000; Rafols et al. 2005; Sutton 1999a,b; Sutton et al. 2000, 2009a,b; Van Hasselt 2012; Xu et al. 2014), which were not part of the scope of this work, but clearly indicate a line of future work. It was understood that for the first approach on the application of Q-learning to the investment decision problem, it was more convenient to use the tabular version of this mechanism, due to the bigger level of research about it, even knowing its limitations.
3. It is possible that the future behavior of asset price could not be predicted based on historical information. This is one of the theories about market prediction [efficient market hypothesis Fama (1965, 1970)].

From this work some lines of future work are proposed:

1. **Model of the environment**: The search for a model that represents enough information to determine the future market behavior, based on discrete or continuous variables, can be considered a line of further research. The sources of information are many, like technical analysis, fundamental analysis, expert opinion, financial news, etc.
2. **Q-learning generalization**: As was already mentioned, the tabular Q-learning algorithm does not allow generalizing between similar states. This constraint could be eliminated if another mechanism of Q-values estimation is considered. One possible solution is using function approximation techniques. In this way, instead of using Q-values stored in a table, there are estimated values based on a function built by approximation from a set of samples. These samples could be the own experience of agent. An interesting example of function approximation technique that could be applied to add generalization to Q-learning is neural networks. Some examples in this line of research can be found in Bertsekas and Yu (2012), Maei et al. (2010), Precup et al. (2000), Rafols et al. (2005),

Sutton (1999a), Sutton (1999b), Sutton et al. (2000), Sutton et al. (2009a), Sutton et al. (2009b), Van Hasselt (2012), and Xu et al. (2014). This line of future work includes the treatment of higher-dimensional models of the environment.

3. **Portfolio selection strategies**: This work was limited to the study of the application of continuous learning and decision-making mechanisms for investment recommendations on a single asset. How to insert a system like this into a portfolio selection strategy, considering recommendations on many assets as well as risk measures, can be also considered a particular line of further research.

4. **Application of reinforcement learning to other financial problems**: Reinforcement learning, and particularly Q-learning, offers a very interesting theoretical framework for the treatment of decision problems oriented to short-term results. Financial problems, in general, have this characteristic. But nevertheless, it was not detected in the context of this work, the existence of an important number of works about the application of this mechanism in this field.

5. **Techniques to avoid overfitting**: When the same set of training data is used to train the system, like it was made in the present work, overfitting is a logical consequence. The use of techniques to avoid overfitting can give to this work a more complete coverage. Some ideas can be found in Whiteson et al. (2011).

References

Bertsekas, D.P., Yu, H.: Q-Learning and enhanced policy iteration in discounted dynamic programming. Math. Oper. Res. **37**(1), 66–94 (2012)

Borodin, A., El-Yaniv, R., Gogan, V.: Can we learn to beat the best stock. J. Artif. Intell. Res. **21**, 579–94 (2004)

Dorigo, M., Gambardella, L.M.: Ant colony system: a cooperative learning approach to the traveling salesman problem. IEEE Trans. Evolutionary Computation **1**(1), 53–66 (1997)

Fama, E.F.: The behavior of stock-market prices. J. Bus. **38**(1), 34–105 (1965)

Fama, E.F.: Efficient capital markets: a review of theory and empirical work. J. Finance **25**(2), 383–417 (1970)

Maei, H.R., Szepesvari, C., Bhatnagar, S., Precup, D., Silver, D., Sutton, R.S.: Convergent temporal-difference learning with arbitrary smooth function approximation. In: Advances in Neural Information Processing Systems. 23rd Annual Conference on Neural Information Processing Systems, Vancouver, 7–10 December 2009, pp. 1204–1212. La Jolla (2010)

Markowitz, H.M.: Portfolio Selection. Yale University Press, New Haven (1959)

Murphy, J.J.: Technical Analysis of the Financial Markets: A Comprehensive Guide to Trading Methods and Applications. New York Institute of Finance, New York (1999)

Precup, D., Sutton, R.S., Dasgupta, S.: Off-policy Temporal-difference Learning with Function Approximation. In: Proceedings of the Eighteenth International Conference on Machine Learning (ICML-2001), June 2001, pp. 417–424. Morgan Kaufmann, San Francisco (2000)

Rafols, E.J., Ring, M.B., Sutton, R.S., Tanner, B.: Using predictive representations to improve generalization in reinforcement learning. In: Proceedings of the Nineteenth International Joint Conference on Artificial Intelligence, Edinburgh, July 2005, pp. 835–840. Professional Book Center, Denver (2005)

Russell, S.J., Norvig, P.: Artificial Intelligence: A Modern Approach. Prentice Hall/Pearson Education, Upper Saddle River (1995)

Russell, S.J., Norvig, P.: Artificial Intelligence: A Modern Approach. Prentice Hall/Pearson Education, Upper Saddle River (2003)

Russell, S.J., Norvig, P.: Artificial Intelligence: A Modern Approach. Pearson, Boston (2010)

Sutton, R.S.: Open theoretical questions in reinforcement learning. In: Computational Learning Theory. Lecture Notes in Computer Science, vol. 1572, pp. 637–638. Springer, Berlin (1999)

Sutton, R.S.: Reinforcement learning: past, present and future. In: Simulated Evolution and Learning. Lecture Notes in Computer Science, vol. 1585, pp. 195–197. Springer, Berlin (1999)

Sutton, R.S., Barto, A.G.: Reinforcement Learning: An Introduction. MIT, Cambridge (2000)

Sutton, R.S., McAllester, D. Singh, S., Mansour, Y.: Policy Gradient Methods for Reinforcement Learning with Function Approximation. In: Advances in Neural Information Processing Systems, 1999 Conference, vol. 12, pp. 1057–1063. MIT, Cambridge (2000)

Sutton, R.S., Maei, H.R., Precup, D., Bhatnagar, S., Silver, D., Szepesvári, C., Wiewiora, E.: Fast gradient-descent methods for temporal-difference learning with linear function approximation. In: Proceedings, Twenty-sixth International Conference on Machine Learning, pp. 993–1000. Omnipress, Madison (2009a)

Sutton, R.S., Szepesvari, C., Maei, H.R.: A convergent o(n) algorithm for off-policy temporal difference learning with linear function approximation. In: Advances in Neural Information Processing Systems. 22nd Annual Conference on Neural Information Processing Systems, Vancouver, 8–10 December 2008, vol. 21 pp. 1609–1616, Curran, Red Hook (2009b)

Van Hasselt, H.: Reinforcement learning in continuous state and action spaces. In: Reinforcement Learning, pp. 207–251. Springer, Berlin (2012)

Watkins, C.J.C.H.: Learning from Delayed Rewards. University of Cambridge, Cambridge (1989)

Whiteson, S., Tanner, B., Taylor, M., Stone, P.: Protecting against evaluation overfitting in empirical reinforcement learning. In: Symposium on Adaptive Dynamic Programming And Reinforcement Learning (ADPRL), pp. 120–127. IEEE, Paris (2011)

Xu, X., Zuo, L., Huang, Z.: Reinforcement learning algorithms with function approximation: recent advances and applications. Inform. Sci. **261**, 1–31 (2014)

Yahoo! Finance - Business Finance, Stock Market, Quotes, News. http://finance.yahoo.com/. Accessed 19 December 2011

Chapter 19
Relative Entropy Criterion and CAPM-Like Pricing

Stylianos Z. Xanthopoulos

Abstract The minimal relative entropy criterion for the selection of an equivalent martingale measure in an incomplete market seems to still hold some mystique in its financial interpretation. In this paper we work toward this interpretation by suggesting and exploring the idea of relating a martingale measure selection criterion to a CAPM-like pricing scheme. We examine this idea in the case of the minimal relative entropy criterion and we present some preliminary results. We work within a one-period financial market and show that the minimal relative entropy pricing criterion is equivalent to some CAPM-like pricing scheme where the classical beta coefficient formula has been replaced by some "entropic beta" and the market portfolio by some "appropriate" reference portfolio. Furthermore, we show that if the assets involved have returns that are jointly normal, then this "entropic beta" formula coincides with the classical beta coefficient. Additionally and for comparison reasons, we briefly illustrate that if our criterion for the choice of the martingale measure was the minimization of the variance of the Radon–Nikodym derivative, then the resulting martingale pricing and the pricing implied by the classical CAPM scheme would be the same.

Keywords Minimal relative entropy criterion • Equivalent martingale measure • Incomplete market • CAPM

19.1 Introduction

It is well known that in an incomplete market, not all contingent claims are replicable as portfolios of traded assets. Therefore, a deeper understanding of how the market prices such claims presents extra challenges. A large number of studies in the area of incomplete markets have already shed some light to a better understanding of the functioning of financial markets [see, e.g., Magill and Quinzii (2002), Delbaen and Schachermayer (2006) for relevant overviews]; however a complete theory on the price selection procedure in incomplete markets is still

S.Z. Xanthopoulos (✉)
Department of Mathematics, University of the Aegean, Karlovassi Samos, 83200, Greece
e-mail: sxantho@aegean.gr

© Springer International Publishing Switzerland 2016
A.A. Pinto et al. (eds.), *Trends in Mathematical Economics*,
DOI 10.1007/978-3-319-32543-9_19

missing. It is well known, for example, that in an incomplete market, there exists an infinity of pricing kernels, leading to a whole interval of "legitimate" non-arbitrage prices for a non-replicable contingent claim. The difficulty to determine one single price stems from the fact that, for a non-replicable claim, one cannot hedge away all of its risk just by trading in the market's assets. Although there is a vast literature focusing on the determination of the upper and lower hedging prices (see, e.g., Davis et al. 2001, Delbaen and Schachermayer 1994, Sircar and Zariphopoulou 2004), it seems that additional criteria are needed if one is to figure out one single price, out of the whole band of the non-arbitrage prices, at which the contingent claim is eventually traded. This in a sense amounts to introducing criteria for the selection of some "appropriate" pricing kernel. One of the most popular and interesting of such criteria, which has been proposed in the literature, amounts to the minimization of an entropy measure and was introduced by Frittelli (1995) and further elaborated in Frittelli (2000).

Typically, a pricing kernel can be interpreted as a probability measure—usually called an "Arrow–Debreu" or equivalent martingale measure—under which the price of a European claim is the expectation of its discounted payoff (equivalently, a measure under which the discounted, under the risk-free rate, price process of each traded asset is a martingale). According to the minimal relative entropy criterion, the pricing kernel that is selected is the one corresponding to the Arrow–Debreu measure Q that is "closer" to the "true" statistical measure P which governs the possible states of the world. In other words, it is this martingale measure Q that minimizes a Kullback–Leibler-like entropy measure $I(Q, P)$. This suggestion has been supported by utility pricing arguments which are related with the relative entropy minimization problem via duality [see Frittelli (1995), Frittelli (2000), Bellini and Frittelli (2002), Föllmer and Schied (2011), and references therein].

Despite the popularity of the minimal relative entropy criterion and its connection to utility pricing via duality, it seems that there is still some mystique in its financial interpretation. In this work we will suggest a research path idea toward this interpretation, and we will present some preliminary results by attempting to relate the minimal relative entropy pricing criterion to a CAPM-like pricing scheme.

One should recall that Black and Scholes, in their seminal paper (Black and Scholes 1973) on option pricing, explain how CAPM could be used as an alternative way to derive their famous differential equation. In fact they claim that this derivation "offers more understanding on the way in which one can discount the value of an option to the present, using a discount rate that depends on both time and the price of the stock." Furthermore they stress the fact that "CAPM provides a general method for discounting under uncertainty."

The capital asset pricing model (CAPM), introduced independently by Treynor (1961b), Sharpe (1964), Lintner (1965a,b) and Mossin (1966), relates the expected return with the risk of an asset when the market is at an equilibrium and was based on the pathbreaking work of Markowitz on portfolio theory (Markowitz 1952).

While the notion of expected return is unambiguous, the concept of risk is rather subtle, and a lot of research has been devoted in order to better understand its nature and explore the ways to define it and measure it. The traditionality of standard deviation as the "appropriate" measure of risk gave its place to wider families of risk measures, like coherent, convex, and spectral risk measures [Heath et al. (1999), Artzner et al. (2002), Föllmer and Schied (2002), Acerbi (2002), etc.]. These in turn allowed for alternative considerations of the Markowitz portfolio theory and furthermore alternative CAPM considerations. For example, Kadan et al. (2014) generalize the concept of systemic risk to a broad class of risk measures, offering thus a whole spectrum of "alternative" CAPMs and extending the traditional beta to capture multiple dimensions of risk.

In this paper we will work within a one-period financial market, and we will show that the minimal relative entropy pricing criterion is equivalent to some CAPM-like pricing scheme. In particular, we will exhibit an "entropic beta" coefficient β^* and an appropriate reference portfolio G so that the pricing of an asset F via the formula $E(R^F) - r = \beta^*(E(R^G) - r)$ and the pricing of F as discounted expectation of its final value, under the minimal relative entropy martingale measure, lead to the same price. Furthermore, we will show that if the assets involved have returns that are jointly normal, then this "entropic beta" coefficient formula coincides with the classical beta coefficient. These results are presented in Sect. 19.3. With regard to the rest of the paper, in Sect. 19.2 we present some necessary preliminaries and we fix notation, while in Sect. 19.4, we illustrate even further, via an example, that this idea of relating a criterion for the choice of a martingale measure to some CAPM-like pricing scheme can work more generally. More precisely, we show that if our criterion for the choice of the martingale measure was the minimization of the variance of the Radon–Nikodym derivative, then the resulting martingale pricing would be the same as the one implied by the classical CAPM pricing scheme.

19.2 Preliminaries and Notation

We consider a one-period financial market with a finite number of final states (although the finite states consideration can be rather easily relaxed under the appropriate technical considerations).

There are two trading days t_0 and T and n possible states of the world at time T, labeled $\omega_1, \ldots, \omega_n$. The uncertainty about the state of the world that will be realized at time T is described by some probability measure $P = (p_1, \ldots, p_n)$ which assigns positive probability $p_i = P(\omega_i)$ to each ω_i, $i = 1, \ldots, n$.

We assume that there is a riskless asset, bearing a risk-free interest rate r, so that an investment of 1 unit at time t_0 is worth $1 + r$ at time T. Furthermore, we assume the existence of m risky traded assets $S^{(1)}, \ldots, S^{(m)}$. We use the following notation.

For each $i = 1, \ldots, n$ and $j = 1, \ldots, m$ we set:

$S_0^{(j)}$: the price of asset $S^{(j)}$ at time t_0.

$S_i^{(j)}$: the price of asset $S^{(j)}$ at time T, at state ω_i.

$S_0 := (S_0^{(1)}, \ldots, S_0^{(m)})^T$

$S^{(j)} := (S_1^{(j)}, \ldots, S_n^{(j)})$

$$S_i := (S_i^{(1)}, \ldots, S_i^{(m)})^T$$

$$S_T := \begin{pmatrix} S^{(1)} \\ \vdots \\ S^{(m)} \end{pmatrix} = \begin{pmatrix} S_1^{(1)} & \ldots & S_n^{(1)} \\ \vdots & \vdots & \vdots \\ S_1^{(m)} & \ldots & S_n^{(m)} \end{pmatrix}$$

$$R_i^{(j)} := \frac{S_i^{(j)}}{S_0^{(j)}} - 1$$

$$R_0 := (R_0^{(1)}, \ldots, R_0^{(m)})^T$$

$$R^{(j)} := (R_1^{(j)}, \ldots, R_n^{(j)})$$

$$R_i := (R_i^{(1)}, \ldots, R_i^{(m)})^T$$

$$R_T := \begin{pmatrix} R^{(1)} \\ \vdots \\ R^{(m)} \end{pmatrix} = \begin{pmatrix} R_1^{(1)} & \ldots & R_n^{(1)} \\ \vdots & \vdots & \vdots \\ R_1^{(m)} & \ldots & R_n^{(m)} \end{pmatrix}$$

$$E(S_T) := (E_P(S^{(1)}), \ldots, E_P(S^{(m)}))^T$$

$$E(R_T) := (E_P(R^{(1)}), \ldots, E_P(R^{(m)}))^T$$

$$\text{Cov}(X, Y) := \text{Cov}_P(X, Y), \quad \text{Var}(X) := \text{Var}_P(X)$$

In the sequel we will also consider a contingent claim denoted by F that at time T takes values $F_T = (F_1, \ldots, F_n)$; we will write F_0 to denote its value at time t_0 and $R^F := \frac{F_T}{F_0} - 1$ to denote its return during the period.

A probability measure Q on Ω is called martingale measure for this market, if and only if $E_Q(S_T) = S_0(1 + r)$. We denote by $\mathcal{M}$ the set of all martingale measures for this market. A martingale measure Q is called equivalent martingale measure with regard to the measure P if moreover the zero probability events coincide under both Q and P. We denote by $\mathcal{M}_e$ the set of all equivalent martingale measures for this market. The existence of equivalent martingale measures is equivalent to nonexistence of arbitrage opportunities in the market.

19.3 The Minimal Relative Entropy Criterion

Let Q, P be two probability measures. The relative entropy of Q with respect to P is defined as

$$I(Q, P) = E_P(\frac{dQ}{dP} \ln \frac{dQ}{dP}) \ ; Q << P + \infty \ ; \text{otherwise}$$

where $\frac{dQ}{dP}$ denotes the Radon–Nikodym derivative of Q with regard to P.

Let $Q = (q_1, \ldots, q_n)$ and $P = (p_1, \ldots, p_n) > 0$ be probability measures on $\Omega = \{\omega_1, \ldots, \omega_n\}$. In this case the Radon–Nikodym derivative is just $\frac{dQ}{dP} = (\frac{q_i}{p_i})_{i=1,\ldots,n}$, and so the relative entropy of Q with respect to P is defined as

$$I(Q, P) = \sum_{i=1}^{n} q_i \ln \left(\frac{q_i}{p_i} \right), \quad \text{with } 0 \ln 0 := 0 \tag{19.1}$$

The minimal relative entropy criterion, as suggested by Frittelli (1995), amounts
to solving the following optimization problem:

$$\min_{Q \in \mathcal{M}} I(Q, P) \tag{19.2}$$

or equivalently:

$$\min_{q_1, \dots, q_n} \sum_{i=1}^{n} q_i \ln\left(\frac{q_i}{p_i}\right) \tag{19.3}$$

under the restrictions

$$q_1, \dots, q_n \geq 0$$
$$\sum_{i=1}^{n} q_i = 1$$
$$\sum_{i=1}^{n} q_i S_i^{(j)} = S_0^{(j)}(1+r) \ \forall j = 1, \dots, m$$

It can then be easily proved [see Frittelli (1995) for the details of the proof] that
the solution to the above problem (19.3) is given by

$$q_i = \frac{p_i \exp(-\gamma_1 R_i^{(1)} - \dots - \gamma_m R_i^{(m)})}{\sum_{j=1}^{n} p_j \exp(-\gamma_1 R_j^{(1)} - \dots - \gamma_m R_j^{(m)})} \tag{19.4}$$

where $\gamma = (\gamma_1, \dots, \gamma_m)$ is the unique solution of the system of equations:

$$r \sum_{i=1}^{n} p_i \exp\left(-\sum_{j=1}^{m} \gamma_j R_i^{(j)}\right) = \sum_{i=1}^{n} p_i R_i^{(h)} \exp\left(-\sum_{j=1}^{m} \gamma_j R_i^{(j)}\right)$$

$$\text{for all } h = 1, \dots, m$$

It is also clear that the so-defined martingale measure $Q = (q_1, \dots, q_n)$ is in fact
an equivalent martingale measure.

It will be convenient to write the expressions (19.4) and (19.5) in a more compact
form. For this we set $\Gamma = \sum_{j=1}^{m} \gamma_j$ and we consider the portfolio G consisting
of positions on the traded assets $S^{(1)}, \dots, S^{(m)}$ with respective weights $g_1, \dots, g_m$
where $g_j = \frac{\gamma_j}{\Gamma}$. Let R^G denote the return of this portfolio and R_i^G the return of
the portfolio at state i. Then $R^G = \sum_{j=1}^{m} g_j R^{(j)}$ and $R_i^G = \sum_{j=1}^{m} g_j R_i^{(j)}$ and the
relations (19.4) and (19.5) can be written as follows:

$$q_i = \frac{p_i \exp(-\Gamma R_i^G)}{E(\exp(-\Gamma R^G))} \tag{19.6}$$

where $\Gamma, g_1, \dots, g_m$ is the unique solution of the system:

$$rE(\exp(-\Gamma R^G)) = E(R^{(h)} \exp(-\Gamma R^G)); \ h = 1, \dots, m \tag{19.7}$$

with $\sum_{j=1}^m g_j = 1$.

Now it is a straightforward exercise to show that this last expression (19.7) is equivalent to

$$r - E(R^{(h)}) = \frac{\text{Cov}(R^{(h)}, \exp(-\Gamma R^G))}{E(\exp(-\Gamma R^G))} \quad \forall h = 1, \ldots, m \tag{19.8}$$

Thus we can give the following definition.

Definition 1. The minimal relative entropy martingale measure $Q = (q_1, \ldots, q_n)$ is given by the relation:

$$\frac{q_i}{p_i} = \frac{\exp(-\Gamma R_i^G)}{E(\exp(-\Gamma R^G))}$$

where $R^G = \sum_{j=1}^m g_j R^{(j)}$ and $(\Gamma, g_1, \ldots, g_m)$ is the unique solution of the martingale equations system:

$$r - E(R^{(h)}) = \frac{\text{Cov}(R^{(h)}, \exp(-\Gamma R^G))}{E(\exp(-\Gamma R^G))} \quad ; \ h = 1, \ldots, m$$

$$\sum_{j=1}^m g_j = 1$$

Remark 1. It is just a matter of calculus of variation technicalities to show that in the case of continuous probability measures, the previous Definition 1 still works but with the first relation written in the appropriate form:

$$\frac{dQ}{dP} = \frac{\exp(-\Gamma R^G)}{E(\exp(-\Gamma R^G))}$$

where $\frac{dQ}{dP}$ denotes the Radon–Nikodym derivative [see also Frittelli (2000)].

Remark 2. Let $R^\Pi = \sum_{j=1}^m w_j R^{(j)}$ denote the return of a portfolio consisting of positions on the risky assets $S^{(1)}, \ldots, S^{(m)}$ with respective weights $w_1, \ldots, w_m$ (i.e., $w_1 + \ldots + w_m = 1$). Then it is clear that

$$r - E(R^\Pi) = \frac{\text{Cov}(R^\Pi, \exp(-\Gamma R^G))}{E(\exp(-\Gamma R^G))} \tag{19.9}$$

Remark 3. If we had further assumed that R^Π and $-\Gamma R^G$ are jointly normal, then the previous Remark 2 combined with Stein's Lemma would imply that

$$\Gamma = \frac{E(R^\Pi) - r}{\mathrm{Cov}(R^\Pi, R^G)} \tag{19.10}$$

In particular, if the portfolio Π is taken to be G itself, then the previous equation implies that

$$\Gamma = \frac{E(R^G) - r}{\mathrm{Var}(R^G)}$$

which is equivalent to

$$\Gamma \,\mathrm{stdv}(R^G) = \frac{E(R^G) - r}{\mathrm{stdv}(R^G)} \tag{19.11}$$

This last equation shows that Γ measures the Sharpe ratio of this particular portfolio G in standard deviation units.

Definition 2. Let Π be some portfolio consisting of positions on the risky traded assets $S^{(1)}, \ldots, S^{(m)}$ and let F be a contingent claim in this market. The entropic beta of F with respect to Π is defined as

$$\beta^*_{(F,\Pi)} := \frac{\mathrm{Cov}(R^F, \exp(-\Gamma R^G))}{\mathrm{Cov}(R^\Pi, \exp(-\Gamma R^G))} \tag{19.12}$$

Remark 4. Suppose R^F and $-\Gamma R^G$ are jointly normal and that R^Π and $-\Gamma R^G$ are jointly normal as well. Then Stein's lemma would imply that

$$\beta^*_{(F,\Pi)} := \frac{\mathrm{Cov}(R^F, R^G)}{\mathrm{Cov}(R^\Pi, R^G)}$$

In particular

$$\beta^*_{(F,G)} := \frac{\mathrm{Cov}(R^F, R^G)}{\mathrm{Var}(R^G)} \tag{19.13}$$

which coincides with the familiar formula for beta in the classical CAPM.

Now we are in the position to show in the next proposition that, independently of the distribution of the various asset returns, the minimal relative entropy pricing criterion is equivalent to a CAPM-like pricing scheme. More precisely, the price obtained when using the minimal relative entropy martingale criterion is the same as the price obtained when using a CAPM-like pricing scheme, where the β coefficient of the classical CAPM has been replaced by a generalized β^* coefficient. Furthermore, the corollary that follows shows that by imposing an additional condition of joint normality of returns, the minimal relative entropy pricing is equivalent to the classical CAPM and highlights the role of the portfolio G, which now plays a role similar to that of the market portfolio.

Proposition 1. *Let Π be a portfolio consisting of positions on $S^{(1)}, \ldots, S^{(m)}$ and let F be a contingent claim. Let $R^F = \frac{F_T}{F_0} - 1$ be the return of F. Consider the equation:*

$$E(R^F) - r = \beta^*_{(F,\Pi)}(E(R^\Pi) - r) \tag{19.14}$$

where

$$\beta^*_{(F,\Pi)} = \frac{Cov(R^F, \exp(-\Gamma R^G))}{Cov(R^\Pi, \exp(-\Gamma R^G))} \tag{19.15}$$

with $R^G = \sum_{j=1}^m g_j R^{(j)}$ and $\Gamma, g_1, \ldots, g_m)$ the unique solution of the system

$$r - E(R^{(h)}) = \frac{Cov(R^{(h)}, \exp(-\Gamma R^G))}{E(\exp(-\Gamma R^G))} \quad ; h = 1, \ldots, m$$

$$\sum_{j=1}^m g_j = 1$$

Then F_0 satisfies Eq. (19.14) if and only if $(1 + r)F_0 = E_Q(F_T)$ where Q is the minimal entropy martingale measure.

Proof. Equation (19.14) is equivalent to

$$(R^F) - r = \frac{Cov(R^F, \exp(-\Gamma R^G))}{Cov(R^\Pi, \exp(-\Gamma R^G))}(E(R^\Pi) - r)$$

$$\Leftrightarrow_{(19.9)} E(R^F) - r = \frac{Cov(R^F, \exp(-\Gamma R^G))}{(r - E(R^\Pi))E(\exp(-\Gamma R^G))}(E(R^\Pi) - r)$$

$$\Leftrightarrow E(R^F)E(\exp(-\Gamma R^G)) + Cov(R^F, \exp(-\Gamma R^G)) = rE(\exp(-\Gamma R^G))$$

$$\Leftrightarrow E(R^F \exp(-\Gamma R^G)) = rE(\exp(-\Gamma R^G)) \Leftrightarrow$$

$$\Leftrightarrow E\left(\left(\frac{F_T}{F_0} - 1\right)\exp(-\Gamma R^G)\right) = rE(\exp(-\Gamma R^G))$$

$$\Leftrightarrow \frac{1}{F_0}E(F_T \exp(-\Gamma R^G)) = (1 + r)E(\exp(-\Gamma R^G))$$

$$\Leftrightarrow (1 + r)F_0 = \frac{E(F_T \exp(-\Gamma R^G))}{E(\exp(-\Gamma R^G))}$$

$$\Leftrightarrow (1 + r)F_0 = \frac{E_Q(\frac{dP}{dQ}F_T \exp(-\Gamma R^G))}{E(\exp(-\Gamma R^G))}$$

which by Remark 1 is equivalent to

$$(1+r)F_0 = \frac{E_Q(\frac{E(\exp(-\Gamma R^G))}{\exp(-\Gamma R^G)} F_T \exp(-\Gamma R^G))}{E(\exp(-\Gamma R^G))} \Leftrightarrow (1+r)F_0 = E_Q(F_T)$$

where Q such that $\frac{dQ}{dP} = \frac{\exp(-\Gamma R^G)}{E(\exp(-\Gamma R^G))}$ is the minimal relative entropy martingale measure according to Definition 1 and Remark 1.   □

Corollary 1. *Suppose that R^F and $-\Gamma R^G$ are jointly normal. Then the minimal relative entropy pricing is equivalent to CAPM pricing, where the role of the market portfolio is played by portfolio G.*

Proof. Remark 4 implies that Eqs. (19.14) and (19.15) of the previous proposition become

$$E(R^F) - r = \beta^*_{(F,G)}(E(R^G) - r)$$

and

$$\beta^*_{(F,G)} := \frac{\text{Cov}(R^F, R^G)}{\text{Var}(R^G)},$$

respectively, and the result follows.   □

19.4 The Minimal Variance Criterion

In this section we will provide an example illustrating that the idea presented in the previous section works effectively for other pricing criteria as well. In Frittelli (1995), Frittelli compares the minimal relative entropy criterion with a method that was initially proposed by Follmer and Sondermann (1986) and which amounts to choosing the signed martingale measure that minimizes the variance of the Radon–Nikodym derivative, i.e., to solve the problem:

$$\min_{Q \in \mathcal{M}} \text{Var}(dQ/dP) \qquad (19.16)$$

For simplicity of exposition, we will consider that our market has only one risky asset (therefore the market portfolio coincides with this asset). In this market example, the solution to Problem 19.16 turns out to be

$$q_i = p_i \left(1 + \frac{(E(S_T) - S_i)(E(S_T) - S_0(1+r))}{\text{Var}(S_T)}\right) \qquad (19.17)$$

which we call the minimal variance martingale measure. It has to be noted that this is not necessarily an equivalent martingale measure (see Frittelli 1995). We will show here that the employment of the minimal variance criterion, as a pricing method, is equivalent to the standard CAPM pricing formula.

Proposition 2. *Consider the equation:*

$$E(R_F) - r = \beta(E(R_S) - r) \tag{19.18}$$

where $R_F = F_T/F_0 - 1$, $R_S = S_T/S_0 - 1$ *and*

$$\beta = \frac{Cov(R_F, R_S)}{Var(R_S)}$$

Then F_0 satisfies (19.18) if and only if $F_0 = E_Q(F_T/(1+r))$, where Q is the minimal variance martingale measure

Proof.

$$E(R_F) - r = \frac{Cov(R_F, R_S)}{Var(R_S)}(E(R_S) - r)$$

$$\Leftrightarrow E(F_T) - (1+r)F_0 = \frac{Cov(F_T, S_T)}{Var(S_T)}(E(S_T) - (1+r)S_0)$$

$$\Leftrightarrow (1+r)F_0 = E(F_T) + \frac{(E(F_T)E(S_T) - E(F_T S_T))(E(S_T) - (1+r)S_0)}{Var(S_T)}$$

$$\Leftrightarrow (1+r)F_0 = \sum p_i F_i + \frac{(\sum p_i F_i E(S_T) - \sum p_i F_i S_i)(E(S_T) - (1+r)S_0)}{Var(S_T)}$$

$$\Leftrightarrow (1+r)F_0 = \sum F_i p_i (1 + \frac{(E(S_T) - S_i)(E(S_T) - S_0(1+r))}{Var(S_T)})$$

$$\Leftrightarrow (1+r)F_0 \sum F_i q_i \Leftrightarrow F_0 = E_Q(F_T/(1+r)) \qquad \square$$

This example illustrates also that if a martingale selection criterion is problematic [as is known to be the case for the minimum variance criterion where arbitrage prices may be produced, (see Frittelli 1995)], then the same may be true for the corresponding equivalent equilibrium model (the CAPM in our case) and vice versa.

References

Acerbi, C.: Spectral measures of risk: a coherent representation of subjective risk aversion. J. Bank. Finance **26**(7), 1505–1518 (2002)

Artzner, P., Delbaen, F., Eber, J.M., Heath, D.: Coherent measures of risk. Math. Financ. **9**(3), 203–228 (1999)

Bellini, F., Frittelli, M.: On the existence of minimax martingale measures. Math. Finance **12**(1), 1–21 (2002)

Black, F., Scholes, M.: The pricing of options and corporate liabilities. J. Polit. Econ. **81**(3), 637–654 (1973)

Davis, M.H.A., Schachermayer, W., Tompkins, R.G.: Pricing, no-arbitrage bounds and robust hedging of instalment options. Quant. Finan. **1**(6), 597–610 (2001)

Delbaen, F., Schachermayer, W.: A general version of the fundamental theorem of asset pricing. Math. Ann. **300**(1), 463–520 (1994)

Delbaen, F., Schachermayer, W.: The Mathematics of Arbitrage. Springer Science & Business Media, Berlin (2006)

Föllmer, H., Schied, A.: Convex measures of risk and trading constraints. Finance Stoch. **6**(4), 429–447 (2002)

Föllmer, H., Schied, A.: Stochastic Finance: An Introduction in Discrete Time. Walter de Gruyter, Berlin (2011)

Follmer, H., Sondermann, D.: Contributions to Mathematical Economics. North Holland, Amsterdam (1986)

Frittelli, M.: Minimal entropy criterion for pricing in one period incomplete markets. Technical Report (1995)

Frittelli, M.: The minimal entropy martingale measure and the valuation problem in incomplete markets. Math. Finance **10**(1), 39–52 (2000)

Heath, D., Delbaen, F., Eber, J.M., Artzner, P.: Coherent measures of risk. Math. Finance **9**, 203–228 (1999)

Kadan, O., Liu, F., Li, S.: Generalized systematic risk. Available at SSRN 2444039 (2014)

Lintner, J.: Security prices, risk, and maximal gains from diversification. J. Finance **20**(4), 587–615 (1965)

Lintner, J.: The valuation of risk assets and the selection of risky investments in stock portfolios and capital budgets. Rev. Econ. Stat. **47**(1), 13-37 (1965)

Magill, M., Quinzi, M.: Theory of Incomplete Markets, vol. 1. MIT, Cambridge (2002)

Markowitz, H.: Portfolio selection. J. Finance **7**(1), 77–91 (1952)

Mossin, J.: Equilibrium in a capital asset market. Econometrica J. Econ. Soc. 768–783 (1966)

Sharpe, W.F.: Capital asset prices: a theory of market equilibrium under conditions of risk. J. Finance **19**(3), 425–442 (1964)

Sircar, R., Zariphopoulou, T.: Bounds and asymptotic approximations for utility prices when volatility is random. SIAM J. Control Optim. **43**(4), 1328–1353 (2004)

Treynor, J.L.: Toward a theory of market value of risky assets. Unpublished manuscript (1961) [Treynor, J.L. (ed.): Toward a theory of market value of risky assets. In: Treynor on Institutional Investing. Wiley, Hoboken (2012). doi:10.1002/9781119196679.ch6]

Index

A

Arbitrage possibilities, 342, 346

B

Bellman equation, 6, 10, 24, 153–155,
157–159, 354, 355
Bitcoin, 73–95
Black–Scholes, 100, 107, 115, 116, 120, 320

C

Capital asset pricing model (CAPM), 191,
370–378
Circular flow, 1–32, 165, 166, 200
Coalitional structures, 209–213
Command economy, 65–68
Competition in sports leagues, 246
Complementarity, 247, 261
Contrarian, 221–223, 228, 241, 348
Cooperative games, 182, 210, 280, 281, 289,
290, 299, 300, 304
Correspondences, 126–130, 132, 133, 139,
142, 264, 268, 271–274
Creative and productive sets, 222, 227
Credibility, 36, 40
Cryptocurrencies, 73, 74, 77, 78, 80, 90, 91,
93, 95

D

Deferred acceptance (DA), 43–50
Derivatives, 11, 30, 59, 60, 66, 100, 103, 107,
119, 145, 175, 177, 178, 183, 185, 195,
196, 204, 256, 324, 332, 338, 340, 344,
371, 372, 374, 377

E

Economic decision making, 317, 318
Economics and finance, 78, 83, 94–95,
317–328
Endogenous growth, 54–57, 70
Equivalent martingale measure, 370, 372, 373,
378
European vs. American sports leagues, 246
Expected utility, 31, 40, 148, 149, 151, 152,
155, 163, 167–168, 172–179, 192, 195,
201, 282, 289
Externalities, 53–70, 123–145, 171, 201,
245–248, 252, 257, 260, 261, 281,
299–314

F

Finance, 2, 5, 12–13, 20, 29, 30, 34, 58, 77, 78,
83, 94–95, 124, 317–328, 332, 361
Forward rate models, 332, 335–336, 343–344
Fractal market hypothesis, 75, 77, 80–81,
92–93

G

Gain and lost games, 210
General equilibrium, 3, 12, 31, 54, 56, 125,
126, 131, 164, 331
Gödel incompleteness, 219, 222, 225–233
Greeks, 99–120

I

Incentives, 38, 69, 79, 219, 281–283, 293
campaigns, 39
Incomplete market, 369, 370

A.A. Pinto et al. (eds.), *Trends in Mathematical Economics*,
DOI 10.1007/978-3-319-32543-9

CPSIA information can be obtained
at www.ICGtesting.com
Printed in the USA
LVOW02*1449061216

516054LV00002B/22/P